Second Edition

Reviewing Intermediate-Level Science

Preparing for Your Eighth-Grade Test

Paul S. Cohen

Former Assistant Principal, Science
Franklin Delano Roosevelt High School
Brooklyn, New York

Anthony V. Sorrentino, D.Ed.

Former Director of Computer Services
Former Earth Science Teacher
Monroe-Woodbury Central School District
Central Valley, New York

Amsco School Publications, Inc.,
a division of Perfection Learning®

We would like to recognize the following teachers who acted as reviewers.

Cheryl Dodes
Science Chair, Retired
Weber Middle School
Port Washington, NY

Martin Solomon
Science Teacher
Daniel Cater Beard JHS 189 Q
Queens, NY

Janet Iadanza
Science Teacher
Rockwood Park School
Howard Beach, NY

Diana K. Harding
Retired
New York State Department of Education

Text Design: A Good Thing, Inc.
Cover Design: Meghan J. Shupe
Composition: Northeastern Graphic, Inc.
Artwork: Hadel Studio
Cartoon page 38 bottom: Courtesy of Alison Solomon
Cover photo: M101 Spiral Galaxy

Please visit our Web sites at: ***www.amscopub.com*** and ***www.perfectionlearning.com***

When ordering this book, please specify:
either **13462** REVIEWING INTERMEDIATE-LEVEL SCIENCE, SECOND EDITION

ISBN: 978-1-56765-931-3

Printed in the United States of America
8 9 10 11 17 16 15 14 13

Contents

To the Student

This book provides a complete review of intermediate-level science to help you prepare for your Grade 8 Intermediate-Level Science Written Test.

The text consists of 13 chapters. These cover topics in life, physical, and Earth science, as well as the scientific method, the history and nature of scientific inquiry, and the interactions of science, technology, and society.

The text presents the major ideas of each topic in a clear manner that is easy to understand. It is aligned and corresponds with the National Standards (as well as those of New York City and New York State). The many illustrations help make clear the concepts presented. Each chapter is divided into several sections by topic. Each section is followed by Review Questions, which are divided into Part I, multiple-choice questions, and Part II, extended-response questions. These questions are similar to those found in the Grade 8 Intermediate-Level Test: Science—Written Test. These questions test and reinforce the main points covered in the section. The questions are designed to test your abilities in *acquiring, processing,* and *extending* scientific knowledge.

In addition, special features called Process Skills and Laboratory Skills appear throughout the text. These features will teach you a particular process-oriented or laboratory-oriented skill, including reading scales, using a compass, and interpreting graphs or diagrams. Each feature guides you through the skill and ends with several follow-up questions that have you apply the skill on your own.

Throughout the book, important vocabulary terms are printed in ***bold italic type***. These terms are defined in the text and appear with formal definitions in the glossary at the back of the book. Terms that are less important or that may be unfamiliar to you are printed in *italic type*. These terms do not appear in the Glossary but are listed in the index.

Finally, the book begins with a Diagnostic Test followed by a Diagnostic Checklist. These will help you identify any weak areas in your science knowledge. There are also two sample tests that are similar to Grade 8 Intermediate-Level Test: Science—Written Test. These tests will help you practice. The authors hope you enjoy using this book and that it helps you do well in science.

Introduction

I. What Is the Grade 8 Intermediate-Level Science Written Test?

Eighth-grade students in New York State must pass the Grade 8 Intermediate-Level Science Written Test. Students will usually take this examination in the spring. You will also take a Performance Test before you take the written test.

The questions on the written test will measure your knowledge and understanding of science. The test has two parts:

Part I consists of 45 multiple-choice questions. You will record your answers to these questions on a separate answer sheet, using a No. 2 pencil.

Part II consists of 28 to 33 open-ended questions. You will write the answers to Part II in the spaces provided in the test booklet.

You may use a calculator to answer the questions on the test if needed. You will have two hours to answer the questions on the test.

II. What Should I Review?

The following are the areas of science you will need to study to be successful on the Grade 8 Intermediate-Level Science Written Test.

A. Key Ideas

You will need to understand the key ideas covered in intermediate-level science.

Physical Setting

- Earth and celestial events can be described by principles of relative motion and frame of reference.
- Many of the events we observe on Earth are the result of interactions among air, water, and land.
- Matter is made up of particles whose properties determine the characteristics of matter and how it reacts.
- Energy exists in many forms, and when these forms change, energy is conserved.
- Energy and matter interact through forces that result in changes in motion.

Living Environment

- Living things are similar to and different from each other and from nonliving things.
- Organisms inherit genetic information in a variety of ways that result in parents and offspring having the same structure and function.
- Individual organisms and species change over time.
- Reproduction and development maintain the continuity of life.
- Organisms maintain a dynamic equilibrium that supports life.
- Plants and animals depend on each other and on their physical environment.
- Human decisions and activities have had a large effect on the physical and living environment.

The content in this review book and the questions in each chapter will help you review these ideas. Any words in ***boldface italics*** may appear on the examination without any explanation. You must know the meaning of these words. Other words in the review book are *italicized* for emphasis if they are important but they don't need to be memorized.

B. The Nature of Scientific Inquiry

What science is and how scientific research is conducted will also be tested. It is important to realize that science is both a body of knowledge and a way of learning about how the world works. Scientific explanations are developed using observations and experimental evidence to add to what has been learned previously. All scientific explanations may change if better ones are found. Good science involves asking questions, observing, experimenting, finding evidence, collecting and organizing data, drawing valid conclusions, and discussing results with other scientists.

C. Laboratory Skills

As an intermediate-level science student, you need certain laboratory skills that may be tested on the written test and the performance test. These skills are:

General Skills

- Follow safety procedures in the classroom and laboratory.
- Safely and accurately use the following measurement tools: metric ruler, balance, stopwatch, graduated cylinder, thermometer, spring scale, and voltmeter.
- Use appropriate units for measured and calculated values.

- Recognize and analyze patterns and trends
- Classify objects according to an established scheme or a scheme you develop.
- Develop and use a dichotomous key.
- Sequence events.
- Identify cause-and-effect relationships.
- Use acid-base indicators and interpret the results.

Living Environment Skills

- Use a compound microscope to view microscopic objects.
- Determine the size of a microscopic object using a compound microscope.
- Prepare a wet-mount slide.
- Use appropriate staining techniques.
- Design and use a Punnett square or a pedigree chart to predict the probability of certain traits.
- Classify living things according to an established scheme and a scheme you develop.
- Interpret and/or illustrate the energy flow in a food chain, energy pyramid, or food web.
- Identify pulse points and pulse rates.
- Identify structure and function relationships in organisms.

Physical Setting Skills

- Given the latitude and longitude of a location, show its position on a map; determine the latitude and longitude of a given location on a map.
- Using identification tests and a flowchart, identify mineral samples.
- Use a diagram of the rock cycle to determine geological processes that led to the formation of a specific rock type.
- Plot the location of recent earthquake and volcanic activity on a map and identify patterns of distribution.
- Use a magnetic compass to find cardinal directions.
- Measure the angular elevation of an object using appropriate instruments.
- Draw and interpret field maps, including topographic and weather maps.

- Predict the characteristics of an air mass based on the source of the air mass.
- Measure weather variables such as wind speed and direction, relative humidity, barometric pressure, etc.
- Determine the density of liquids, and regular- and irregular-shaped solids.
- Determine the volume of a regular- and irregular-shaped solid, using water displacement.
- Using the Periodic Table, identify an element as a metal, nonmetal, or noble gas.
- Determine the identity of an unknown element using physical and chemical properties.
- Using appropriate resources, separate the parts of a mixture.
- Determine the electrical conductivity of a material using a simple circuit.
- Determine the speed and acceleration of a moving object.

III. How Should I Study?

A. Basic Strategies for Study

Make a regular time and place for study at home and plan a daily schedule. As a guideline, on a regular school night, you should spend 20 minutes *per subject* on homework and study. Success will come—if you work hard!

Look up any words you don't understand, even nonscience words. Keep a dictionary nearby.

Prepare your own notes about important ideas and terms as you read. This helps to keep your mind concentrated on the subject while you are reading.

Write your own questions about what you read, close the book, and then try to answer your own questions. After completing this task, open the book again and review and correct your answers. If there is a question to which you cannot find an answer, ask your teacher.

Study when you are feeling alert and try to avoid distractions (such as the Internet, telephone, radio, and television).

Take five-minute breaks every hour to help you stay alert and retain what you have studied.

B. Test-Taking Strategies

Multiple-Choice Questions

For multiple-choice questions, **read** the statement or question. If a diagram is included, look at it carefully. Do not look at the answers until you have read the question. Reread the question. It is helpful to underline key words as you read. Now read each of the answers. Do not stop when you think you have found the right one! Look at all the answers. You are looking for the word or expression that **best** completes the statement or answers the question. There will be only one answer that is the best choice.

As you read the answers for a multiple-choice question, cross out the answers that you know are not correct. It is helpful to think about why answers are wrong as you cross them out. This will help you choose an answer. Every choice you eliminate increases your chance of choosing the correct answer. Go back to the statement or question as often as you need to as you read the answers. After crossing out any answers you know are wrong, choose wisely, or in other words, guess from the remaining answers. **Make a choice. Do not leave a blank.** A wrong answer is better than none at all.

If you have time at the end of the test, and you are reviewing your answers for multiple-choice questions, be very careful before making any changes. After following the thinking process described above, the answer you chose the first time is more likely to be correct than another choice made later. In other words, be careful not to "second-guess" yourself.

Example:

What makes it possible for an astronaut to see Earth from space?

1 Earth emits light from its surface.
2 The moon emits light, and this light is reflected from Earth.
3 The sun emits light, and this light is reflected from Earth.
4 The sun emits light, and this light is reflected from the moon.

Analysis:

The key words are "see Earth from space." Now we know that the question is about how we see things. You know that we see most things by the light they reflect. We see the sun and stars because they emit light. At this point, we look at the answers. We see that Answer 1—"Earth emits light from its surface"—is wrong because Earth does not emit light. Answer 2—"The moon emits light, and this light is reflected from Earth"—is wrong, too. Although the Earth reflects light, the moon does not emit light. Even when part of an answer is correct, if another part is incorrect, the answer is wrong. Answer 3—"The sun emits light, and this light is reflected from Earth"—fits what we think is the answer to the question. The sun does emit light and Earth does reflect light. We continue by looking at Answer 4—"The sun emits light, and this light is reflected from the moon." Although this statement it true, it does not answer how the astronaut sees Earth from space. It is not the correct answer. Answer 3 is the correct choice.

You should follow this process for each multiple-choice question. While this takes time, it is the only method that will allow you to make use of what you have learned and avoid selecting the wrong answer.

Reading Questions

Part II of the Grade 8 Intermediate-Level Written Test may have questions based on reading passages. You will usually have to write your answers to questions based on the reading passages. Prepare yourself for the reading passage by giving the passage your **full attention**:

- The reading passage **will** contain interesting information, even if the topic is something you have not learned before. Understanding reading material is much greater when you are interested in what you are reading.
- Focus on the passage. Do not let what is going on around you distract you from your task.
- Stop looking at your watch to check the time.

The following are suggested strategies for questions based on reading passages. You may wish to use one or more of these strategies, based on what you think will work best for you:

- Look over the questions before reading the passage. This may give you an idea of what to look for while you are reading. However, once you start reading, don't stop for individual answers because this will break your concentration.
- As you read each paragraph, ask yourself what the paragraph is about. If a reading passage has more than one paragraph, this means it has more than one important idea. Each paragraph has one important idea.
- Look carefully at the first sentence of the first paragraph; it usually states the topic for the reading passage.
- Underline the sentence that states the main idea of the reading passage and of each paragraph.
- Circle specific facts.
- Reread the whole passage before working on the questions.

Work on one question for the reading passage at a time. A number in square brackets ([ ]) following the question tells you how many points a correct answer is worth. Note this number. If the question is worth two or more points, be certain that you write a complete answer to earn all these points. Underline the **key words** in the question and **include these words in your written answer**. This will make it much more likely that the teacher who reads your answer and marks your test will understand what you are trying to say. Also, write very neatly. If this is not possible, then print your answer. A teacher marking the test cannot give points for the answers that he or she cannot read!

Data Interpretation and Graphing

The Grade 8 Intermediate-Level Science Written Test may include questions that have you read about an experiment, study a data table showing the results of the experiment, graph the data, and answer one or more questions about the data. The data for a typical experiment is placed in a data table with at least two columns. The first column is usually the independent variable. This is the condition or variable that the researcher sets. The numbers in this column show each level at which the independent variable was set. The other columns are for the dependent variables. The values in these columns are the data that was measured or collected at each level of the independent variable. These values depend on the independent variable.

As you read the passage about the experiment, study the data table. Use the reading passage to understand the data table. If you are asked to construct a line graph for the data, follow the directions carefully. The title for a graph should refer to **both** the independent and the dependent variables.

The most important part of a graph is the marking of an **appropriate scale**. The scale begins at the lower left, with the values increasing going to the right or going up. The scale for the independent variable is on the horizontal axis (*x*-axis) and the scale for the dependent variable is on the vertical axis (*y*-axis). You must use a scale that includes the lowest and the highest values for the variable. Label the axes, stating what quantity each measures and the units. Each space on the graph paper must represent the same quantity. To plot the data points, refer to the data table. Each point on the line graph is located by two pieces of data from the data table, that is, the value for each level of the independent variable with the corresponding value for the dependent variable.

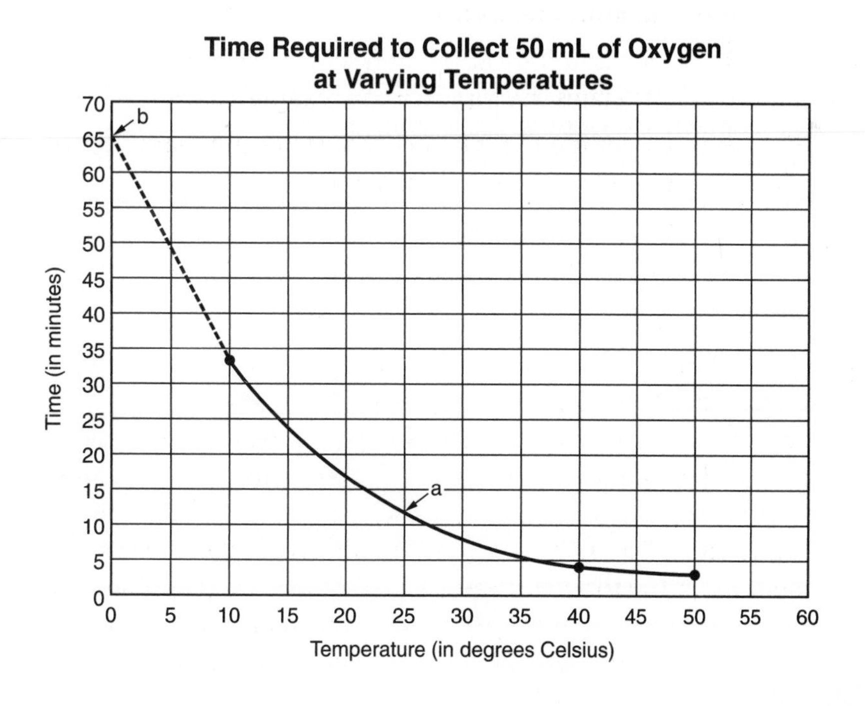

A bar graph compares data rather than showing relationships between variables. A bar graph has no scale along the horizontal axis. Rather, labels for each set of data are placed on the horizontal axis, with a scale for the data on the vertical axis.

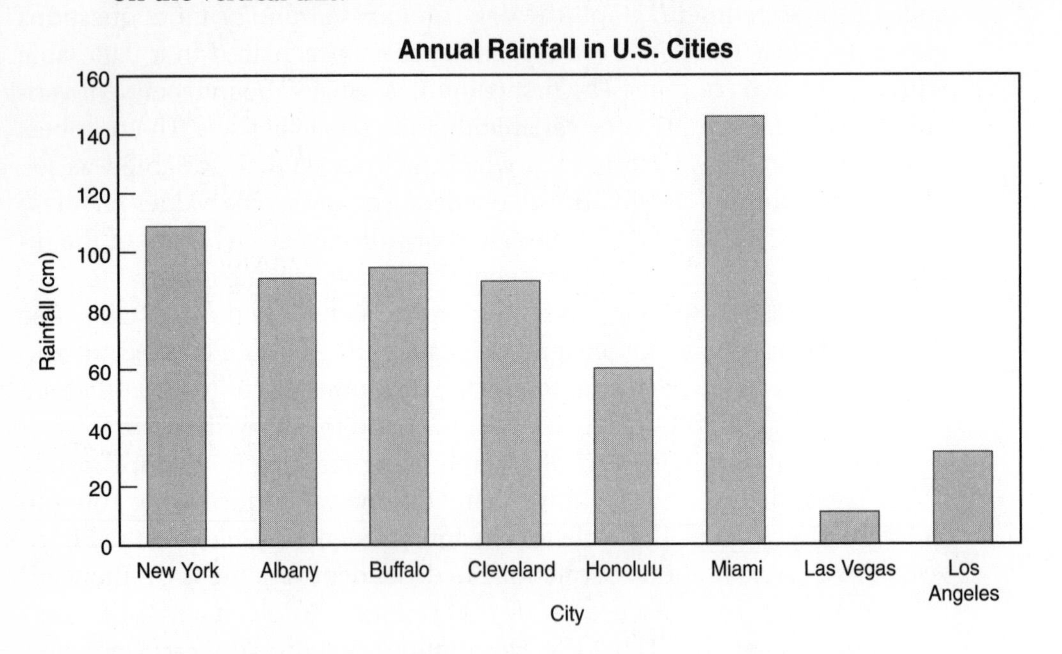

C. Final Test Preparation

During the review period before you take the test, you should set aside time to take a complete practice exam. Pretend that you are taking the real exam and do all the questions on the test. Record your answers according to the directions given in the practice test. Once you have taken the complete test, mark your answers. Note what topics were particularly difficult for you and return to the review book to study those topics again. Then take one of the other practice tests provided in this review book. Finally, don't try to cram in all your studying on the night before you take the test. It is important that you be well rested when you take the Grade 8 Intermediate-Level Science Test.

Remember: Hard work increases your abilities and will improve your test results.

Nonscience Words Students Should Understand

abnormal
abrupt
abundant
accomplish
acquire
adjustment
administer
advantage
alter
alteration
application
apply
appropriate
arrangement
associated
beneficial
benefit
boundary
camouflage
capable
capacity
captivity
characteristic
communicate
composed (of)
concerning
contaminate
contamination
continuously
contribute(d)
conversion
coordinated
current (adj.)
decreased
defective
delivery
deplete
deprive
desirable
develop
differentiate
direct
disadvantage
disorder
disrupt
distribution
duration
effectiveness
eliminate
emit
engulf
entirely
ethical
expel
expenditure
exposure
formation
function
grant (v.)
identical
impact (v.)
independently
influenced
intensity
interaction
maintenance
malfunction
maturation
migrate
minimize
modification
mount (v.)
occurrence
optimum
outbreak
overview
possess
potential
precaution
predominant(ly)
presence
prior
procedure
production
promote
propose
provide
quantity
rate
rebound
reduced
regulate/regulation
rely
remain
replenish
revise
short-term
significant
situation
specialization
specifically
stable (adj.)
stimulate
structure
summarize
synthesize
thrive
treat (v.)
trigger (v.)
unaware
unintended
unstable
valid
variation
variety
well-accepted

Diagnostic Test

Part I

1. Which statement correctly describes the relationship between minerals and rocks?

   (1) Rocks are composed of minerals.
   (2) Minerals are composed of rocks.
   (3) A rock wears away to form a mineral.
   (4) A mineral wears away to form a rock.

2. The streak color of a mineral is determined by scratching a mineral on the unglazed side of a ceramic tile. The streak color is

   (1) the same as the color of a mineral
   (2) a chemical property of a mineral
   (3) a physical property of a mineral
   (4) related to a mineral's luster

3. The geologic cross section below shows rock layers and a type of fossil found in each rock layer. The rock layers have not been displaced.

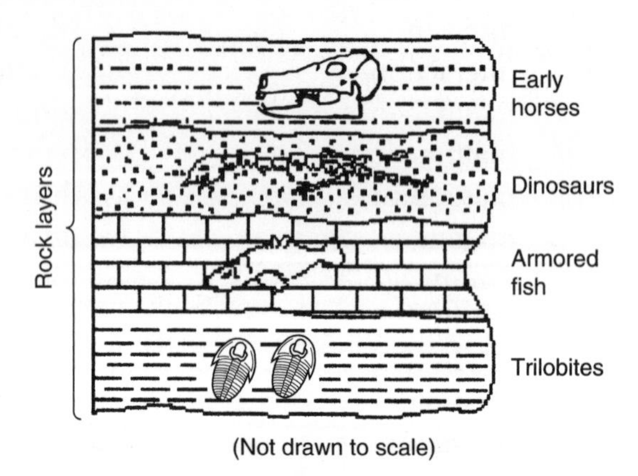

(Not drawn to scale)

   Which fossil is the oldest in the cross section shown?

   (1) early horses
   (2) dinosaurs
   (3) armored fish
   (4) trilobites

4. Fossils are most commonly found in

   (1) igneous rocks
   (2) sedimentary rocks
   (3) metamorphic rocks
   (4) volcanic rocks

5. Rocks, minerals, and mountains are part of Earth's

(1) atmosphere
(2) hydrosphere
(3) lithosphere
(4) hemisphere

6. Earthquakes are most commonly associated with

(1) ocean currents
(2) faults
(3) magma and lava
(4) weathering and erosion

7. Which type of wave provides information about Earth's internal structure?

(1) ocean waves
(2) earthquake waves
(3) light waves
(4) sound waves

8. On a hot sunny day a puddle on the sidewalk disappears within an hour after it rains. The puddle disappears because water molecules

(1) gain heat energy and evaporate
(2) gain heat energy and condense
(3) lose heat energy and evaporate
(4) lose heat energy and condense

9. North America has the second highest amount of carbon dioxide emissions (25%). Which sector of the pie chart represents North America?

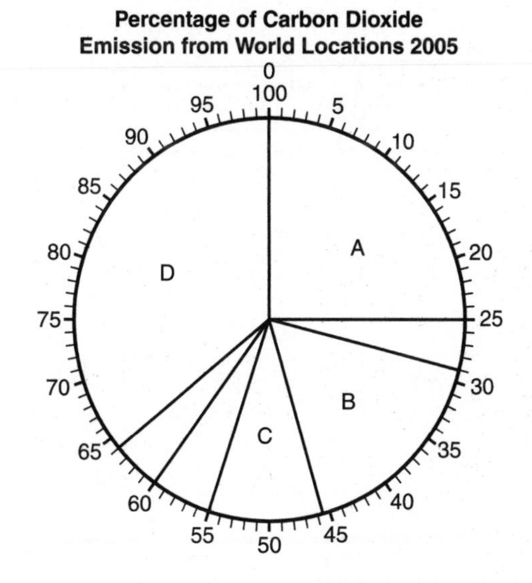

(1) Sector A
(2) Sector B
(3) Sector C
(4) Sector D

10. The map below shows an air mass that formed over Mexico and moved from location A to location B.

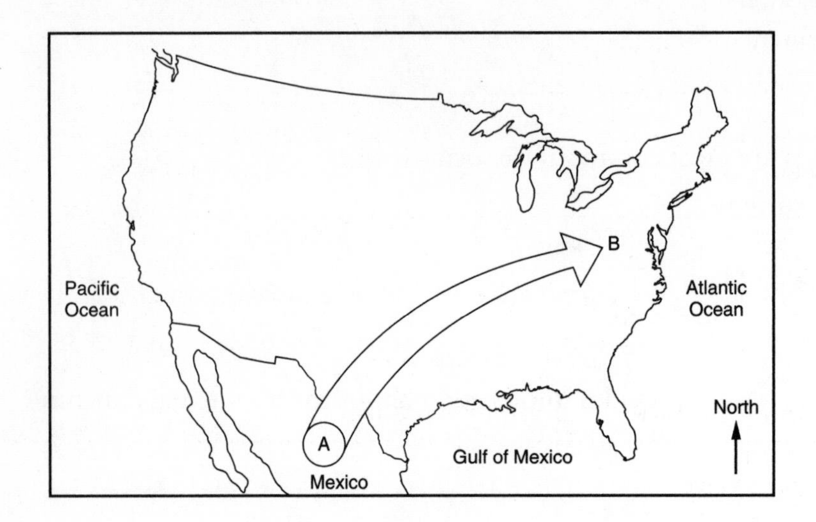

How did the air mass affect the weather conditions at location B?

(1) warmer temperature and drier air
(2) warmer temperature and more humid air
(3) colder temperature and drier air
(4) colder temperature and more humid air

11. How long does it take Earth to complete one revolution around the sun?

(1) one day
(2) one week
(3) one month
(4) one year

12. Which is a *chemical* change?

(1) iron rusts
(2) an eggshell cracks
(3) a car is dented
(4) ice melts

13. Which statement best describes the process called evaporation?

(1) Evaporation is a phase change that occurs fastest at low temperatures.
(2) Evaporation is a phase change that occurs fastest at high temperatures.
(3) Evaporation is a change in composition that occurs fastest at low temperatures.
(4) Evaporation is a change in composition that occurs fastest at high temperatures.

Questions 14 and 15 are based on the diagram below, which represents a water molecule.

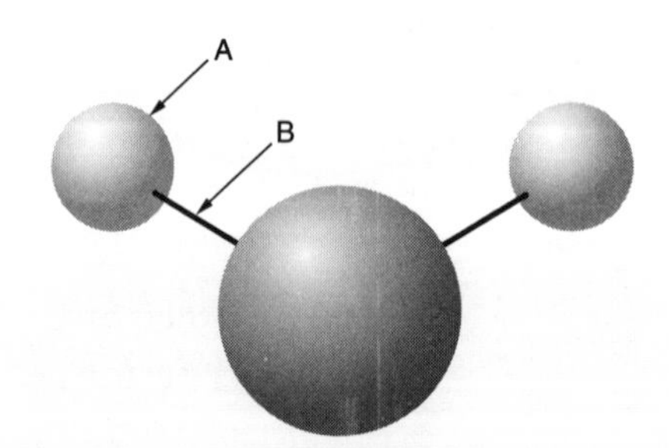

14. Arrow A is pointing to a representation of

(1) a hydrogen atom
(2) a nitrogen atom
(3) an oxygen atom
(4) a chemical bond

15. Arrow B is pointing to a representation of

(1) a hydrogen atom
(2) a nitrogen atom
(3) an oxygen atom
(4) a chemical bond

---

16. The data below shows the temperature of a sample of warm water that has been placed in a freezer.

Time (in minutes)	Temperature (in °C)
0	70
10	65
20	60
30	55
40	50

At the same rate of cooling, how much *more* time would be needed for the water to reach a temperature of 30°C?

(1) 10 minutes
(2) 20 minutes
(3) 30 minutes
(4) 40 minutes

Base your answers to questions 17 and 18 on the table below and on your knowledge of science. The table shows the results of four machines that convert electrical energy into mechanical energy.

Machine	Units of Energy Input	Units of Energy Output
A	110	100
B	125	100
C	135	100
D	150	100

17. Which machine was most efficient in its use of energy?

(1) machine A
(2) machine B
(3) machine C
(4) machine D

18. Which machine most likely produced the most undesired heat energy?

(1) machine A
(2) machine B
(3) machine C
(4) machine D

19 The diagram below shows two forces applied to a cart in opposite directions. If the forces are applied to the cart at the same time, what will happen?

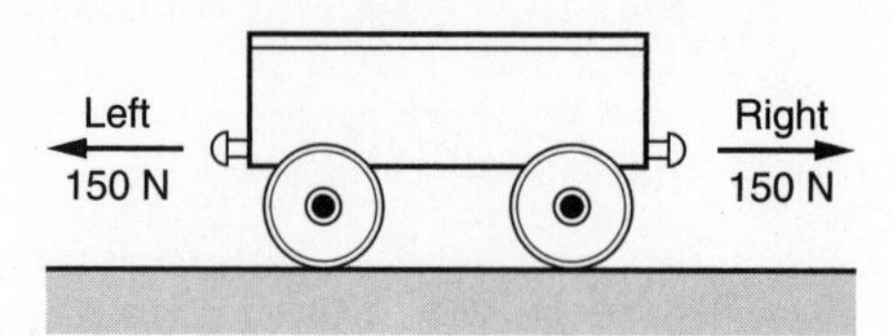

(1) The cart will move to the right.
(2) The cart will move to the left.
(3) The cart will not move.
(4) The cart will be lifted off the ground.

20. The diagram below illustrates what property of light?

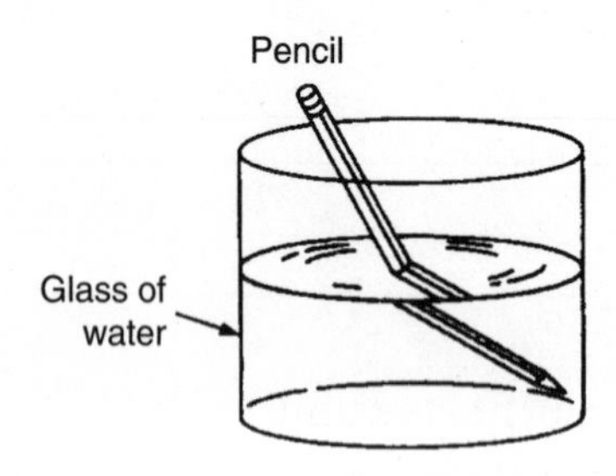

(1) absorption of light waves
(2) refraction of light waves
(3) reflection of light waves
(4) production of light waves

21. The illustration below shows four positions on the path of a person riding a sled in the snow. At which position does the person have the greatest potential energy?

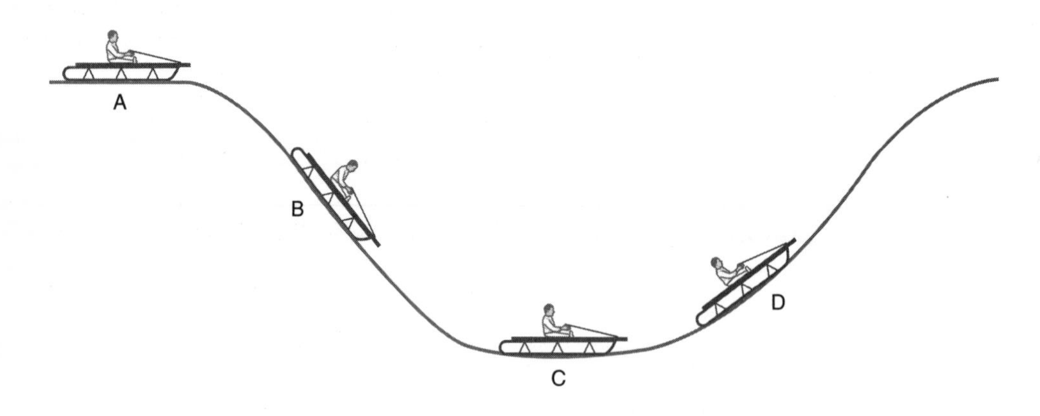

(1) position A
(2) position B
(3) position C
(4) position D

22. Which process occurs in *all* of the cells shown below?

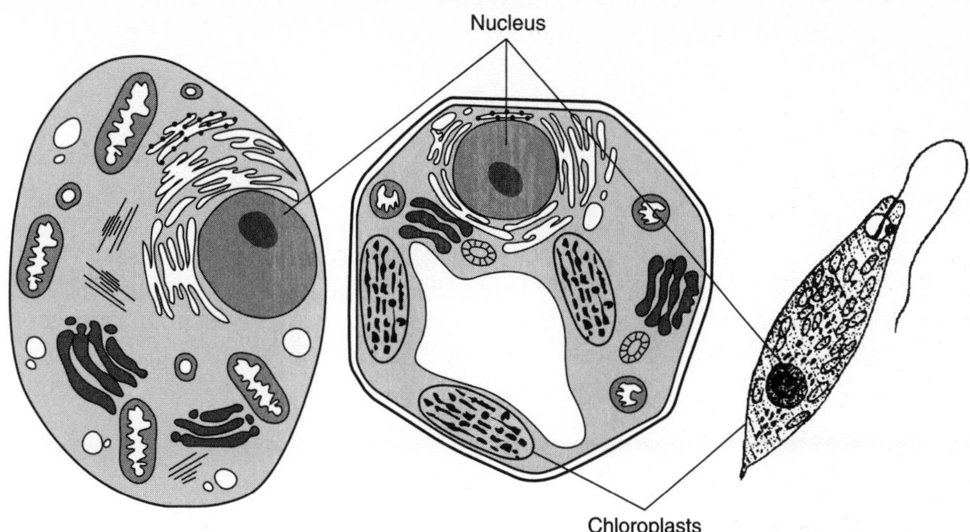

(1) photosynthesis
(2) respiration
(3) locomotion
(4) meiosis

23. The thick fur and white color of the polar bear are examples of

(1) responses to stimuli
(2) adaptations for survival in a cold climate
(3) learned behavior
(4) instinctive behavior

24. Offspring that have characteristics different from those of the parents are produced through

(1) binary fission
(2) asexual reproduction
(3) sexual reproduction
(4) vegetative propagation

25. Choose the two terms that correctly complete the following statement: A group of similar cells working together is called a(n) ________, while a group of organs working together is called a(n) ________.

(1) organ, tissue
(2) organ system, tissue
(3) tissue, organ system
(4) tissue, chromosome

26. Human body cells contain 46 chromosomes. How many chromosomes are contained in a human sperm cell?

(1) 23 (2) 46 (3) 92 (4) 184

27. Butterflies and mosquitoes belong to the same class of living things—*insecta*. They must also be classified in the same

(1) order
(2) genus
(3) kingdom
(4) family

28. Tulip plants produce enlarged underground stems called bulbs, which store food for the plants. These bulbs can be removed and planted separately, to produce new tulip plants. This process is most correctly described as

(1) asexual reproduction
(2) sexual reproduction
(3) grafting
(4) evolution

29. The owners of female race horses sometimes pay thousands of dollars to other owners so that they can mate their horses with the fastest, most successful male horses. This procedure is an example of

(1) selective breeding
(2) natural selection
(3) survival of the fittest
(4) mutation

30. Four different reproductive processes are illustrated below:

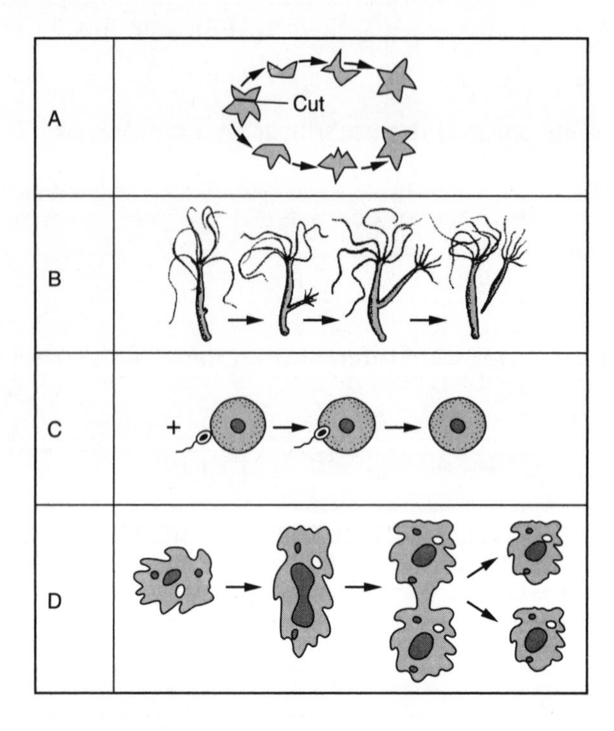

Which of these is an example of *sexual* reproduction?

(1) A (2) B (3) C (4) D

31. Sexual reproduction in plants may take place after a

(1) flower is pollinated
(2) long stem bends to the ground, grows roots, and forms a new plant
(3) new plant is grafted onto the stem of an old plant
(4) plant is cut down, and its stem grows new roots in a new location

Base your answers to questions 32 and 33 on the diagram below and your knowledge of science. The diagram shows the life cycle of the frog.

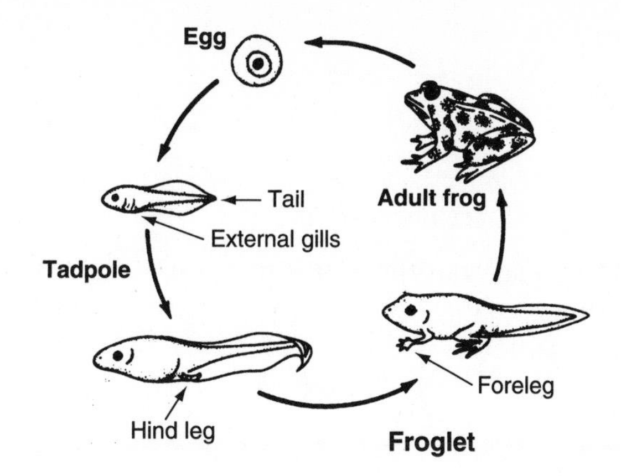

32. This diagram illustrates a process called

(1) evolution
(2) metamorphosis
(3) adaptation
(4) asexual reproduction

33. A frog is an amphibian. The word "amphibian" comes from a Greek word meaning "both lives." The word describes the life of the frog because the frog lives

(1) part of its life as a male, and part of its life as a female
(2) part of its life aboveground, and part of its life belowground
(3) part of its life in the air, and part of its life on the ground
(4) part of its life in water, and part of its life on land

---

34. A calorie is a unit that is used to measure

(1) mass
(2) weight
(3) temperature
(4) energy

35. Complete the analogy with the best term from the four choices below:

Nucleus is to cell as ____ is to human being.

(1) muscle
(2) blood
(3) lung
(4) brain

36. Evolution causes animal species to change over very long periods of time. Scientists often study changes in bacteria, which occur much more rapidly. One reason that change can occur comparatively quickly in bacteria is that bacteria

(1) can cause disease
(2) reproduce very rapidly
(3) grow very slowly
(4) contain DNA

37. Competition for food is most likely to occur between

(1) butterflies and owls
(2) apple trees and bees
(3) hawks and snakes
(4) lions and zebras

38. The energy that every cell needs to carry out its life functions is produced through

(1) photosynthesis
(2) cellular respiration
(3) decomposition
(4) excretion

39. A maple tree is called a producer because it

(1) produces red-colored leaves in the autumn
(2) produces its own food
(3) produces carbon dioxide
(4) produces water

40. Organisms that absorb food from the remains of dead organisms are called

(1) producers
(2) herbivores
(3) decomposers
(4) omnivores

41. The exchange of gases between the air and the blood takes place in the

(1) nose
(2) trachea
(3) bronchi
(4) air sacs

42. One purpose of the respiratory system is to supply the cells with

(1) carbon dioxide
(2) nutrients
(3) oxygen
(4) water

43. Which behavior could be considered an instinct?

(1) A child eats breakfast.
(2) A crow learns how to open a garbage can to get food.
(3) A spider builds a web to catch insects.
(4) A blue jay returns to the same bird feeder every morning.

44. Hereditary information is contained in genes. Genes are composed of

(1) RNA
(2) DNA
(3) ADP
(4) ATP

45. Which of the following is most likely to cause the extinction of a species?

(1) evolution
(2) selective breeding
(3) migration
(4) environmental changes

Part II

Base your answers to questions 46 through 48 on the diagram below and on your knowledge of science. The diagram shows the rock cycle.

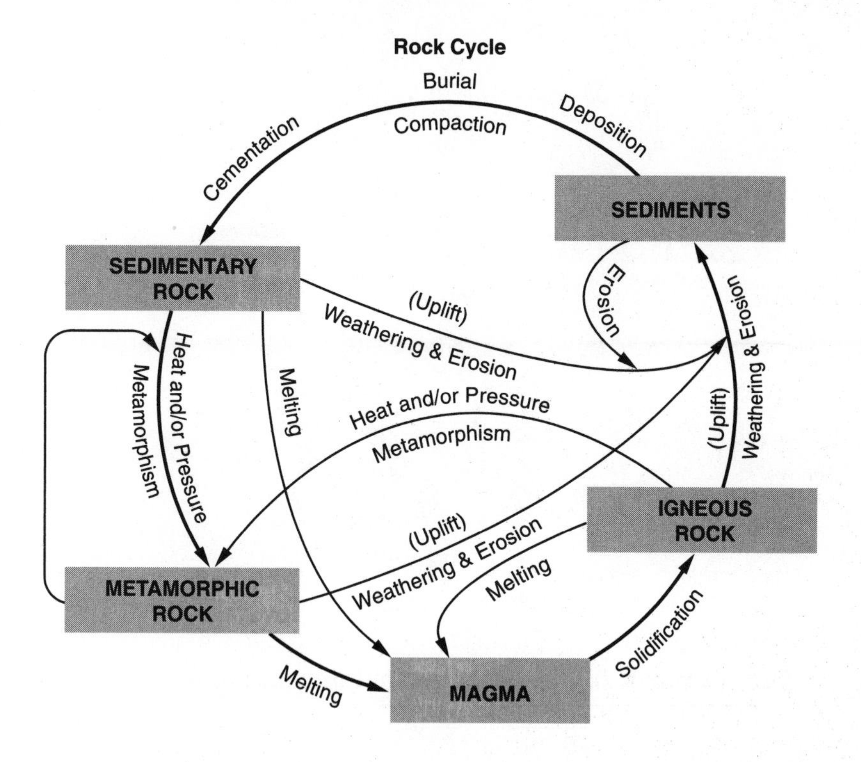

46. Identify two processes that can cause a sedimentary rock to become a metamorphic rock. [2]

47. What type of rock forms from particles that are deposited, compacted, and cemented? [1]

48. What term describes liquid rock that reaches Earth's surface? [1]

---

49. Base your answer to the question on the properties of gypsum and feldspar listed below and your knowledge of science.

    Gypsum—light colored, white streak, hardness 2, does not fizz or bubble in weak hydrochloric acid

    Feldspar—light colored, no streak, hardness 6, does not fizz or bubble in weak hydrochloric acid

    Given only a specimen of gypsum and feldspar, how can you determine which is gypsum and which is feldspar? [2]

Base your answers to questions 50 through 52 on the map below and your knowledge of science. The map shows the position of the continents 240 million years ago.

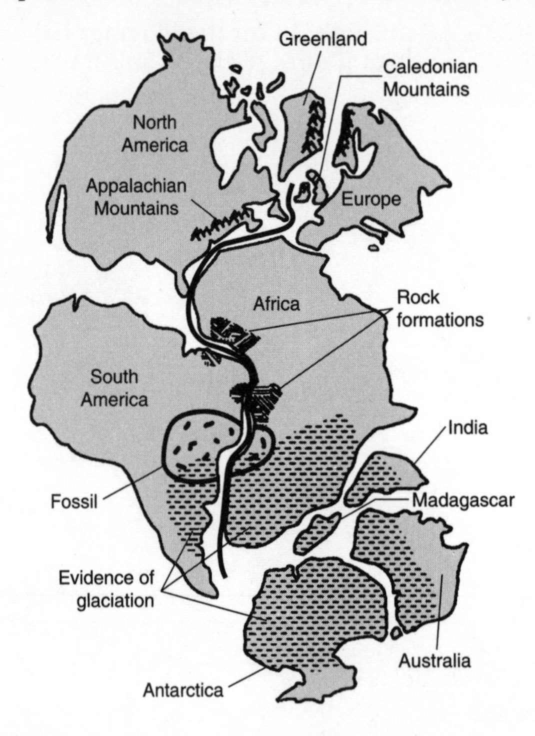

50. What ocean floor feature formed along the bold line on the map? [1]

51. What force is responsible for the separation of the continents? [1]

52. Give two forms of evidence supporting the idea that the continents were once connected. [2]

Base your answers to questions 53 through 55 on the map below and your knowledge of science. The map shows the size and path of a certain weather event.

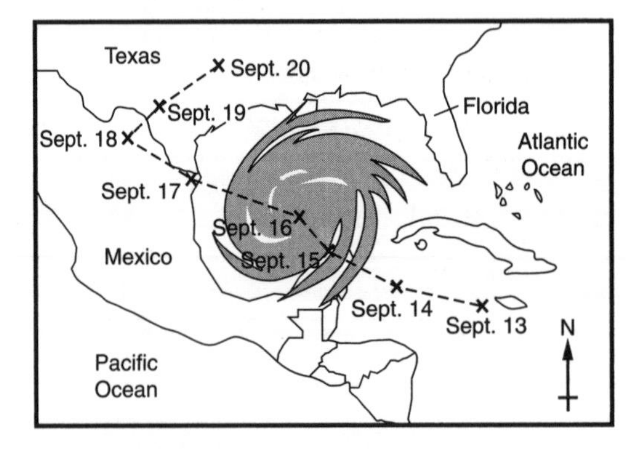

53. The path of what weather event is shown on the map? [1]

54. Between which two dates did the event move the greatest distance? [1]

55. Describe two characteristics of this weather event. [2]

Base your answers to questions 56 through 58 on the diagram below and your knowledge of science. The diagram shows four positions of the moon in its orbit around Earth.

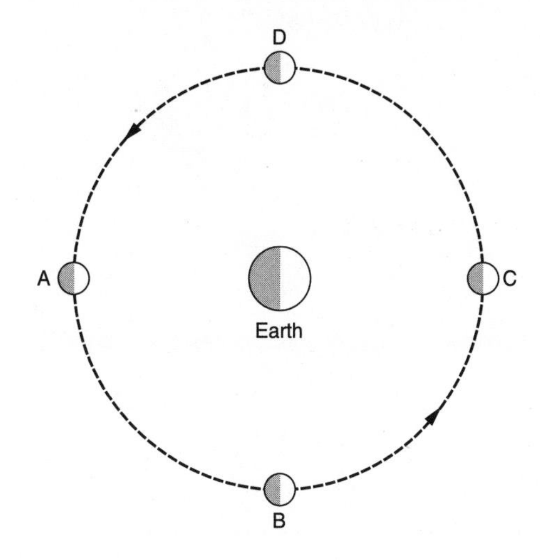

56. At what position would the moon be when a person on Earth would see a new moon? [1]

57. About how much time does it take for the moon to travel from position A back to position A? [1]

58. On your answer sheet draw how the moon would look from Earth at position D. [1]

Base your answers to questions 59 and 60 on the diagram below and on your knowledge of science. The diagram shows a glass containing water and ice.

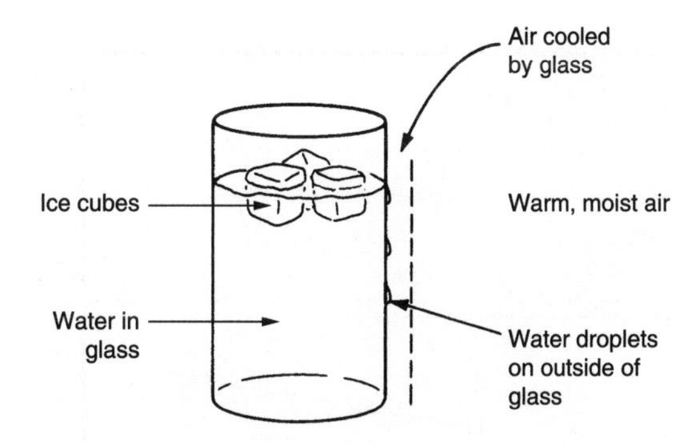

59. Describe the heat exchange between the water and the ice cubes. [1]

60. Explain why water droplets formed on the outside of the glass. [1]

Base your answers to questions 61 through 63 on the diagram below and on your knowledge of science. The diagram shows an activity done by a teacher in the science laboratory.

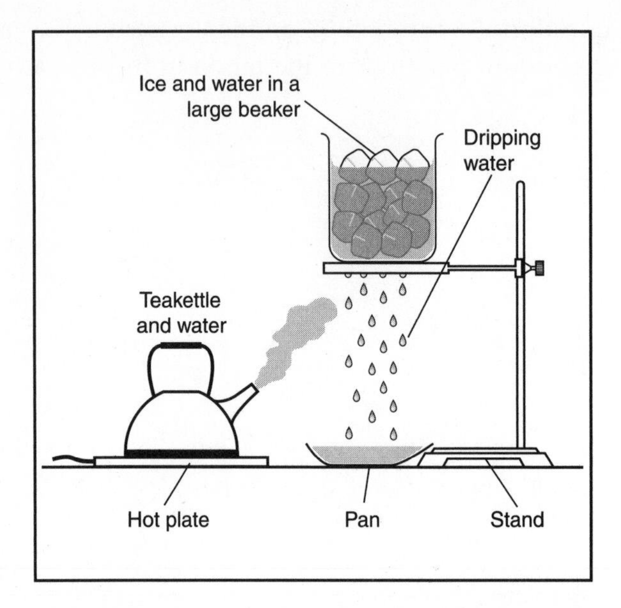

61. Describe and name the phase change occurring in the teakettle. [2]

62. Water droplets are forming on the bottom of the beaker. What process is causing the water vapor to change into water droplets? [1]

63. What heat-transfer process is occurring between the hot plate and the bottom of the teakettle? [1]

Base your answers to questions 64 through 66 on the graph below and your knowledge of science. The graph shows the amounts of three solid solutes that will dissolve in 100 grams of water at temperatures between 0 and 100°C

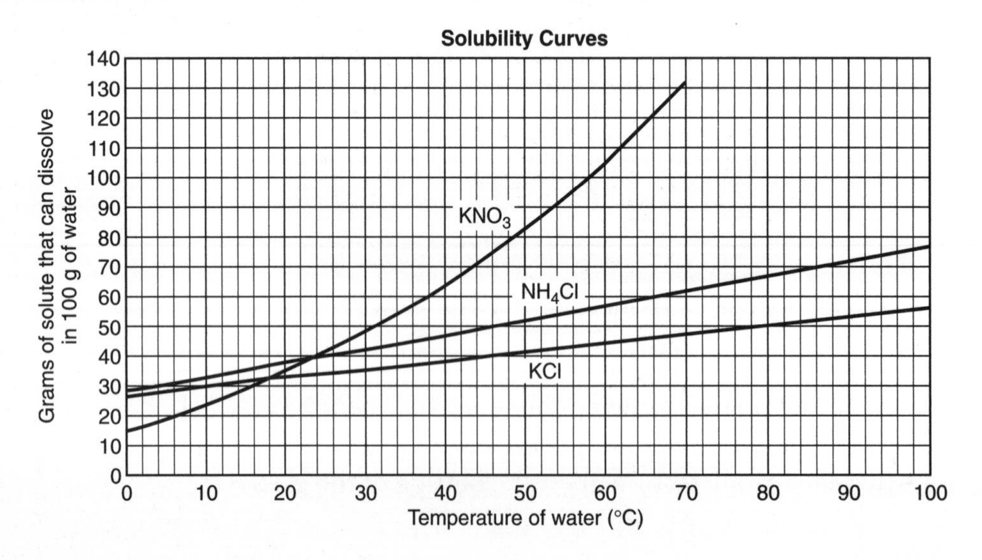

64. Based on the graph, what effect does the temperature of the water have on the amount of solute that can dissolve? [1]

65. Which of the three solids shows the greatest *change* in solubility between 0 and 15°C? [1]

66. How much KNO_3 can dissolve in 100 grams of water at 44°C? [1]

Base your answers to questions 67 through 69 on the information below and your knowledge of science.

A student wishes to separate a mixture of sand, sugar, and iron filings. He first places the mixture in a beaker, adds water, and stirs thoroughly. He then filters his mixture by pouring it through filter paper.

67. Which materials remain on the filter paper? [1]

68. How could he separate the substances that remain on the filter paper? [2]

69. What *two* substances passed through the filter paper? [1]

To answer questions 70 and 71, use the Punnett square below that represents the result of a cross between two fruit flies, one with red eyes and one with white eyes.

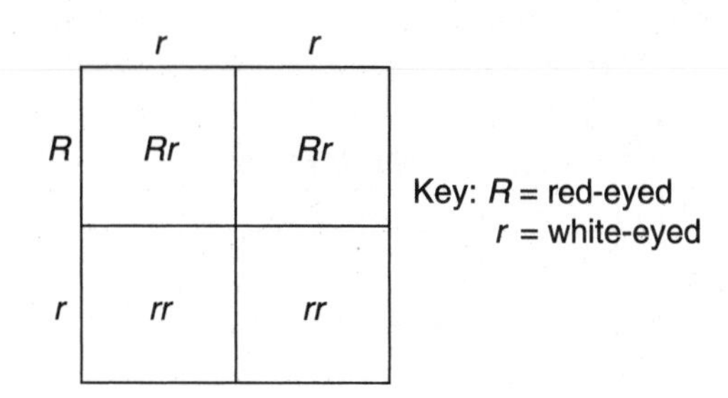

70. Which eye color is the dominant trait? Explain your answer. [2]

71. If the two red-eyed offspring were crossed, what percentage of the next generation is likely to be red-eyed? Using a Punnett square, show how you arrived at your answer. [2]

---

72. Scientists often collect data on the number of birds of various species found in a specific area. Every June, scientists count the number of house finches, downy woodpeckers, and mockingbirds in a woody area of Long Island. During their study, some of the trees were removed to make way for new homes.

Year	Number of House Finches	Number of Downy Woodpeckers	Number of Mockingbirds	Number of Homes
1998	4	8	12	0
1999	10	8	12	4
2000	16	6	10	12
2001	22	5	12	20
2002	28	4	11	25
2003	30	2	11	32

(a) Which bird population was affected *least* by the use of the land for home building? [1]

(b) Explain why the number of downy woodpeckers is declining. [2]

(c) Why might the number of house finches be increasing? [2]

Use the following information to answer questions 73 and 74. A balanced diet provides protein, carbohydrates, vitamins, and minerals.

73. Marathon runners have to run just over 26 miles. Before running a marathon, these athletes sometimes load up on one of these nutrients. Which one? Explain your answer. [2]

74. Which of these nutrients is mainly responsible for growth and repair in the body? [1]

Diagnostic Checklist

Use this checklist to evaluate your Diagnostic Test answers. The checklist is designed so that you can determine which skills you need to improve in your preparation for the Intermediate Level Science test. Answer columns are provided so that you can compare your answers with those given by your teacher. If you miss an answer, check the box next to the number of that answer. Once you have checked all your answers, note which sections you need to review by referring to the sections listed in the chapter column. Page numbers in the last column provide the location of those sections in this book.

Answers		Check if Missed	MST Standard, NY Performance Indicator (Area)	Chapter	Pages
Yours	Key				
Part I					
1.		☐	4-2.1e (Physical Setting)	9	244
2.		☐	4-2.1e (Physical Setting)	9	244
3.		☐	4-2.2d (Physical Setting)	9	257–259
4.		☐	4-2.1f (Physical Setting)	9	258–259
5.		☐	4-2.1c (Physical Setting)	9	244
6.		☐	4-2.2a,f (Physical Setting)	10	280–281
7.		☐	4-2.2b (Physical Setting)	10	281, 283
8.		☐	4-4.2c (Physical Setting)	1	35
9.		☐	4-7.2d (Living Environment), 4-2.2r (Physical Setting)	8 13	 325, 376
10.		☐	4-2.2p (Physical Setting)	11	309–310
11.		☐	4-1.1e (Physical Setting)	12	333
12.		☐	4-3.2c (Physical Setting)	1	52–54
13.		☐	4-4.2c (Physical Setting)	1	35
14.		☐	4-3.3c (Physical Setting)	1	32
15.		☐	4-3.3d (Physical Setting)	1	32
16.		☐	1-M 2.1 (Mathematics)	1	36
17.		☐	4-5.2c (Physical Setting)	4	146–147
18.		☐	4-4.5b (Physical Setting)	4	146–147
19.		☐	4-5.1c (Physical Setting)	4	125
20.		☐	4-4.4b (Physical Setting)	2	94–95
21.		☐	4-4.1e (Physical Setting)	2	62–63
22.		☐	4-1.1b (Living Environment)	5	153–155
23.		☐	4-5.1b (Living Environment)	8	222–223
24.		☐	4-2.1e (Living Environment)	6	180–183
25.		☐	4-1.1e (Living Environment)	7	197–198
26.		☐	4-4.2b (Living Environment)	6	180
27.		☐	4-1.1h (Living Environment)	6	177
28.		☐	4-2.1d (Living Environment)	6	180
29.		☐	4-3.1c (Living Environment)	6	192
30.		☐	4-4.1a (Living Environment)	6	180–181
31.		☐	4-4.1c (Living Environment)	6	183–184
32.		☐	4-4.3d (Living Environment)	5	158–159
33.		☐	4-4.3c (Living Environment)	5	159
34.		☐	4-5.2d (Living Environment)	3	102
35.		☐	4-1.1b (Living Environment)	5	153

Answers		Check if Missed	MST Standard, NY Performance Indicator (Area)	Chapter	Pages
Yours	Key				
36.		☐	4-3.2d (Living Environment)	6	192
37.		☐	4-3.2a (Living Environment)	8	233–234
38.		☐	4-5.2a (Living Environment)	5	157–158
39.		☐	4-6.1b (Living Environment)	8	229
40.		☐	4-6.1c (Living Environment)	8	231
41.		☐	4-1.1g (Living Environment)	7	212–213
42.		☐	4-1.2d (Living Environment)	7	212–213
43.		☐	4-1.2h (Living Environment)	7	203–204
44.		☐	4-2.1a (Living Environment)	6	186–187
45.		☐	4-5.1g (Living Environment)	8	224–225
Part II					
46.		☐	4-2.2g (Physical Setting)	9	252–253
47.		☐	4-2.2h (Physical Setting)	9	247
48.		☐	4-2.2a (Physical Setting)	9	246
49.		☐	4-2.1e (Physical Setting)	9	244–246
50.		☐	4-2.2e (Physical Setting)	10	286
51.		☐	4-2.2a (Physical Setting)	10	285
52.		☐	4-2.2d (Physical Setting)	10	283
53.		☐	4-2.2q (Physical Setting)	11	321
54.		☐	1-M 3.1 (Mathematics)	11	322–323
55.		☐	4-2.2q (Physical Setting)	11	321–322
56.		☐	4-1.1g (Physical Setting)	12	346–347
57.		☐	4-1.1g (Physical Setting)	12	346
58.		☐	6-5.2 (Patterns of Change)	12	347
59.		☐	4-4.2a (Physical Setting)	1	35
60.		☐	4-4.2b (Physical Setting)	1	35
61.		☐	4-4.2c (Physical Setting)	1	36
62.		☐	4-4.2c (Physical Setting)	1	35
63.		☐	4-4.2b (Physical Setting)	2	71–73
64.		☐	1-S3.2h (Scientific Inquiry)	1	41
65.		☐	4-3.1b (Physical Setting)	1	41–42
66.		☐	1-M 1.1c (Mathematics)	1	41
67.		☐	1-S3.2f (Scientific Inquiry)	1	52
68.		☐	4-3.2b (Physical Setting)	1	52
69.		☐	4-3.1g (Physical Setting)	1	52
70.		☐	4-2.2b (Living Environment)	6	188–189
71.		☐	4-2.2c (Living Environment)	6	188–189
72.		☐	4-7.1b (Living Environment)	8	222–223
73.		☐	4-5.1d (Living Environment)	5	155
74.		☐	4-5.2b (Living Environment)	5	155

Chapter 1 Matter

All matter is made of atoms and molecules.

Vocabulary at a Glance

atom
boiling
boiling point
buoyancy
buoyant
chemical bond
chemical change
compound
concentrated
concentration
condensation
corrosion
density
dilute
element
erosion
evaporation
freezing
freezing point
insoluble
Law of Conservation of Matter
mass
matter
melting
melting point
metal
molecule
noble gas
nonmetal
nucleus
physical change
physical property
solubility
soluble
solute
solution
solvent
volume

Major Concepts

- Matter is anything that has mass and occupies space. Matter can exist as a solid, liquid, or gas.
- Elements are substances that cannot be broken down into simpler substances. Two or more elements can combine to form a compound.
- Atoms are composed of protons, neutrons, and electrons.
- Changes in matter may be chemical or physical. During a chemical change, new substances are formed.
- Compounds are groups of two or more elements that are bonded together. Compounds cannot be separated by physcial means.
- During a physical change, the identity of substances is unchanged.
- Mixtures do not form new substances. Solutions are mixtures.
- $\text{Density} = \frac{\text{mass}}{\text{volume}}$.
- Floating and sinking depend on differences in density compared with water.
- Matter cannot be created or destroyed in a chemical reaction. Energy is absorbed or released during a chemical change. The rates of chemical and physical changes are influenced by such factors as temperature and particle size.

Defining Matter

Look around you. The objects you see, such as this book, your desk and chair, and the walls and ceiling, are all made of matter. The air (a mixture of gases) that surrounds you is also made of matter. In fact, every solid, liquid, and gas is a form of matter.

Matter is defined as anything that has mass and takes up space. ***Mass*** is the total amount of material in an object. We measure mass with a triple-beam balance or an equal-arm balance, as shown in Figure 1-1. Notice that a balloon filled with air has a greater mass than an empty balloon, because air has mass. The amount of space an object occupies is called its ***volume.*** The air in the filled balloon in Figure 1-1 takes up space, giving that balloon a greater volume than the empty balloon.

Is there anything that is not made of matter? Is there anything that has no mass and takes up no space? Figure 1-2 shows that shining a light on a balance has no effect on the balance. This is because light is a form of energy. Energy is not matter, since it has no mass and no volume. Some other forms of energy are heat and sound.

Elements

The basic building blocks of matter are called ***elements***. All substances are made up of one or more elements. Oxygen, hydrogen, gold, and iron are examples of elements. Each element is represented by a symbol made up of one or two letters. For example, the symbol for hydrogen is H, oxygen is O, and gold is Au. The first letter of the symbol is always capitalized while the second letter, if any, is always lowercase. At the time this book was written, 118 elements had been discovered. By the time you read it there may be more. However, less than half of all elements occur commonly in nature. (Note: The number of elements changes as new elements are created in research laboratories.) Table 1-1 lists the most common elements found in Earth's crust.

The smallest particle of an element that has the properties of that element is called an ***atom***. All atoms of a particular element are alike, but they are different from the atoms of any other element. For instance, all hydrogen atoms are alike, but they differ from oxygen atoms. If there are 118 different elements, there are 118 different kinds of atoms.

Figure 1-1. *The air-filled balloon is heavier and takes up more space than the empty balloon because air is matter.*

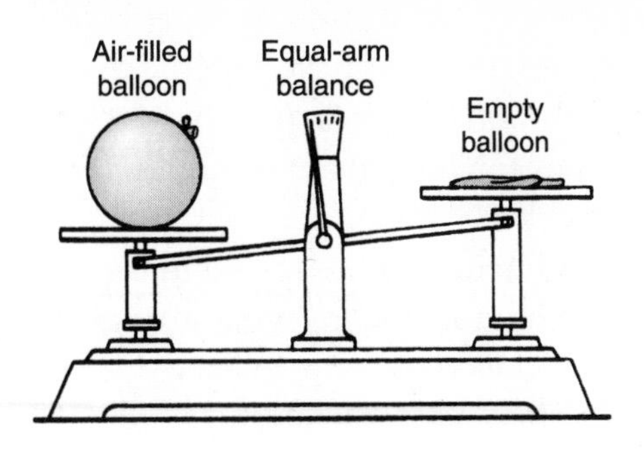

Figure 1-2. *The balance is not affected by the light shining on it because light is not matter.*

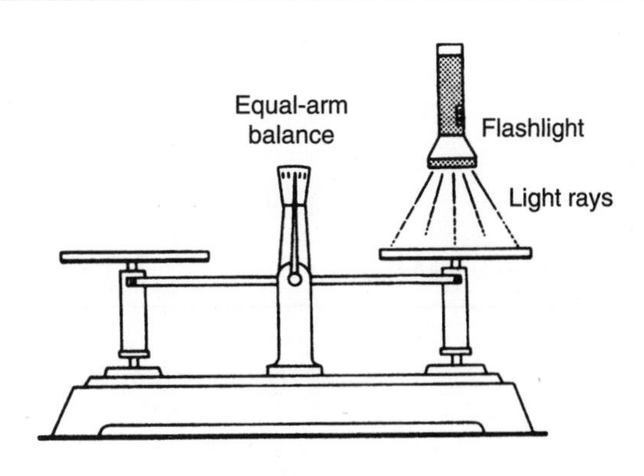

Table 1-1. *Most Common Elements in Earth's Crust*

Element	**Chemical Symbol**
Oxygen	O
Silicon	Si
Aluminum	Al
Iron	Fe
Calcium	Ca
Sodium	Na
Potassium	K
Magnesium	Mg

Atomic Structure

All atoms are composed of smaller *subatomic* particles called *protons, neutrons,* and *electrons.* They differ in their mass, electrical charge, and location in the atom. Protons and neutrons have roughly the same mass, while electrons are much lighter. Protons have a positive (+) charge, and electrons have a negative (–) charge. Neutrons have no electrical charge: they are electrically neutral.

Protons and neutrons are found in the center, or ***nucleus***, of the atom. Electrons orbit the nucleus, moving around the nucleus very rapidly. The negatively charged electrons are attracted to the positively charged protons in the nucleus because oppositely charged particles attract each other. Since like charges repel and all electrons are negatively charged, electrons repel other electrons. Table 1-2 summarizes the properties of the subatomic particles.

The atoms of different elements have a different number of protons in their nuclei. Oxygen has 8 protons, carbon has 6, and uranium has 92. The number of protons in the nucleus is called the *atomic number. The Periodic Table of the Elements* arranges the elements according to their atomic number.

Table 1-2. *Properties of the Subatomic Particles*

Particle	Mass (AMU)*	Charge	Location
Proton	1	+	Nucleus
Neutron	1	0	Nucleus
Electron	0.00054 (often rounded to 0)	–	Outside the nucleus

*The atomic mass unit (AMU) is a special unit created for measuring the mass of very small particles.

The Periodic Table of the Elements

Scientists organize the elements based on their properties on *The Periodic Table of the Elements* (see Figure 1-3). In this table, elements with similar properties are placed in the same vertical column, called a group. These groups are numbered from 1 through 18. The Periodic Table is a useful model for classifying elements. It is also used to predict properties of elements (metals, nonmetals, and noble gases).

The majority of elements are shiny solids that conduct electricity. These elements are called ***metals***. (Mercury, which is a liquid at room temperature, is also considered a metal.) A smaller number of elements are poor conductors of electricity and lack the shine of the metals. These are called ***nonmetals***. A still smaller group of elements, which are all gases at room temperature, seldom react with other elements. These are called the ***noble gases***.

Figure 1-3. *The Periodic Table of the Elements.*

Key: 6 — Atomic number; C — Symbol; 12.01 — Atomic mass

1	2	3	4	5	6	7	8	9	10	11	12	13	14	15	16	17	18
1 **H** 1.008																	2 **He** 4.003
3 **Li** 6.941	4 **Be** 9.012											5 **B** 10.81	6 **C** 12.01	7 **N** 14.01	8 **O** 16.00	9 **F** 19.00	10 **Ne** 20.18
11 **Na** 22.99	12 **Mg** 24.31											13 **Al** 26.98	14 **Si** 28.09	15 **P** 30.97	16 **S** 32.07	17 **Cl** 35.45	18 **Ar** 39.95
19 **K** 39.10	20 **Ca** 40.08	21 **Sc** 44.96	22 **Ti** 47.88	23 **V** 50.94	24 **Cr** 52.00	25 **Mn** 54.94	26 **Fe** 55.85	27 **Co** 58.93	28 **Ni** 58.69	29 **Cu** 63.55	30 **Zn** 65.39	31 **Ga** 69.72	32 **Ge** 72.61	33 **As** 74.92	34 **Se** 78.96	35 **Br** 79.90	36 **Kr** 83.80
37 **Rb** 85.47	38 **Sr** 87.62	39 **Y** 88.91	40 **Zr** 91.22	41 **Nb** 92.91	42 **Mo** 95.94	43 **Tc** (98)	44 **Ru** 101.1	45 **Rh** 102.9	46 **Pd** 106.4	47 **Ag** 107.9	48 **Cd** 112.4	49 **In** 114.8	50 **Sn** 118.7	51 **Sb** 121.8	52 **Te** 127.6	53 **I** 126.9	54 **Xe** 131.3
55 **Cs** 132.9	56 **Ba** 137.3	57 **La** 138.9	72 **Hf** 178.5	73 **Ta** 181.0	74 **W** 183.8	75 **Re** 186.2	76 **Os** 190.2	77 **Ir** 192.2	78 **Pt** 195.1	79 **Au** 197.0	80 **Hg** 200.6	81 **Tl** 204.4	82 **Pb** 207.2	83 **Bi** 209.0	84 **Po** (209)	85 **At** (210)	86 **Rn** (222)
87 **Fr** (223)	88 **Ra** 226.0	89 **Ac** 227.0	104 **Rf** (261)	105 **Db** (262)	106 **Sg** (263)	107 **Bh** (262)	108 **Hs** (265)	109 **Mt** (268)	110 **Ds** (281)	111 **Rg** (272)	112 **Uub** (285)	113 **Uut** (284)	114 **Uuq** (289)	115 **Uup** (288)	116 **Uuh** (292)	117	118 **Uuo** (294)

58 **Ce** 140.1	59 **Pr** 140.9	60 **Nd** 144.2	61 **Pm** (145)	62 **Sm** 150.4	63 **Eu** 152.0	64 **Gd** 157.3	65 **Tb** 158.9	66 **Dy** 162.5	67 **Ho** 164.9	68 **Er** 167.3	69 **Tm** 168.9	70 **Yb** 173.0	71 **Lu** 175.0
90 **Th** 232.0	91 **Pa** 231.0	92 **U** 238.0	93 **Np** 237.0	94 **Pu** (244)	95 **Am** (243)	96 **Cm** (247)	97 **Bk** (247)	98 **Cf** (251)	99 **Es** (252)	100 **Fm** (257)	101 **Md** (258)	102 **No** (259)	103 **Lr** (260)

The Periodic Table contains a zigzag line that separates the metals to the left from the nonmetals to the right. The last group of elements, Group 18, contains the noble gases.

Compounds

Scientists know there are millions of different substances. How is this possible if there are only about 118 elements? Elements combine to form new substances. A substance formed when two or more different elements combine chemically is called a ***compound***. Since many different combinations of elements are possible, many different compounds can exist. Water is the compound formed when the elements hydrogen and oxygen combine.

A compound is represented by a chemical formula that indicates which elements have combined, and in what proportions. The chemical formula for water, H_2O, indicates that water contains two atoms of hydrogen to every atom of oxygen. Table 1-3 lists some common compounds and their chemical formulas.

Table 1-3. *Some Common Compounds and Their Chemical Formulas*

Compound	Formula	Elements
Table salt	$NaCl$	Sodium, chlorine
Water	H_2O	Hydrogen, oxygen
Sugar (sucrose)	$C_{12}H_{22}O_{11}$	Carbon, hydrogen, oxygen
Quartz	SiO_2	Silicon, oxygen
Ammonia	NH_3	Nitrogen, hydrogen

The smallest particle of a compound is called a ***molecule***. A water molecule is composed of two hydrogen atoms and one oxygen atom, as shown in Figure 1-4. Atoms of the same element can also combine to form molecules. For example, two oxygen atoms combine to form a molecule of oxygen gas, O_2.

Figure 1-4. *The arrangement of atoms in a molecule of water.*

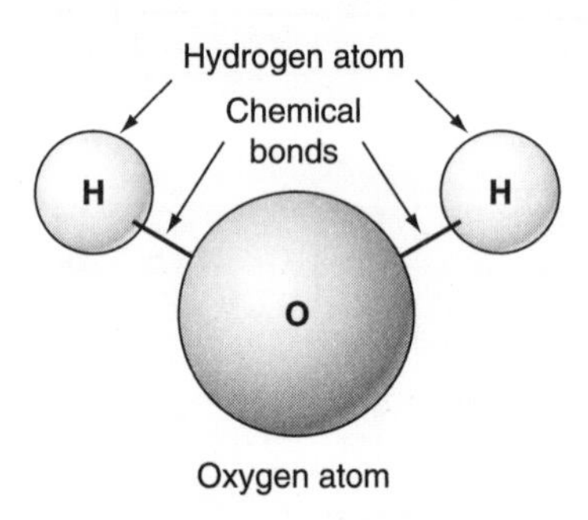

Atoms and molecules are extremely small. To get an idea of just how small, consider that 1 teaspoonful of water contains about 175 *sextillion* water molecules. (That would be written as 1.75×10^{23}, or 175 followed by 21 zeros!)

Chemical Bonds

Atoms in a molecule are joined by a special link called a ***chemical bond*** (see Figure 1-4). These bonds store chemical energy. Sometimes this energy can be released by a chemical reaction. Burning is one type of chemical reaction that releases energy. When wood is burned, energy stored in the chemical bonds within the wood is released as heat and light. Respiration (See Chapter 5) is another chemical reaction that releases energy from chemical bonds.

Mixtures

When two or more materials are put together and do not make a new substance, a *mixture* is formed. Salt water, for example, is a mixture of salt and water. Sand is a mixture of minerals. Blood is a mixture of different cells, water, and other nutrients. Air is a mixture of several gases.

Unlike compounds, mixtures cannot be represented with a chemical formula. Table salt—a compound—is always NaCl. However, salt water—a mixture—can be more or less salty and still be salt water. For example, salt (NaCl) in Utah is exactly the same as salt in New York, but the salt water in the Great Salt Lake in Utah is quite different from the salt water in the Atlantic Ocean off New York.

Physical Properties

Have you ever mistaken salt for sugar? They look very much alike. How might you tell them apart? Scientists faced with similar problems identify substances by examining their *properties.*

A difference in taste helps you distinguish salt from sugar. (**Note**: You should never use taste as a property, unless your teacher says that it is safe. Some substances may be poisonous.) A difference in color (as well as taste) helps you distinguish salt from pepper. Taste and color are ***physical properties***—properties that can be determined without changing the identity of a substance. All substances have unique physical properties we can use to identify them. Table 1-4 lists some physical properties often used to identify substances.

Table 1-4. *Examples of Physical Properties of Substances*

Property	**Example**
Phase	Mercury is a liquid at room temperature.
Color	Sulfur is yellow.
Odor	Hydrogen sulfide smells like rotten eggs.
Density	Lead is much denser than aluminum.
Solubility	Salt dissolves in water.
Melting point	Ice melts at 0°C.
Boiling point	Water boils at 100°C.

Phases

One obvious physical property of a substance is whether it is a solid, a liquid, or a gas. These three forms, or states, of matter are called *phases.* The arrangement and motion of the molecules within a substance determine its phase.

1. In *solids,* the molecules are close together, move relatively slowly, and remain in fixed (unchanging) positions. A solid has a definite shape and a definite volume; that is, its shape and size do not depend on the container it is in.
2. In *liquids,* the molecules are usually farther apart and moving faster than the molecules in solids. The molecules in a liquid can change

position and flow past each other. A liquid has no definite shape; it takes on the shape of its container. However, liquids do have a definite volume. A given quantity of a liquid takes up the same amount of space regardless of the shape and size of its container.

3. In *gases*, the molecules are much farther apart and move even faster than in liquids. The molecules of a gas can move anywhere within their container. A gas has no definite shape or volume but expands or contracts to fill whatever container it is in.

Figure 1-5 shows how molecules are typically arranged in solids, liquids, and gases.

Figure 1-5. *The three phases of matter: solid, liquid, and gas.*

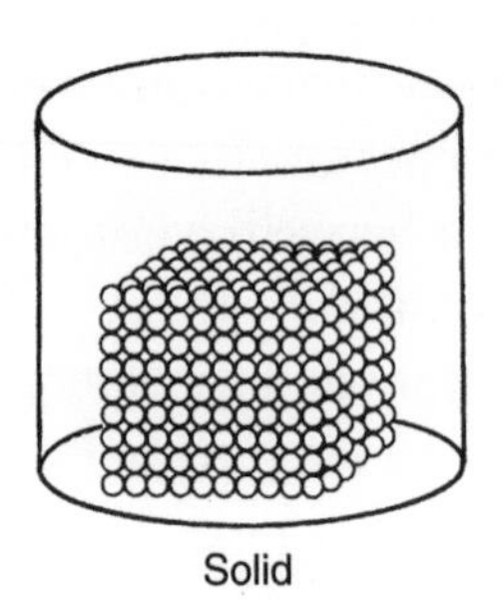

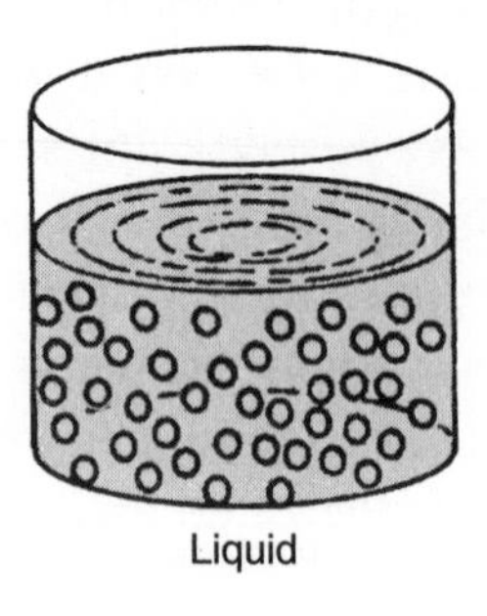

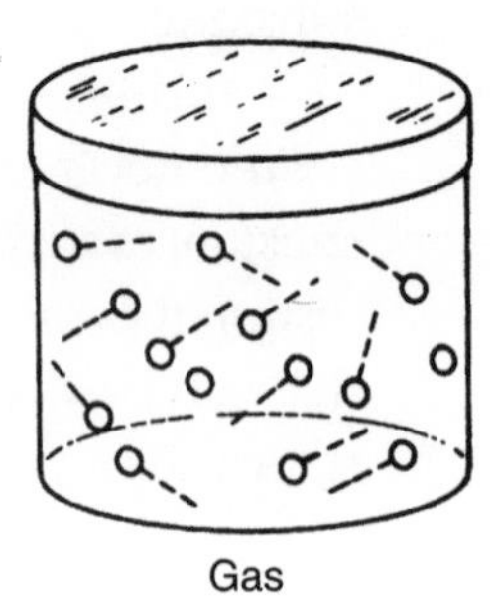

Process Skill 1

Reading for Understanding

A Fourth Phase of Matter

Three phases of matter may not be enough. Scientists have identified types of matter that do not fit the definitions of solids, liquids, and gases. One example is glass.

Glass has a definite shape, but the particles are not arranged in a regular pattern. Most solids, as discussed in the next section, will change from solid to liquid at one particular temperature called the melting point. When glass is heated, it gradually gets softer and softer until it begins to flow freely. Is glass a solid or a liquid? Many scientists say "Neither." They identify glass as another phase of matter.

Questions

1. State one way in which glass resembles a solid.
2. How is glass different from solids?
3. Glass is made mostly of a substance with the formula SiO_2. However, it contains small amounts of other substances. Would you consider glass an element, a compound, or a mixture? Explain your answer.

Changes in Phase

Since the phase of a substance depends on the arrangement of its molecules, a change in this arrangement can cause a change in phase.

1. To change a solid into a liquid, the molecules must generally move farther apart, out of their fixed positions. This is called ***melting***. Heat energy must be added to a substance to separate its molecules, so energy is absorbed during melting. (See Table 1-5.)

Table 1-5. *Examples of Water Phase Changes*

Phase Change	Heat Flow	Examples
Liquid to gas	Water absorbs heat energy.	Puddle evaporates, boiling water
Solid to liquid	Water absorbs heat energy.	Melting snow, ice melting in soda
Gas to liquid	Heat energy is released.	Cloud condensation, water on kitchen window in winter
Liquid to solid	Heat energy is released.	Ice cubes form in freezer, ice forming on lake

2. ***Freezing*** is the opposite of melting. When a liquid freezes into a solid, the molecules come together and bond more tightly into fixed positions. This process releases energy.
3. Changing a liquid into a gas, by ***boiling*** or ***evaporation***, requires that the molecules of the liquid be separated even farther. Energy is therefore absorbed when a liquid changes into a gas.
4. The change from a gas to a liquid is called ***condensation***. During condensation, molecules of a gas move closer together to form a liquid, and energy is released. Figure 1-6 illustrates the energy changes associated with changes in phase.

Figure 1-6. *Energy changes occur during changes in phase.*

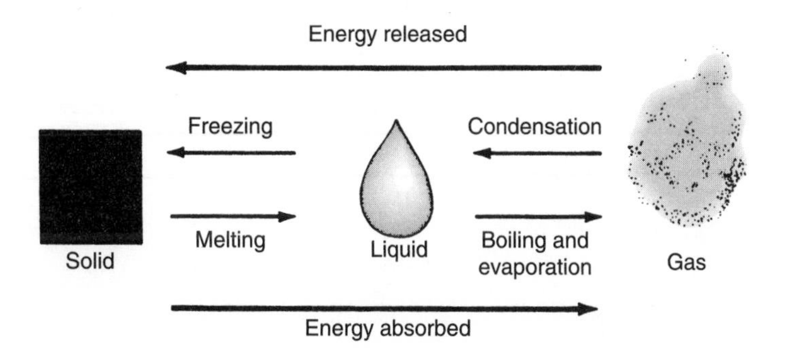

For each substance, the change in phase from a solid to a liquid occurs at a particular temperature called its ***melting point***. The melting point of ice (the solid form of water) is 0 °C. The temperature at which a pure liquid freezes into a solid is called its ***freezing point***. The freezing point of pure water is 0 °C. The freezing point and melting point of a substance are the same.

The temperature at which a liquid boils and changes rapidly into a gas is called its ***boiling point***. The boiling point of water is 100 °C. This is also the temperature at which water vapor, cooling from above 100 °C, begins to condense into liquid water.

While a substance is changing phase, its temperature remains constant. For example, while you are boiling water, the temperature remains at 100 °C even though you are constantly supplying heat. This additional heat causes a phase change rather than a change in temperature. After the phase change is completed, the temperature will rise above 100 °C, as heat is added.

Process Skill 2

Interpreting Data in a Table

By using melting-point and boiling-point information, we can determine what phase a substance will be in at a given temperature. If the temperature of a substance is below its melting point, the substance is a solid. If the temperature is above its boiling point, the substance is a gas. If the temperature is between its melting and boiling points, the substance is a liquid. For example, at room temperature (20°C), water is a liquid, because 20°C is between the melting point (0°) and boiling point (100°) of water.

The table below lists the melting point and boiling point of some common substances. What phase would table salt be in at a temperature of 1000°C? Since 1000°C is above the melting point of salt but below its boiling point, table salt would be a liquid at that temperature. Use the same kind of reasoning, based on the table below, to answer the following questions.

Melting Points and Boiling Points of Some Common Substances

Substance	*Melting Point (°C)*	*Boiling Point (°C)*	*Phase at 20°C (room temperature)*
Water	0	100	Liquid
Alcohol	−117	78	Liquid
Table salt	801	1413	Solid
Oxygen	−218	−183	Gas

Questions

1. At a temperature of −190°C, oxygen is in the form of a
   (1) gas (2) liquid (3) solid
2. Alcohol would be a liquid at all of the following temperatures except
   (1) −100°C (2) 32°C (3) 100°C (4) 77°C
3. The only substance listed that could be a liquid at a temperature of 90°C is
   (1) table salt (2) water (3) alcohol (4) oxygen

Review Questions

Part I

1. Which choice is considered matter?

   (1) sound (3) heat
   (2) air (4) light

2. Complete the sentence by selecting the best choices to fill in the blanks.

   Matter is anything that has _____ and takes up _____.

   (1) space, air (3) mass, space
   (2) weight, energy (4) air, solids

3. The particles generally are closest together when matter is

   (1) a gas (3) a solid
   (2) a liquid (4) energy

4. What property of matter can be measured using the tool shown below?

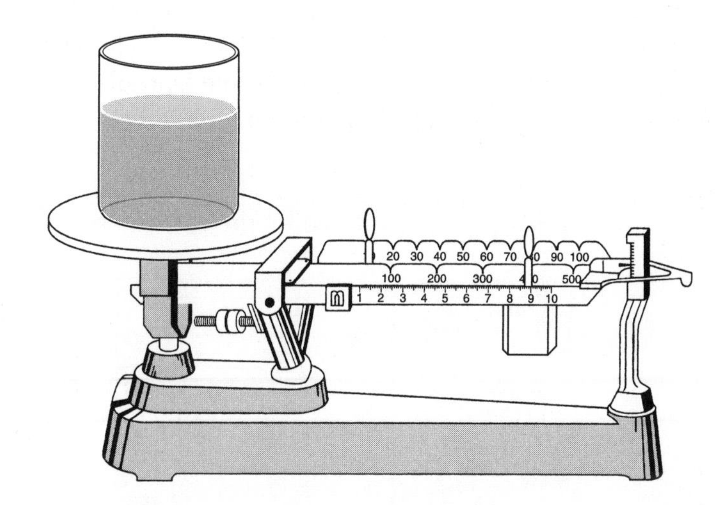

   (1) volume (3) phase
   (2) temperature (4) mass

5. Which particle is positively charged?

(1) proton
(2) electron
(3) neutron
(4) molecule

6. Which substance is an element?

(1) water
(2) air
(3) iron
(4) table salt

7. Which process is illustrated in the cartoon?

(1) melting, with a release of energy
(2) freezing, with a release of energy
(3) melting, with an absorption of energy
(4) freezing, with an absorption of energy

8. Which substance is classified as a compound?

(1) hydrogen (H_2)
(2) water (H_2O)
(3) iron (Fe)
(4) oxygen (O_2)

9. When water is heated from 20°C to 40°C, its molecules

(1) move faster and get farther apart
(2) move faster and get closer together
(3) move slower and get farther apart
(4) move slower and get closer together

10. Which principle is illustrated by the cartoon below?

(1) Like charges attract.

(2) Opposite charges attract.

(3) Like charges repel.

(4) Opposite charges repel.

11. The process of changing a liquid to a gas is called

(1) condensation

(2) evaporation

(3) melting

(4) freezing

12. Scientists organize elements based on their properties using a model called a

(1) dichotomous key

(2) periodic table

(3) graphical organizer

(4) pedigree chart

Part II

Base your answers to questions 13 through 15 on the diagrams below, which show three sealed containers, labeled A, B, and C.

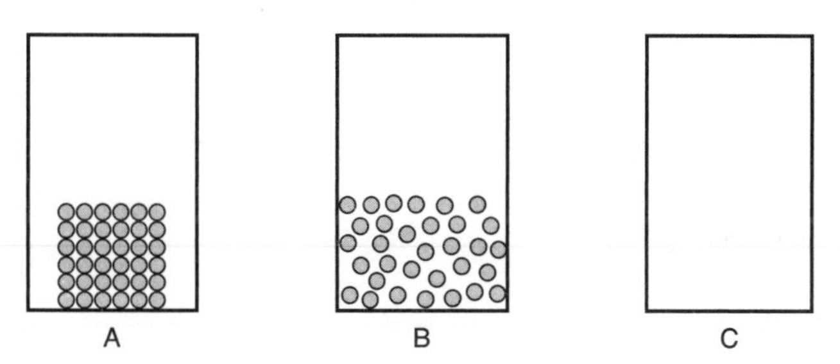

The circles in container A represent the molecules of a substance in the solid state.

The circles in container B represent the molecules of the same substance after it has been melted.

13. In your notebook, draw container C showing the molecules of this substance in the gaseous state.

14. What must be done to the molecules in container B to convert them to the gaseous state?

15. Changing a substance from one state to another is considered a physical change. Explain why this is a physical change and not a chemical change.

Base your answers to questions 16 and 17 on the diagram on page 40, top, and on your knowledge of science. The diagram shows a phase change represented by letter A.

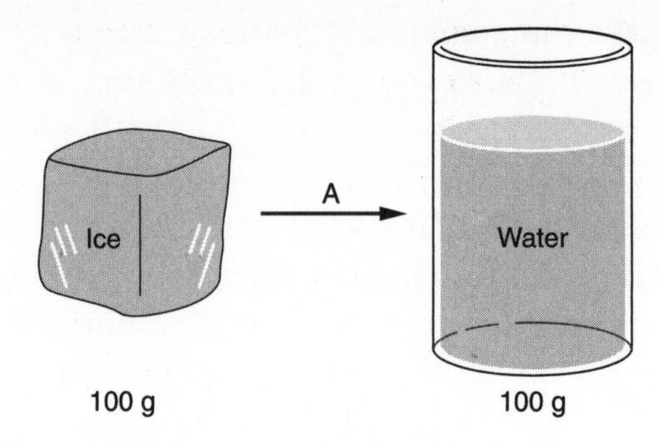

16. What is the name of the phase change at A?

17. Would the phase change at A be considered a physical change or a chemical change? Explain your answer.

18. The top row of the table below represents a compound in its three phases (solid, liquid, and gas). Copy this table into your notebook. Complete the table by filling in the answers that correspond to the drawing at the top of each column and the question in each row.

What is the name of this phase?			
Does this phase have a *definite volume* or *no definite volume*?			
Does this phase have a *definite shape* or *no definite shape*?			
Describe the particles of this phase as being *very close*, *close*, or *very far apart*.			

Solutions

How does a mixture of table salt and water differ from a mixture of sand and water? When table salt is placed in water, the particles of salt disappear, yet they can still be detected when tasting the mixture. A ***solution*** is a mixture in which the components (parts) remain evenly distributed. Salt water is a solution. In a mixture of sand and water, the sand remains clearly visible and settles at the bottom of the container. Sand and water is not a solution. The physical property that distinguishes the sand from table salt is called ***solubility***. We say that salt dissolves in water to form a solution, but sand does not dissolve in water. Table salt is ***soluble*** in water, but sand is ***insoluble*** in water.

A solution generally has two parts, the ***solute*** and the ***solvent***. The solvent does the dissolving, while the solute gets dissolved. For example, when a solid dissolves in a liquid, the solid is the solute and the liquid is the solvent. Gases or other liquids may also dissolve in liquids to form a solution. Table 1-6 shows some common solutions with water as the solvent.

Table 1-6. *Some Solutes That Dissolve in Water (Solvent)*

Solution	***Solute***	***Phase of Solute***
Seltzer	Carbon dioxide	Gas
Tea	Tea	Solid
Vinegar	Acetic acid	Liquid

Do you like your tea strong or weak? When you brew tea, the solid tea is the solute, while the water is the solvent. You can make a stronger tea solution by using more tea, or by using less water. Weaker tea would contain less tea, or more water. The strength of a solution is called its ***concentration***. A strong solution is called a ***concentrated*** solution and a weak solution is called a ***dilute*** solution.

Rate of Dissolving

What do you do after you add sugar to a cup of tea? You probably stir the mixture. Why? Stirring is one method of increasing the rate of dissolving. Which would dissolve faster, a sugar cube or a packet of granulated sugar? The smaller the particle size is, the faster the dissolving process will be. Therefore, granulated sugar dissolves faster than a lump of sugar. Even granulated sugar dissolves quite slowly in iced tea. It dissolves much faster in hot tea. An increase in temperature usually increases the rate of dissolving.

Solubility

Not only does an increase in temperature dissolve sugar faster, it also allows more sugar to dissolve. In fact, it is possible to dissolve two cups of sugar in one cup of water if the water is hot enough (100°C).

In general, raising the temperature increases the amount of solid solute that can dissolve in a liquid. The maximum amount of solute that can dissolve in a given amount of solvent is called the solubility. Generally, the solubility of a solid in a liquid increases as temperature increases. Gases, however, behave differently. The solubility of a gas in a liquid decreases when temperature increases.

The solubility of a gas is also affected by pressure; the higher the pressure, the more soluble the gas. When you open a bottle of soda, you decrease the pressure on the solution. The carbon dioxide gas that was dissolved at the higher pressure comes out of the solution, forming bubbles.

Choosing a Solvent

You have probably heard the expression, "Oil and water don't mix." A chemist might say instead, "Oil is not soluble in water." Oil is soluble in other solvents. We often need to choose a suitable solvent for a given solute. For example, nail polish does not dissolve in water. It does dissolve in acetone, a liquid often used as a nail polish remover. Grease and oil, which often stain clothing, do not dissolve in water. The "dry cleaners" use a liquid that dissolves the grease without harming the fabric. The solubility of a given solute in a given solvent depends on the chemical bonds in the two substances.

Process Skill 3

Using a Dichotomous Key

A dichotomous key organizes information by asking a series of yes-no questions. It helps you to identify matter by offering two choices at each step. One way to make a dichotomous key is in the form of a tree—a graphic organizer that leads you through the questions and, finally, to a correct conclusion.

The figure below uses a tree to classify matter into one of several categories—elements, compounds, and two types of mixtures. Copy the chart into your notebook and complete it by writing the name of the correct types of matter in each oval.

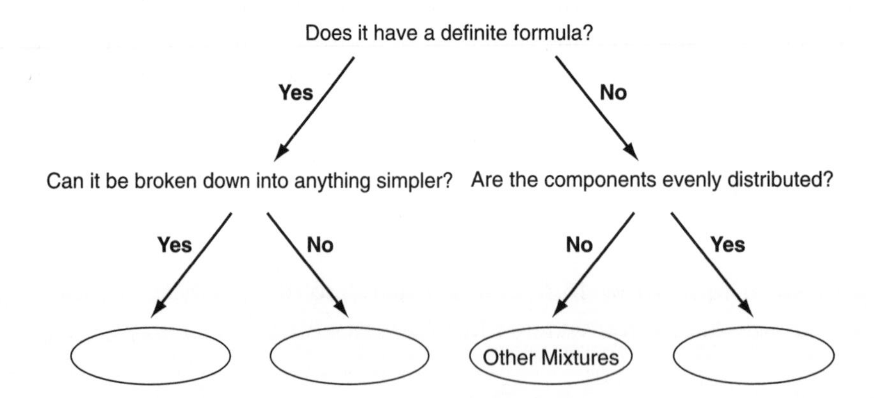

A dichotomous key can also be written as a table of answers to yes or no questions. Copy the table below into your notebook and complete the table based on the tree you have created.

A Key to Identifying Matter		
Couplet	Description	
1a 1b	Does have a definite formula Does not have definite formula	Go to 2 Go to 3
2a 2b	It can be broken down into something simpler It cannot be broken down into something simpler	
3a 3b	Components are evenly distributed Components are not evenly distributed	 Other mixtures

Density

Why are airplanes made of aluminum, and fishing sinkers made of lead? You might answer that aluminum is a light metal, while lead is a heavy metal. Yet an aluminum airplane has a much larger mass than a lead fishing sinker. When we say that lead is heavier than aluminum, we really mean that when these two pieces of metal are the same size, the lead piece will be heavier. (If the pieces are not the same size, we need another way to compare them.) The quantity that compares the mass of an object to its size (***volume***) is called density. ***Density*** is defined as the mass of an object divided by its volume:

$$\text{density} = \frac{\text{mass}}{\text{volume}}$$

While the mass and volume of a piece of metal depend on the size of the piece, the density depends on only the nature of the metal and its temperature. (We will discuss the effect of temperature later on.) Let us compare the densities of lead and aluminum. At room temperature, the density of aluminum is 2.7 grams per cubic centimeter, which we abbreviate as 2.7 g/cm^3. Since density is mass divided by volume, the unit of density contains a mass unit, grams, divided by a volume unit, cubic centimeters. (Another unit of volume, the milliliter, abbreviated mL, is often used when measuring the volume of a liquid.) When we say that aluminum has a density of 2.7 g/cm^3, this means that a piece of aluminum with a volume of 1 cubic centimeter has a mass of 2.7 grams. The density of lead at room temperature is 11.3 g/cm^3. Lead is about four times as dense as aluminum.

Table 1-7. *Performing Calculations with Density*

Finding mass from volume and density	$\textbf{density} = \dfrac{\textbf{mass}}{\textbf{volume}}$
To get mass by itself, multiply both sides of the equation by volume.	$\text{volume} \times \text{density} = \dfrac{\text{mass}}{\cancel{\text{volume}}} \times \cancel{\text{volume}}$
Volume cancels on the right side, so we get:	$\text{mass} = \text{density} \times \text{volume}$
Finding volume from mass and density	$\textbf{density} = \dfrac{\textbf{mass}}{\textbf{volume}}$
To get volume out of the denominator, multiply both sides by volume.	$\text{volume} \times \text{density} = \dfrac{\text{mass}}{\cancel{\text{volume}}} \times \cancel{\text{volume}}$
Volume cancels on the right side, so:	$\text{volume} \times \text{density} = \text{mass}$
To get volume by itself, divide both sides by density.	$\text{volume} \times \dfrac{\cancel{\text{density}}}{\cancel{\text{density}}} = \dfrac{\text{mass}}{\text{density}}$
Density cancels on the left side, so:	$\text{volume} = \dfrac{\text{mass}}{\text{density}}$

Which is heavier, 10 grams of lead or 10 grams of aluminum? Of course, this is a trick question. Since both are 10 grams, they are equally heavy. Which has a larger volume, 10 grams of lead or 10 grams of aluminum? This question requires some thinking. The density of aluminum is 2.7 g/cm^3, so 10 grams of aluminum has a volume greater than 1 cubic centimeter. Table 1-7 shows how the volume can be calculated. The density of lead is 11.3 g/cm^3, so a 10-gram piece of lead has a volume of less than 1 cubic centimeter. In comparing objects of equal mass, the denser object is the one with the smaller volume. Figure 1-7 compares pieces of lead and aluminum of the same mass.

Figure 1-7. *These metal cubes are of equal mass; the cube with the smaller volume has the greater density.*

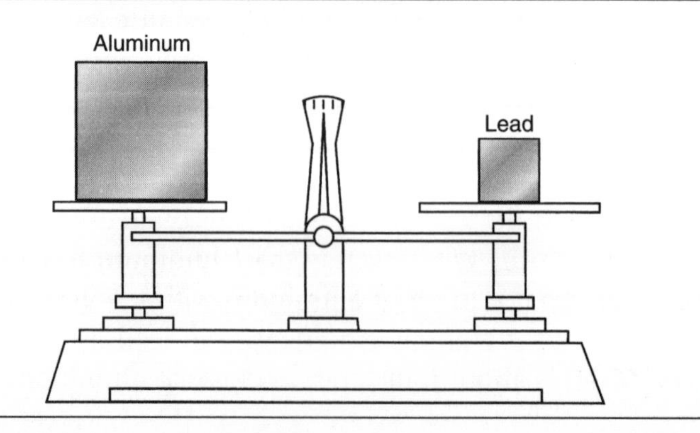

Which is heavier, 10 cm^3 of aluminum, or 10 cm^3 of lead? When comparing objects of equal volume, the object with the greater density has the greater mass. Thus, 10 cm^3 of lead weighs about four times as much as the same volume of aluminum, since lead is about four times denser than aluminum. Figure 1-8 compares pieces of lead and aluminum that have the same volume.

Figure 1-8. *These metal cubes are of equal volume; the cube with the greater density has the greater mass.*

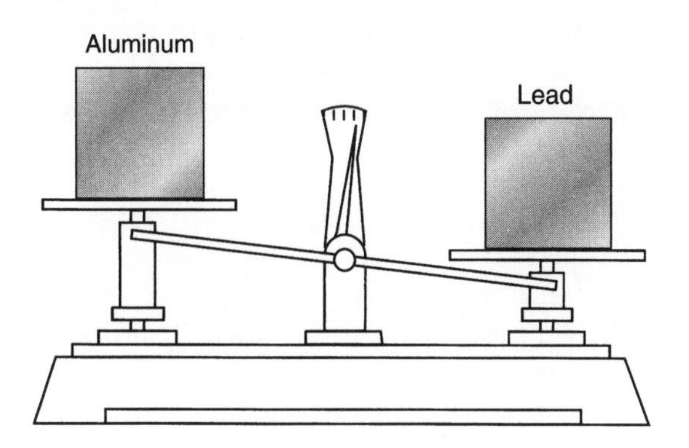

The density of a 10 cm^3 piece of lead is 11.3 g/cm^3. What is the density of a 20 cm^3 piece of lead? The answer: still 11.3 g/cm^3! Table 1-8 gives the mass, volume, and density of several pieces of lead. As you can see, the density remains the same no matter what size the piece of metal is. The density of a material does not depend on the size of the object. This makes density a very useful property for identifying materials.

Table 1-8. *Mass, Volume, and Densities of Several Pieces of Lead*

Mass	**Volume**	**Density (Mass/Volume)**
113 g	10 cm^3	11.3 g/cm^3
226 g	20 cm^3	11.3 g/cm^3
1130 g	100 cm^3	11.3 g/cm^3

Using Density to Identify a Metal

Tin foil and aluminum foil look very much alike. Aluminum has a density of 2.7 g/cm^3, while tin has a density of 7.3 g/cm^3. A chemist measures the volume of a sample of metal foil and finds it to be 5.0 cm^3. He then weighs it and finds its mass to be 36.5 grams. Is it aluminum or tin? Remember that density is mass/volume. To find the density of the metal, divide the mass, 36.5 grams, by the volume, 5.0 cm^3. Thus,

$$\text{density} = \frac{36.5\ \text{g}}{5.0\ \text{cm}^3} = 7.3\ \text{g/cm}^3$$

The metal is tin.

Why Do Objects Float?

A wooden log floats in water, while an iron nail sinks. Why? The answer to this question lies in the density of these materials. A material will float if it is less dense than the liquid in which it is placed. From this information, we can conclude that iron is more dense than water, while wood is less dense than water. Water has a density of 1 g/cm^3. Any object with a density greater than 1 g/cm^3 will sink in water. The density of iron is 7.9 g/cm^3, and therefore it sinks in water.

Some liquids do not mix with each other. For example, the oil and vinegar in salad dressing form separate layers, as shown in Figure 1-9. What can you conclude about the density of oil compared with the density of vinegar? Since the oil floats over the vinegar, the oil must be less dense.

Figure 1-9. *The less dense liquid (oil) floats on top of the more dense liquid (vinegar).*

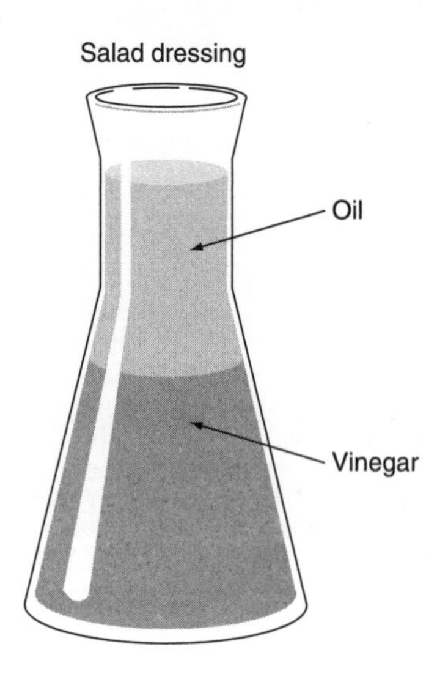

Helium balloons float in air in much the same way that wood floats in water. The density of a helium balloon is much less than the density of air. Some balloons use hot air instead of helium. Since hot-air balloons float, hot air must be less dense than cold air. When most materials are heated, they expand. This means that their volume increases while their mass stays the same. Since density is mass/volume, an increase in volume will cause a decrease in a material's density. In general, an increase in temperature causes a decrease in density.

Objects that float are said to be ***buoyant***. The ***buoyancy*** of an object in water depends on the density of the object relative to the density of the water. The less dense the object is and the denser the fluid is, the greater the buoyancy of the object.

Review Questions

Part I

Base your answers to questions 19 and 20 on the graph below, which shows the amount of potassium nitrate (KNO_3) that will dissolve in 100 grams of water at various temperatures.

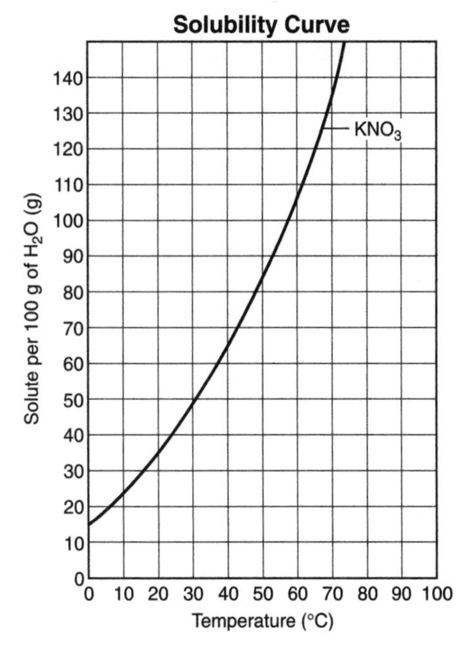

19. Which statement best describes the solubility of KNO_3 in water?

(1) As the temperature of the water increases, the solubility of KNO_3 increases.

(2) As the temperature of the water increases, the solubility of KNO_3 decreases.

(3) As the temperature of the water increases, the solubility of KNO_3 increases and then decreases.

(4) As the temperature of the water increases, the solubility of KNO_3 remains the same.

20. As the water temperature increases from 30°C to 60°C, about how many more grams of KNO_3 will dissolve in 100 grams of water?

(1) 30 g (3) 60 g

(2) 50 g (4) 110 g

21. A solution is made by placing 5 grams of salt in 100 grams of water. In this solution the

(1) water is the solute and the salt is the solute

(2) water is the solvent the salt is the solvent

(3) water is the solute and the salt is the solvent

(4) water is the solvent and the salt is the solute

22. As the temperature of 100 mL of water increases, the amount of sugar that can dissolve in it

(1) increases and the solution becomes more concentrated

(2) decreases and the solution becomes more concentrated

(3) increases and the solution becomes less concentrated

(4) decreases and the solution becomes less concentrated

23. Which substance would dissolve fastest?

(1) powdered sugar in cold water

(2) sugar cubes in cold water

(3) powdered sugar in hot water

(4) sugar cubes in hot water

24. What is the density of an object that has a mass of 20. grams and a volume of 4.0 cm^3?

(1) 0.20 g/cm^3

(2) 4.0 g/cm^3

(3) 5.0 g/cm^3

(4) 80.0 g/cm^3

25. The diagram below shows a tall container with four liquids and their densities.

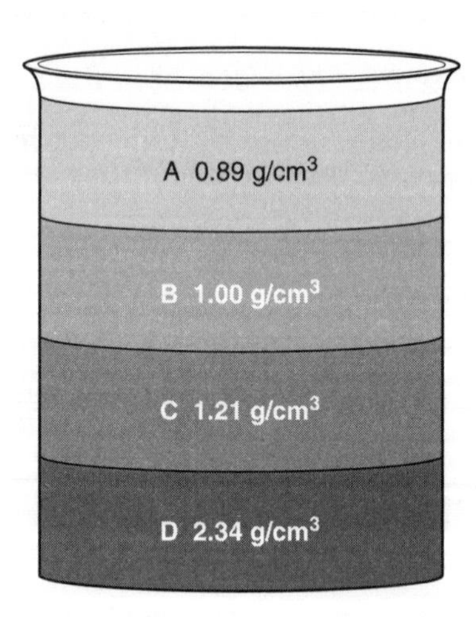

If a ball that has a density of 1.73 g/cm^3 is placed in the container, where will the ball come to rest?

(1) on top of liquid A

(2) between liquids B and C

(3) between liquids C and D

(4) on the bottom of the container

Part II

Base your answers to questions 26 through 28 on the information below.

The mass and volume of an aluminum block were measured and recorded in the Table A below. Table B gives the density of three liquids.

Table A
Mass = 27.0g
Volume = 10.0 cm^3

Table B	
Substance	Density
Water	1.0 g/cm^3
Ethanol	0.8 g/cm^3
Mercury	13.6 g/cm^3

26. What is the density of this aluminum block? (Show your work. Express your answer to the nearest tenth of a unit.)
27. In which of the liquids in Table B would this aluminum block float?
28. How would the density of a 50.0 cm^3 aluminum block compare with that of the block described in Table A above. Explain your reasoning.
29. In two or more sentences, explain the differences between a mixture and a compound. Give two examples of each.
30. Jorge wants to find the density of a glass marble. The tools used by Jorge to measure the glass marble are shown below. Use the information from the diagrams to calculate the density of the marble. Show your work. (Express your answer to the nearest tenth of a unit.)

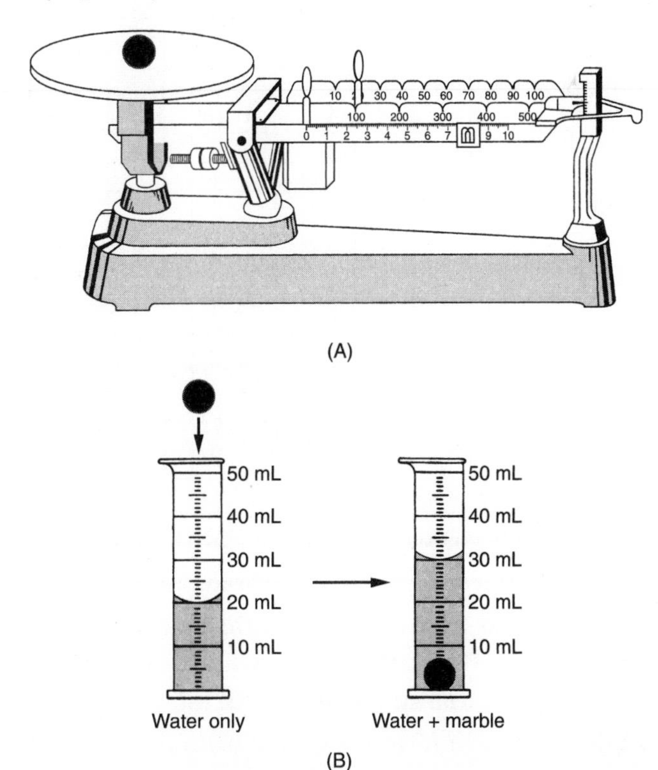

(A)

(B)

Base your answers to questions 31 through 35 on the information below.

Midge performed an experiment to determine what concentration of salt water is best for growing a saltwater plant. She prepared five solutions of salt water in different concentrations as indicated in the data table below. She also used pure water in one of her experimental groups. All the plants were the same height when she started. The plants were allowed to grow for 30 days under the same conditions of light and temperature. At the end of the 30 days, Midge measured the height of each plant and recorded it in the data table below.

Concentration of Salt (%)	*Height (cm)*
0 (pure water)	2
1	9
2	14
3	17
4	17
5	9
6	1

31. Copy the grid below into your notebook and plot the data for this experiment. Use an X to mark each point, and connect the points with a solid line.

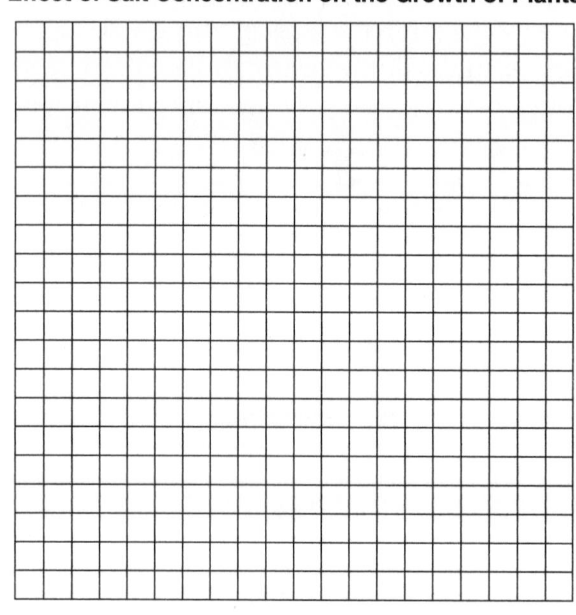

32. Label the x-axis with an appropriate title and proper units.

33. Label the y-axis with an appropriate title and proper units.

34. Midge did research and found that the average concentration of salt in the ocean is 3.5%. What conclusion could you draw about the growth of this plant at a concentration of 3.5%?

35. What conclusion can be drawn from the results of this experiment?

Changes in Matter

We see change around us all the time. There are changes in us. There are changes in the weather. There are changes on Earth. Some of these changes involve physical aspects of matter. Other changes involve chemical aspects.

Physical Changes

As you know, every water molecule is made up of two atoms of hydrogen and one atom of oxygen. Therefore, the chemical formula for water is H_2O. What is the formula for ice? When water freezes, the arrangement of its molecules changes, but the molecules themselves do not change. They are still H_2O. A change of phase, such as freezing or melting, does not produce any new substances. What do you think would be the formula for water vapor? Yes, it would still be H_2O. A change that does not form any new substances is a ***physical change***. All changes of phase are physical changes. Crushing ice cubes into small pieces is also a physical change, since both crushed ice and ice cubes are made of the same substance, H_2O, water.

Similarly, when you dissolve sugar in water, the sugar still tastes sweet and the water is still wet. No new substances have been formed, so dissolving is a physical change. Figure 1-10 shows why boiling, melting, and dissolving are physical changes.

Figure 1-10. *During physical changes, no new substances are formed.*

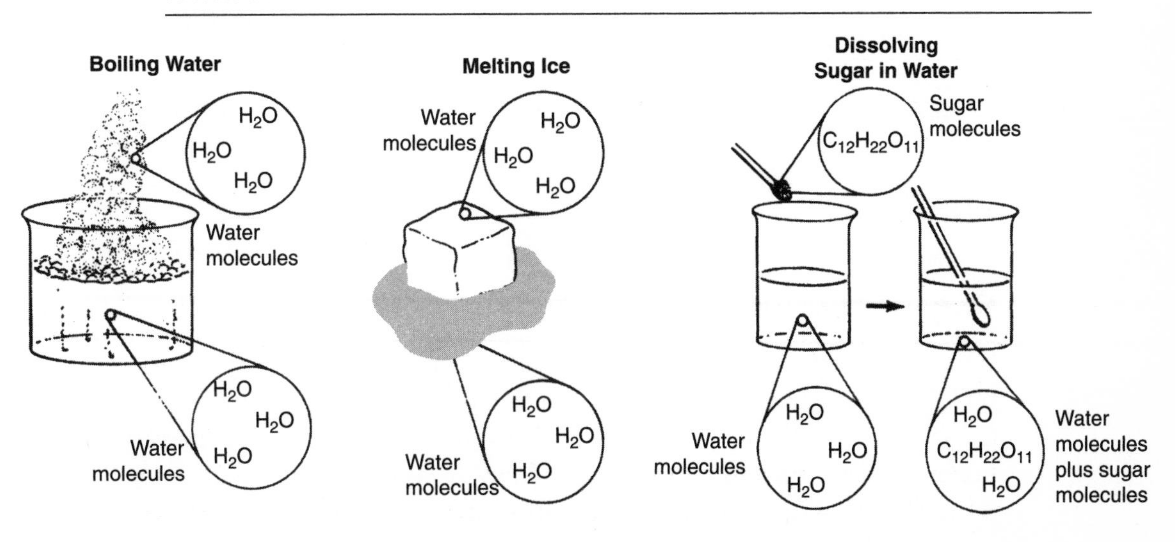

Separating the Parts of a Mixture

A mixture of two or more substances can be separated by differences in their physical properties. For example, a mixture of salt and water can be separated by evaporating (a physical change) the water, which leaves the salt behind. A mixture of iron and silver can be separated with a magnet. The iron will be attracted (a physical property) to the magnet while the silver will not. How might a mixture of salt and sand be separated? This would involve a series of physical changes. First, the mixture could be added to water. The salt would dissolve, while the sand would not. This mixture could then be filtered, separating the undissolved sand from the saltwater solution. The dissolved salt particles are too small to be trapped on the filter paper and pass through the tiny openings in the paper. The sand particles are too large to pass through the openings in the filter paper, so they are separated from the salt water. Finally, we can boil off the water from the saltwater mixture leaving the salt behind. The sequence of steps is illustrated in Figure 1-11.

Figure 1-11. *Separating the parts of a mixture: (A) A mixture of salt and sand. (B) Dissolving the salt in water. (C) Separating the sand with a filter. (D) Evaporating the water.*

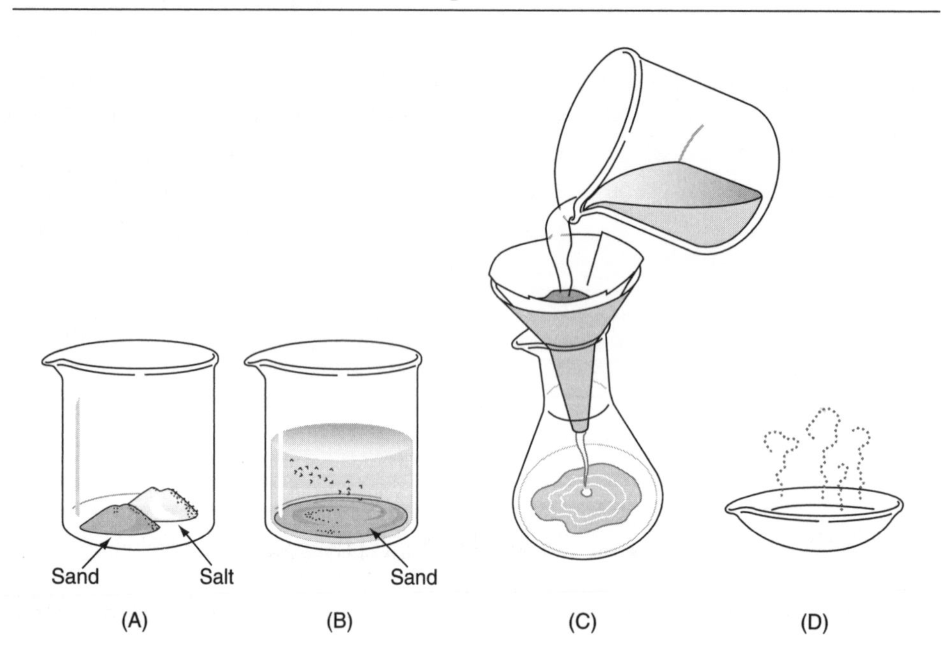

Chemical Changes

What happens if you forget to put a carton of milk back into the refrigerator? First, the milk gets warm. This is a physical change. However, if you leave the milk out too long, it turns sour. The sour taste is caused by the production of a new substance called *lactic acid.* A change that produces one or more new substances is called a ***chemical change.*** When a chemical change occurs, we say there was a chemical reaction. Burning paper produces smoke and ash, both of which are new products. Burning is always a chemical change.

Forming a compound always involves a chemical change, but forming a mixture involves only physical changes. Similarly, it requires a chemical change to break apart a compound. Mixtures, however, can be separated through physical means. For example, salt water can be boiled, leaving the salt behind; and blood can be spun in circles (centrifuged), separating it into its various components.

Chemical changes can be represented by chemical equations. A chemical equation uses formulas and numbers to keep track of a chemical change. The starting materials, called the *reactants,* are listed on the left side of the equation. The final materials, called the *products,* are listed on the right side. An arrow separates the two sides. The equation for the burning of coal, which is mostly carbon, would be written as

$$C + O_2 \rightarrow CO_2$$

A chemist reads this equation as, "Carbon plus oxygen yields carbon dioxide." In this reaction, carbon and oxygen are the reactants, and carbon dioxide is the product. Table 1-9 gives some examples of chemical changes.

Table 1-9. *Examples of Chemical Changes*

Chemical Change	Reactants	Products	Equation
Burning coal	Carbon (C) + oxygen gas (O_2)	Carbon dioxide gas (CO_2)	$C + O_2 \rightarrow CO_2$
Rusting of iron	Iron (Fe) + oxygen gas (O_2)	Rust (Fe_2O_3)	$4Fe + 3O_2 \rightarrow 2Fe_2O_3$
Tarnishing of silver	Silver (Ag) + sulfur (S)	Tarnish (Ag_2S)	$2Ag + S \rightarrow Ag_2S$
Photosynthesis	Carbon dioxide gas (CO_2) + water (H_2O)	Glucose ($C_6H_{12}O_6$) + oxygen gas (O_2)	$6CO_2 + 6H_2O \rightarrow C_6H_{12}O_6 + 6O_2$

Properties and Chemical Changes

The new substances produced by a chemical change have their own set of properties. These properties are different from those of the original substances that reacted, since those substances are no longer present as separate substances. For example, the element sodium is a soft metal that explodes on contact with water. The element chlorine is a poisonous, green gas. When sodium and chlorine combine in a chemical reaction, they produce sodium chloride, commonly known as table salt. The new substance formed has completely different properties from those of the original materials, which no longer exist as separate substances. During a chemical reaction, the atoms are rearranged to form new substances. This involves the breaking of existing chemical bonds and the formation of new bonds.

Both physical and chemical changes occur in nature. The wearing away of a mountain by streams is an example of a physical change called ***erosion***. Erosion is the physical wearing away and movement of rock material at Earth's surface. The Grand Canyon in Arizona was formed over millions of years by this physical change.

The Statue of Liberty in New York City is made of copper but does not look copper-colored (like a new penny). This is due to a chemical reaction between the copper and the air, which produces a new, green-colored substance (patina). The chemical wearing away of a metal is called ***corrosion***. Corrosion, which forms a new substance, is a chemical change. Erosion, which only moves substances around, is a physical change.

Conservation of Matter

In a chemical change, no atoms are created and no atoms are destroyed. Every atom that is present before a reaction takes place is still there after the reaction takes place. What has changed is the way the atoms are arranged. Chemical reactions change only the way atoms are bonded to one another.

Figure 1-12 shows what happens when hydrogen and oxygen combine to form water in a chemical reaction. How many atoms of hydrogen are there before the reaction takes place? How does this compare with the number of hydrogen atoms after the reaction takes place? There are four hydrogen atoms before and after the reaction. How are the starting substances (the reactants) different from the substances formed (the products)?

Figure 1-12. *The Law of Conservation of Matter: In a chemical reaction, such as the formation of water, there are the same numbers of atoms before and after the reaction.*

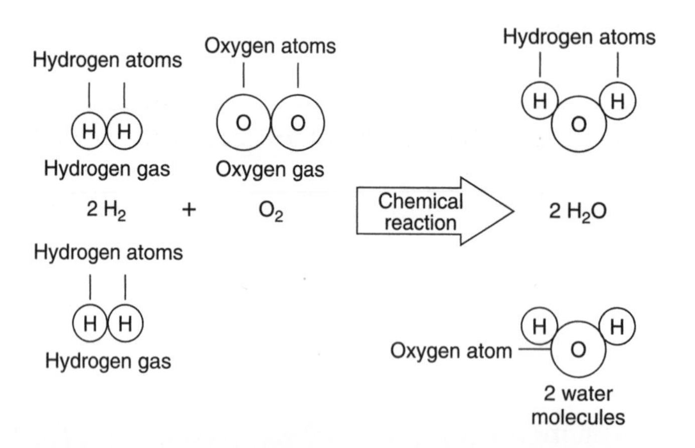

Before the reaction, each hydrogen atom was bonded to one other hydrogen atom; after the reaction, each hydrogen atom was bonded instead to an oxygen atom. How would the mass of the starting materials compare with the mass of the materials formed? The mass remains the same, since no atoms were created or destroyed. This is an example of the ***Law of Conservation of Matter***, which states that matter can be neither created nor destroyed in a chemical reaction. It can, however, be changed from one form to another. It is important to remember to account for all the substances before and after a reaction. In particular, it may be easy to forget about gases, which escape in the air, since it is difficult to capture and weigh them.

Energy and Chemical Changes

As you have learned, new substances are formed during a chemical change. An example is making table salt from sodium and chlorine. However, simply mixing sodium and chlorine together does not produce table salt. Energy is needed to start the chemical reaction. Likewise, a match does not start to burn until you strike it. The friction caused by striking the match provides the heat energy needed to start the chemical reaction of burning.

All chemical changes must be started by the addition of energy, in the form of heat, light, or electricity. Some chemical changes need so little additional energy they can absorb enough energy from their surroundings to get them started. The rusting of iron and the tarnishing of silver are examples of such reactions.

As a chemical reaction proceeds, energy is either absorbed or released. For example, the burning of a match releases energy in the form of heat and light. The chemical reaction that occurs in a battery releases electrical energy. On the other hand, when food is cooked, heat energy is absorbed as the chemical changes take place. Table 1-10 gives some examples of chemical changes that absorb energy and chemical changes that release energy.

We can use chemical reactions to supply us with heat when we need it. For example, campers often use chemical hand warmers in cold weather. When they open the packet, the chemicals in the hand warmer react with oxygen in the air to release heat.

Reactions that absorb heat are also useful. A cold pack contains two chemicals that absorb heat when they react with each other. To start the reaction, you simply break the seal that separates the two chemicals.

Table 1-10. *Energy and Chemical Changes*

Chemical Changes That Release Energy	Type of Energy Released	Chemical Changes That Absorb Energy	Type of Energy Absorbed
Burning of wood	Light, heat	Cooking an egg	Heat
Battery starting a car	Electricity	Recharging a battery	Electricity
Decomposing of organic matter	Heat	Photosynthesis	Light

Process Skill 4

Making Predictions, Determining a Quantitative Relationship, Graphing Data

Nancy learned in science class that temperature is a major factor in determining the rate of a chemical reaction. To investigate the effects of temperature on reaction rate, she decided to time a chemical reaction at several temperatures. Nancy's results are presented in the table below. Study the table and answer the following questions.

Temperature and Reaction Rate

Trial Number	*Time for Completion of Reaction*	*Temperature*
1	120 seconds	10°C
2	60 seconds	20°C
3	30 seconds	30°C

Questions

1. The data seems to indicate a trend: For each 10°C increase in temperature, the time needed to complete the reaction was
(1) doubled (3) decreased by 20 seconds
(2) cut in half (4) increased by 20 seconds

2. Assuming that the observed pattern remains constant for all temperatures, how long would the reaction take at 40°C?
(1) 5 seconds (2) 10 seconds (3) 15 seconds (4) 20 seconds

3. Construct a graph by copying the numbered axes provided below onto a separate sheet of graph paper. Then plot the data from the table, as well as your answer to question 2, on the graph. To do this, mark a point at the intersection of a temperature line and a time line for the result of each trial. (The result of Trial 1 has been plotted as an example to guide you.) Finally, connect the points with a smooth curve.

Effect of Temperature on Time Needed to Complete a Chemical Reaction

Reversible Reactions

Many physical changes, such as melting and dissolving, can be reversed. Water can be turned into ice by cooling it, and ice can be turned back into water by warming it. Sugar can be dissolved in water, and the water can be evaporated to separate it from the sugar. These are examples of changes in which substances can be returned to their original form.

Most chemical changes, however, are very difficult to reverse. It is impossible to "unburn" a match. However, some chemical reactions are reversible. An important example is the reaction that takes place in a rechargeable battery. When you recharge a battery, you are reversing the chemical reaction used by the battery to produce electrical energy. Reversing a chemical change usually requires much more energy than does reversing a physical change.

Rate of Reactions

As discussed earlier, milk turns sour when left out of the refrigerator. However, milk that is refrigerated eventually turns sour, too, though it takes longer to occur. The chemical reaction of souring, which happens rapidly at room temperature, occurs much more slowly in a cold refrigerator. In fact, most chemical reactions take place faster at higher temperatures. Frying an egg takes less time on a high flame than on a low flame. An increase in temperature increases the rate (speed) of a reaction.

Another factor that affects the rate of a chemical reaction is the size of the reacting particles. In general, the smaller the particles are, the faster the reaction is. For instance, a log burns more slowly than does an equal amount of sawdust.

The same factors that influence the rate of a chemical change also affect the rate of many physical changes. Recall that granulated sugar dissolves more rapidly than a cube of sugar does, and sugar dissolves more quickly in hot tea than it does in iced tea.

Part I

36. Which process is a physical change?

    (1) burning of wood

    (2) boiling of water

    (3) rusting of iron

    (4) cooking an egg

37. A mixture of salt, sand, and water is thoroughly stirred. It is then filtered through a filter paper as shown in the diagram below. Which substance(s) pass through the filter?

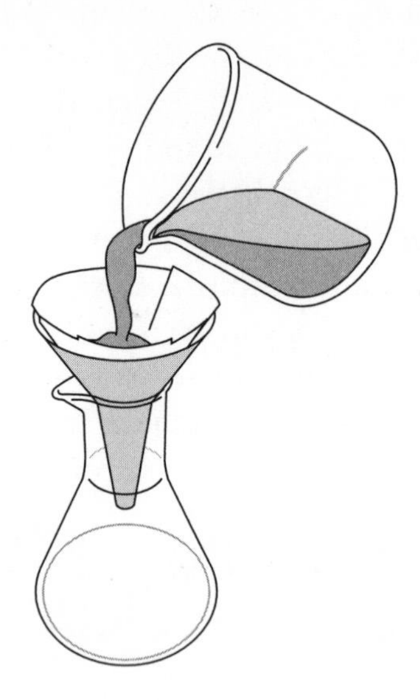

(1) only the water

(2) only the water and the sand

(3) only the water and the salt

(4) the sand, the salt, and the water.

38. When vinegar, a liquid, is added to baking soda, a solid, bubbles of carbon dioxide gas are produced. This is best described as

(1) a physical change
(2) a chemical change
(3) boiling
(4) a mixture

39. Which of the following processes can separate dissolved salt from water?

(1) filtration
(2) stirring
(3) evaporation
(4) sedimentation

40. Which process is a chemical change?

(1) melting
(2) evaporation
(3) condensation
(4) burning

Base your answers to questions 41 and 42 on the information below and your knowledge of science.

When zinc is placed in a test tube of hydrochloric acid, bubbles are observed and the test tube gets hot.

41. Which combination of zinc and hydrochloric acid would react the fastest?

    (1) a zinc strip and hydrochloric acid at 20°C

    (2) a zinc strip and hydrochloric acid at 10°C

    (3) zinc powder and hydrochloric acid at 20°C

    (4) zinc powder and hydrochloric acid at 10°C

42. The reaction between zinc and hydrochloric acid can best be described as

    (1) a physical change with energy released

    (2) a physical change with energy absorbed

    (3) a chemical change with energy released

    (4) a chemical change with energy absorbed

---

43. When sodium combines chemically with chlorine it produces sodium chloride, also known as table salt. The mass of the sodium chloride will be

    (1) greater than the mass of the sodium plus the mass of the chlorine that have combined

    (2) equal to the mass of the sodium plus the mass of the chlorine that have combined

    (3) equal to the mass of the sodium that has reacted

    (4) less than the mass of the sodium that has reacted

Part II

Base your answers to questions 44 through 48 on the following information.

After learning that zinc metal reacts with hydrochloric acid to produce hydrogen gas, the students in Mr. Jackson's science class designed and performed the following experiment.

The students placed equal masses of zinc into each of four separate test tubes that were labeled A, B, C, and D. To test tube A, they added 10.0 mL of a 4% solution of hydrochloric acid. Test tube B received 10.0 mL of an 8% solution of the acid. Test tube C received 10.0 mL of a 12% solution, and test tube D received 10. mL of a 20% solution of the acid.

The students measured the amount of time needed to collect 25.0 mL of the hydrogen gas produced from each test tube in the experiment. It took 64 seconds to collect the hydrogen from test tube A, 33 seconds from B, 16 seconds from C, and 4.0 seconds to collect the hydrogen from test tube D.

44. Create a data table that clearly and logically summarizes the results of this experiment. Be sure to include a proper heading for each column of your table. Fill in your table with the data from the experiment.

45. What problem was this experiment designed to investigate?

46. Mr. Jackson suggested that the students take the temperature of each of the acid samples before using them. Why did he make this suggestion?

47. Unfortunately, there was no 16% acid solution available to be tested. Based on the data in your table, predict the time it would take to collect the 25.0 mL of hydrogen using a 16% solution of hydrochloric acid, under the same conditions.

48. What is the relationship between concentration of acid and the speed of this reaction?

Chapter 2 Energy

When a roller coaster goes downhill, potential energy is converted to kinetic energy. Going uphill, kinetic energy is converted to potential energy.

Major Concepts

- Energy is the ability to do work. There are two states of energy: potential energy and kinetic energy. Some forms of energy include mechanical, chemical, nuclear, sound, heat, electrical, and light.
- The Law of Conservation of Energy states that energy can be neither created nor destroyed; it can only be changed into other forms of energy.
- An electric circuit is a complete path for the flow of electricity. A circuit contains an electric source, a conducting path, and a device that uses the electricity.
- Sound and light travel in waves. Sound is a form of mechanical energy, and light is a form of radiant energy.

Vocabulary at a Glance

amplitude
chemical energy
circuit breaker
conduction
conductor
convection
Doppler effect
electric circuit
electric energy
electromagnetic spectrum
energy
frequency
fuse
heat
heat energy
insulator
kinetic energy
Law of Conservation of Energy
lens
light
mechanical energy
nuclear energy
parallel circuit
potential energy
radiation
series circuit
sound
sound wave
wavelength

Energy

Energy is the ability to do work. Work is done when a force moves an object over a distance. A flowing river has the ability to move a boat. A car moving down the street can carry people from one place to another. Therefore, the river and the car have some form of energy.

States of Energy

There are two basic states of energy: potential and kinetic.

1. *Potential energy* is stored energy that can be used at a later time. An object has potential energy because of its relative position or because of its chemical composition. A rock on top of a cliff has potential energy because of its position above ground level. Gasoline contains potential energy in its chemical bonds.
2. *Kinetic energy* is energy that an object has because it is moving. A rock falling off a cliff has kinetic energy. The heat given off by burning gasoline is also a form of kinetic energy. The faster an object moves, the more kinetic energy it has. Figure 2-1 shows examples of potential and kinetic energy.

Figure 2-1. *Some examples of potential energy and kinetic energy.*

Potential energy may be changed into kinetic energy when an object begins to move. Water held back by a dam has potential energy but no kinetic energy. Releasing the water and letting it flow changes its potential energy into kinetic energy.

Kinetic energy can also be changed into potential energy. When a ball is thrown straight up into the air, its kinetic energy of motion is changed into potential energy as the ball rises higher above the ground. At the highest point of the ball's flight, it is motionless and has only potential energy. As the ball falls back to the ground, the potential energy changes back into kinetic energy. (See Figure 2-2.)

Figure 2-2. *A ball thrown up in the air demonstrates the conversion of kinetic energy into potential energy and back into kinetic energy.*

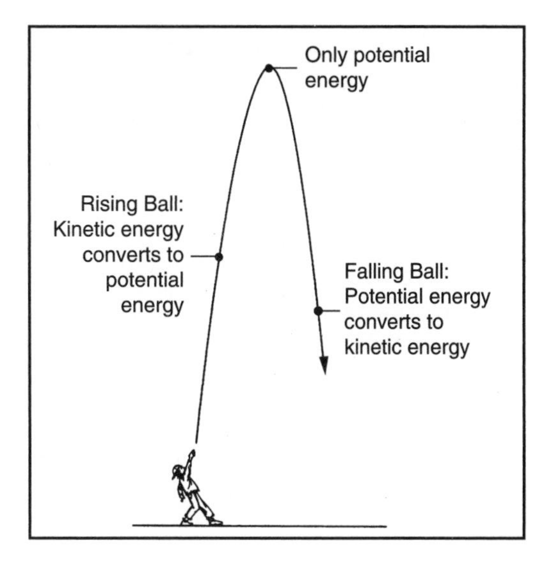

Forms of Energy

Potential and kinetic energy exist in many forms. Table 2-1 shows some examples of different forms of energy.

Table 2-1. *Different Forms of Energy*

Form of Energy	Example
Sound	Bell ringing
Mechanical	Fan turning
Chemical	Candle burning
Nuclear	Fission of radioactive elements
Heat	Toaster heating bread
Electric	Generator producing electricity
Light	Lamp (bulb) glowing

Mechanical energy is a form of kinetic energy in a moving object that is doing work. A hammer striking a nail, a jack lifting a car, and pedals turning the wheels of a bicycle are examples of things using mechanical energy. Sound is one type of mechanical energy.

Chemical energy is a form of potential energy stored in the bonds of atoms and molecules. The energy is released when the bonds are broken, such as when a substance is burned. Coal, oil, propane gas, and foods are examples of substances that contain chemical energy.

Nuclear energy is a form of potential energy stored within the nucleus (center) of an atom. This energy can be released by joining small nuclei together or by splitting large nuclei apart.

Heat energy is a form of kinetic energy associated with vibrating molecules. All matter contains heat energy. Rubbing your hands together, cellular respiration, or burning fuel oil in a home heating system can produce heat energy.

Electric energy is a form of kinetic energy produced by the flow of electrons through a conductor, such as a wire. Computers, lightbulbs, and washing machines are all operated with electric energy. A generator converts mechanical energy into electric energy.

Light is a form of kinetic energy produced when radiant energy moves in waves. Using a magnifying glass to burn a hole in a leaf, or using a laser beam to burn a hole in a steel plate can demonstrate light as a form of energy.

Conservation of Energy and Matter

The *Law of Conservation of Energy* states that energy can be neither created nor destroyed. However, energy can be transformed, or changed, from one type of energy into one or more other types of energy. The *Law of Conservation of Matter* states that matter can be neither created nor destroyed. Energy and matter are related since they are interchangeable. This means that the total amount of energy and matter in the universe is constant, and each can be converted into the other. For example, in the sun, large amounts of matter are being converted into light and heat energy. In addition, scientists have been able to change matter into energy in nuclear reactors and energy into matter under special laboratory conditions.

Energy Transformations

Most of our daily activities involve the transformation of energy. For instance, when you take a bus to school each morning, chemical energy in gasoline is changed into mechanical energy that turns the wheels of the bus. At school, when the bell rings between classes, electric energy is transformed into sound energy. And at night, when you turn on a light, electric energy is changed into light energy. Figure 2-3 shows two energy transformations.

Figure 2-3. *This hand-operated generator transforms mechanical energy into electrical energy, which is then transformed into light energy.*

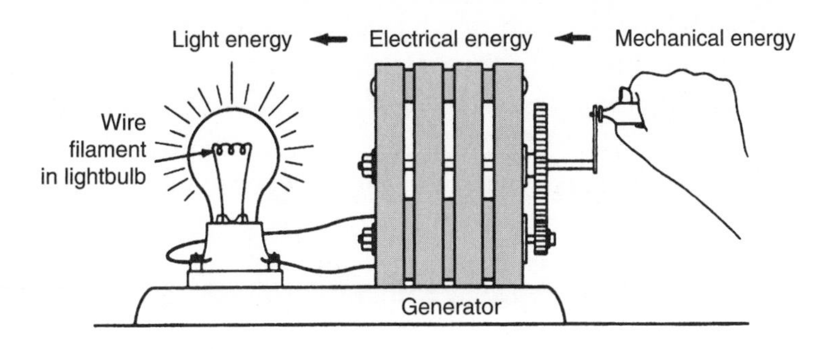

Unusable Energy

Very often during an energy transformation, some heat energy is produced that is not useful; it is waste energy. For example, a car's motor is designed to change chemical energy in the gasoline into mechanical energy to move the car. However, a running motor eventually becomes hot, due to the burning of fuel and the friction of the motor's moving parts rubbing against one another. In other words, some of the chemical energy is transformed into wasted heat energy instead of mechanical energy.

A vacuum cleaner is another example of a machine producing unusable heat energy. A vacuum cleaner has a motor that changes electric energy into mechanical energy. Run a vacuum cleaner for a few minutes and you can feel that it gets warm. The electric energy entering the motor produces mechanical energy to operate the appliance and an amount of wasted heat energy. This is illustrated in Figure 2-4.

Figure 2-4. *The Law of Conservation of Energy: Unusable heat energy is produced when a vacuum cleaner converts electrical energy into mechanical energy.*

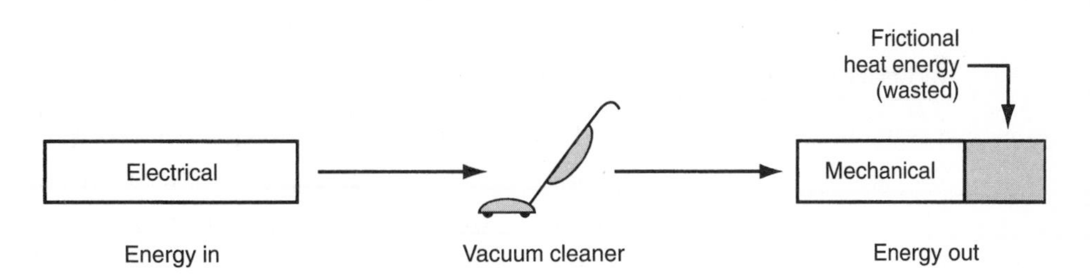

Wave Properties

Sound and light are two forms of energy that travel in the form of waves. The waves produced by sound and light energy are similar to the waves produced when a pebble is tossed into a pool of calm water.

The pebble hitting the water is a source of energy. When the pebble hits the water, it produces a series of waves. The waves travel outward in all directions on the surface of the water. The substance through which the wave travels is the *medium*. For instance, water is the medium for the waves produced by the pebble.

A curved line, as shown in Figure 2-5, can represent these waves. The top of a wave is the crest, and the bottom of the wave is the *trough*. The distance from one wave *crest* to the next wave crest, or one trough to the next, is the ***wavelength***. The height of the crest or the depth of the trough of the wave measured from the undisturbed surface is the ***amplitude*** of the wave. The number of waves that pass by a fixed point in a given amount of time is the wave's ***frequency***.

Figure 2-5. *A representation of sound and light waves.*

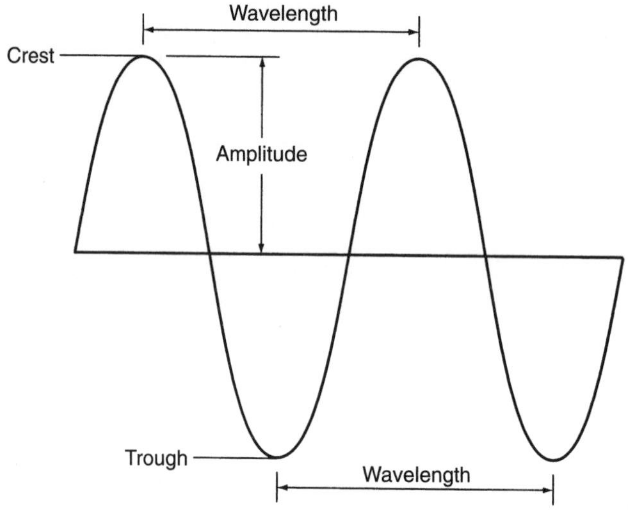

Types of Waves

A wave that vibrates at a right angle (up and down) to the direction in which the wave is traveling is a *transverse wave*. You can form a transverse wave using a rope tied to the back of a chair and shaking the untied end of the rope up and down. Although you see the wave moving along the rope, the material of the rope does not move forward; it moves up and down with each passing crest and trough. (See Figure 2-6.) Light is a transverse wave.

A wave that vibrates back and forth (push and pull) in the same direction as its direction of travel is a *longitudinal wave*. You can form a longitudinal wave using a long coiled spring. Attach the spring to the back of a chair and stretch out the spring, then push it in. You will see a series of "push-and-pull" waves pass through the spring. The area where the spring coils push close together is called a *compression*, and the area where the coils pull apart, or spread out, is called a *rarefaction*. The wavelength is measured from compression to compression or rarefaction to rarefaction. Sound waves are longitudinal waves. (See Figure 2-6.)

Figure 2-6. *A rope can be used to demonstrate a transverse wave, and a coiled spring can be used to demonstrate a longitudinal wave.*

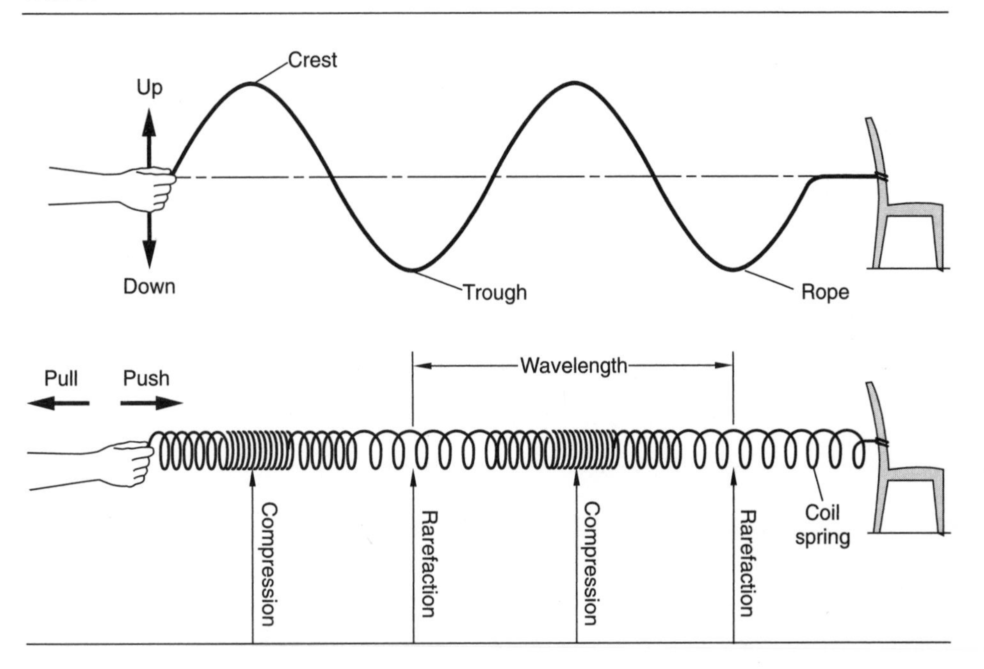

Review Questions

Part I

1. What state of energy is demonstrated by a skier speeding downhill?
   (1) kinetic
   (2) potential
   (3) nuclear
   (4) light

Questions 2 and 3 refer to the following activity.

During his workout Mitch lifts a heavy weight up to his chest and holds it there for 5 seconds. When he releases the weight it falls back to the floor.

2. When the weight is held at Mitch's chest, the weight has
   (1) potential energy
   (2) kinetic energy
   (3) no energy
   (4) motion energy

3. As the weight falls to the floor it has
   (1) potential energy
   (2) kinetic energy
   (3) no energy
   (4) position energy
4. What type of energy transformation is represented in the diagram below?

   (1) chemical energy to sound energy
   (2) chemical energy to heat energy
   (3) light energy to chemical energy
   (4) heat energy to chemical energy
5. Which device converts electrical energy into light and sound energy?
   (1) telephone
   (2) radio
   (3) television
   (4) hair dryer
6. A standard lightbulb uses electricity to produce light. However, much of the electricity is wasted because it produces a large amount of energy in the form of
   (1) heat energy
   (2) sound energy
   (3) chemical energy
   (4) mechanical energy
7. You can hear people talking because their vocal cords vibrate, producing sound waves that travel through the air to your ear. The substance through which sound waves travel is called the
   (1) vibration source
   (2) medium

(3) frequency

(4) energy source

8. The amplitude of a sound wave indicates the loudness of sound. The greater the amplitude is, the louder the sound. The wave that represents the loudest sound is shown in

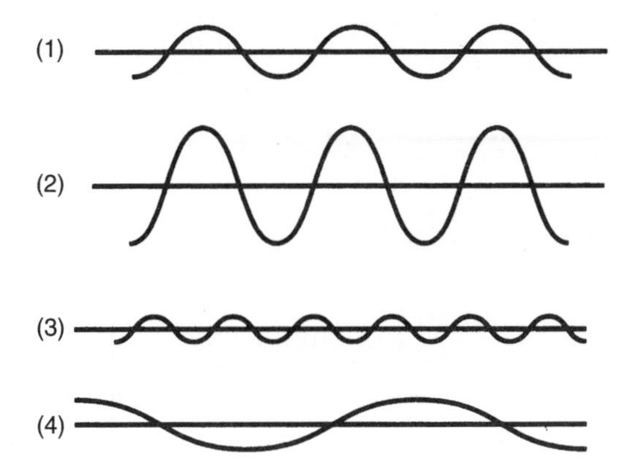

9. Which group of three appliances transforms electrical energy into mechanical energy?

(1) vacuum cleaner, table saw, fan

(2) television, fan, refrigerator

(3) radio, clock, fan

(4) lawn mower, hair dryer, generator

10. To break a strip of aluminum metal, Jason bent the aluminum strip back and forth many times. He was surprised to notice the aluminum getting warm where it bent. What type of energy transformation was happening?

(1) heat energy to mechanical energy

(2) chemical energy to mechanical energy

(3) chemical energy to heat energy

(4) mechanical energy to heat energy

11. The primary purpose of a car engine is to convert chemical energy into mechanical energy. That is, the energy in the gasoline is transformed into the motion of the car. After you drive a car a short distance, the engine gets hot. The heat

(1) is a gain of mechanical energy

(2) is the creation of heat energy

(3) is the destruction of heat energy

(4) is a form of wasted energy

Part II

Base your answers to questions 12 through 14 on the diagram below and your knowledge of science.

12. Explain how the diagram shows potential energy and kinetic energy.
13. Describe the type of energy transformation that occurs when you strike the match on the rough strip of the matchbook cover.
14. Describe the type of energy transformation that occurs when the match is burning.

---

15. Describe the type of energy transformation that takes place when you operate each of the following appliances.
    (a) gas stove
    (b) fan
    (c) hair dryer

Heat

Heat Energy

Heat is a form of energy produced by vibrating molecules. All matter is composed of molecules that vibrate constantly. When heat is added to a substance, the molecules move faster and farther apart. When heat is removed from a substance, the molecules move slower and come closer together.

Heating causes most substances to expand; cooling, the removal of heat, causes most substances to contract. Adding heat to a substance causes its molecules to move farther apart, which causes it to increase its size. Bridges, railroad tracks, and sidewalks have expansion spaces. This allows them to expand and contract freely in response to the great temperature changes that occur between summer and winter. Some thermometers contain alcohol that expands and contracts inside a glass tube, providing a way to measure temperature. (See Figure 2-7.)

Figure 2-7. *When the alcohol in a thermometer is heated, it expands and indicates a higher temperature. When heat is removed, the alcohol contracts and indicates a lower temperature.*

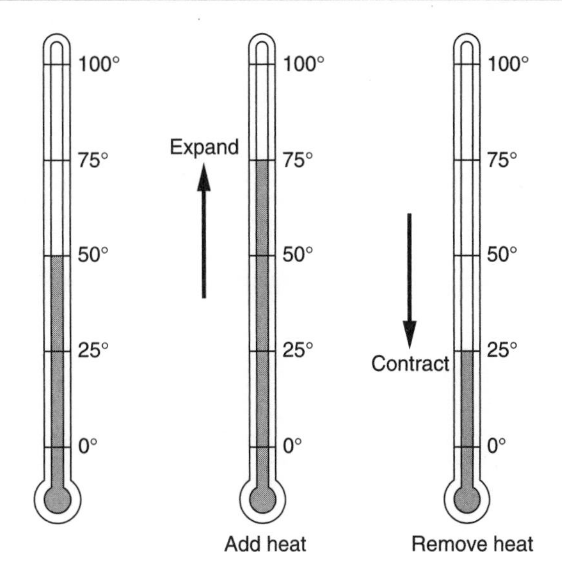

Although most substances expand when heated and contract when cooled, water is an exception. In liquid form, when water cools, it contracts. However, when water is cooled to 4°C, it starts to expand and continues expanding until it becomes ice at 0°C. This is why an unopened bottle of water or soda will crack if left outdoors during freezing weather. The force generated by the expansion of water changing to ice is so powerful that it can crack glass, rocks, concrete, and even steel.

Heat Transfer

Where there is a difference in temperature, heat moves from warmer places to cooler places. The tendency is for heat to become evenly distributed. Place a cold drink in a warm room; after a while, the drink and air will be at the same temperature. The heat in the air moved into the cold drink and caused the temperature of the drink to rise. Similarly, a bowl of hot soup transfers heat to its surroundings until the soup reaches room temperature. Heat can move by conduction, convection, or radiation.

Conduction is the transfer of heat by direct molecular contact. Metal objects conduct heat well. A metal spoon placed in a cup of hot tea gets warm because heat is easily transferred from the tea to the spoon and from molecule to molecule within the spoon. A metal pot on a hot stove quickly distributes the heat throughout the pot by conduction. On the other hand, materials that do not transfer heat well can be used to insulate or reduce the flow of heat. A pot holder reduces heat flow so you can grasp the metal handle of a hot frying pan without being burned. (See Figure 2-8 on page 72.)

Figure 2-8. *In a heated metal bar, heat is transferred by conduction.*

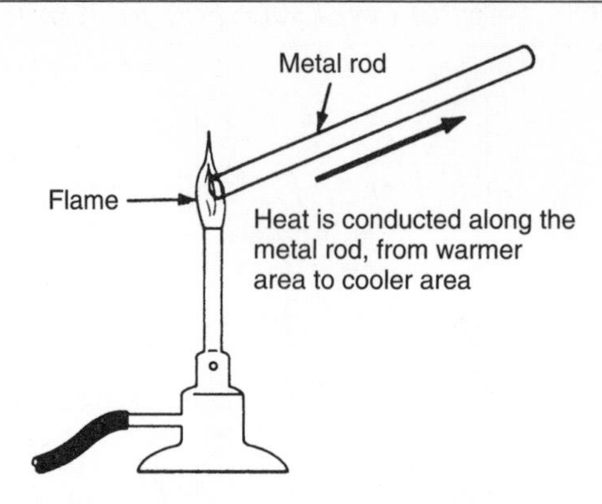

Convection is the transfer of heat by the flowing action within a liquid or gas. Warm air added to a room rises, and the cooler air sinks because of density differences. This causes a circular flow called a convection current that distributes the heat within a room. Figure 2-9 shows how a convection current occurs in a room and eventually equalizes the temperature.

Figure 2-9. *Home heating systems use convection currents to distribute heat in a room.*

Radiation is the transfer of heat in the form of waves. The heat from the sun reaches Earth in the form of radiation waves. Radiation waves do not need a medium; they are capable of traveling through the vacuum of space.

For example, light from the sun reaches Earth by traveling through 150,000,000 kilometers of empty space. Carefully place your hand several inches below a glowing lightbulb and you can feel the warmth of the bulb. The heat from the bulb reaches your hand by radiation. (See Figure 2-10.)

Figure 2-10. *Heat from the bulb reaches your hand by radiation.*

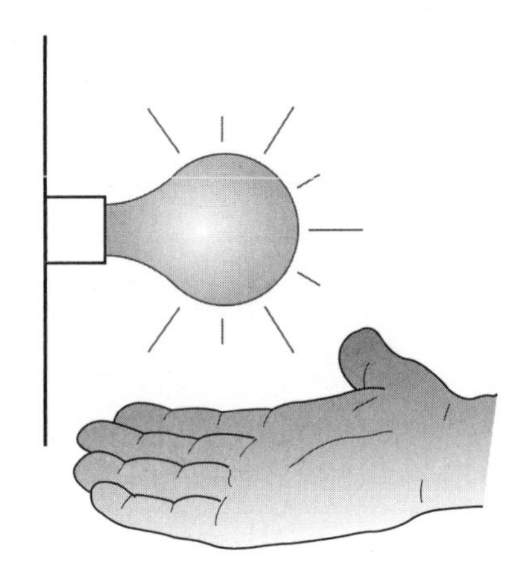

Review Questions

Part I

16. A form of energy produced by the vibrating motion of molecules is called
    (1) sound (3) mechanical
    (2) heat (4) nuclear

17. Removing heat from a substance causes the molecules in the substance to
    (1) stop moving (3) move slower
    (2) move faster (4) continue to move at the same rate

18. When the temperature rises, the alcohol in a thermometer
    (1) contracts (3) remains unchanged
    (2) expands (4) changes from liquid to gas

19. A picnic chest keeps warm objects warm or cold objects cold. A picnic chest must be made from
    (1) heat-conducting material (3) heat-radiating material
    (2) heat-insulating material (4) heat-expanding material

20. The end of a metal bar is placed in a flame. After a few minutes, thermometers at four points on the bar measure the temperature. Thermometer C reads 50°C. Which temperature is most likely the temperature of thermometer B?

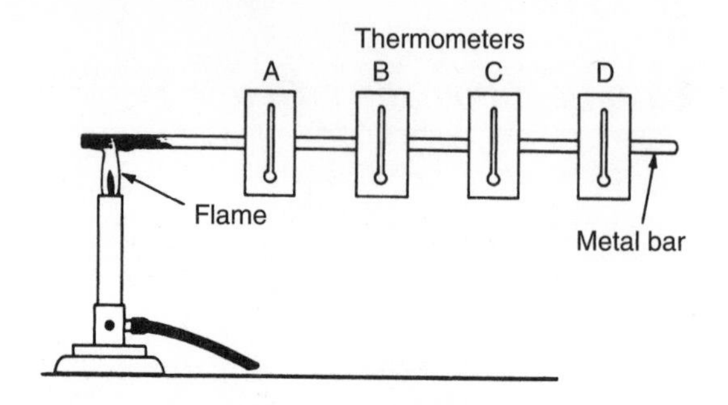

(1) 20°C
(2) 40°C
(3) 50°C
(4) 60°C

21. Earth receives heat from the sun by the process of

(1) conduction
(2) convection
(3) radiation
(4) air circulation

22. A closed bottle of water was left outdoors on a very cold night when the temperature was −5°C. The next morning the water was frozen, and the bottle was cracked. Why was the bottle cracked?

(1) Water contracts when it freezes.
(2) Water expands when it freezes.
(3) Water absorbs heat energy when it freezes.
(4) Water releases heat energy when it freezes.

23. Which graph best shows the relationship between the temperature of a substance and the motion of the particles in the substance when it is heated?

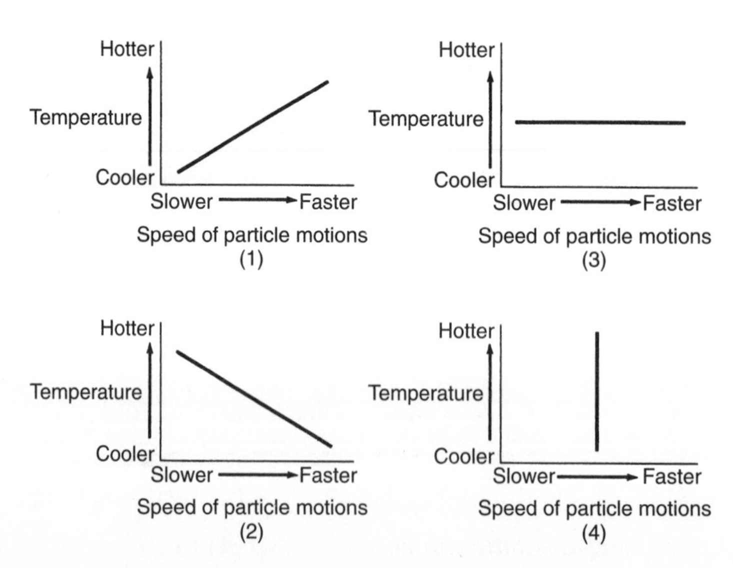

Part II

Base your answers to questions 24 through 26 on the diagram below and your knowledge of science. The diagram shows a 20°C metal spoon in a cup of 90°C hot water.

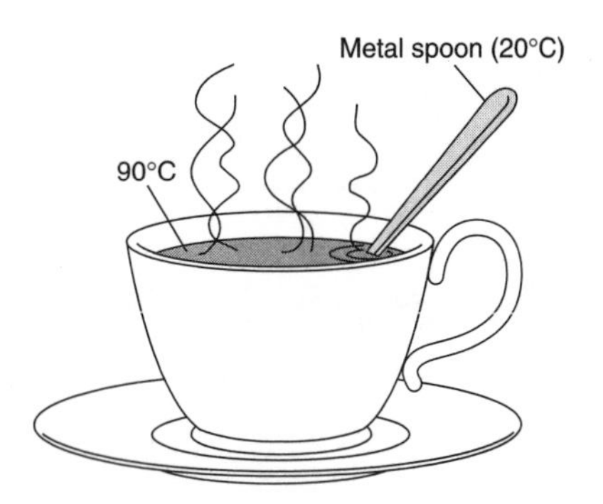

24. What heat transfer process will cause the spoon to get warmer?
25. Make a drawing of a spoon in a cup and place an arrow in the drawing to show the direction of the heat transfer process.
26. Predict what will happen to the temperature of the spoon and hot water over time.

Base your answers to questions 27 and 28 on the diagram below and your knowledge of science. The diagram shows the setup for a demonstration done in science class.

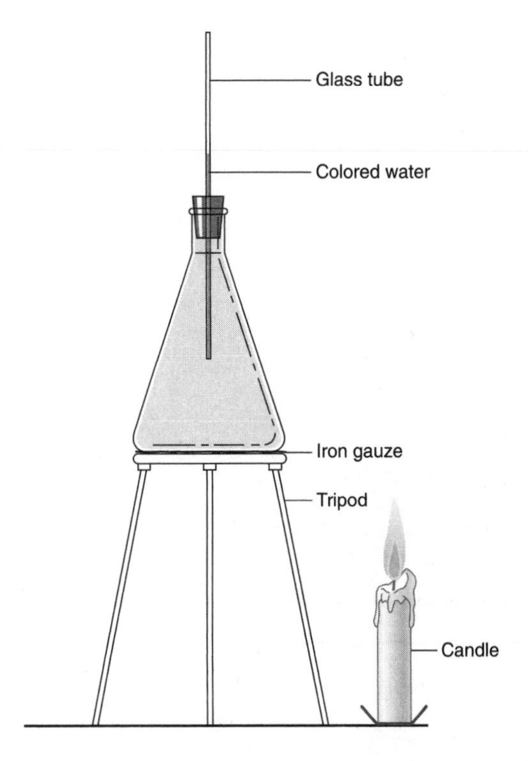

27. Describe what will happen to the liquid in the glass tube if the burning candle is placed under the flask.

28. What instrument works on the principle shown in this demonstration?

Base your answers to questions 29 and 30 on the diagram below and your knowledge of science.

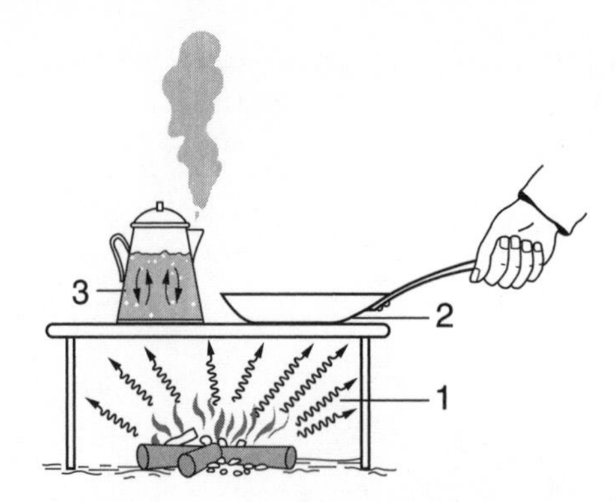

29. What type of heat transfer is occurring at each of the locations: 1, 2, and 3?

30. Describe the energy transformation taking place as the logs burn.

Magnetism and Electricity

Magnetism

Magnets have a north (N) pole and a south (S) pole of equal strength. If you cut a magnet in half, each half will have a north and a south pole. Hang a bar magnet from a string tied around its middle, and it will align itself with Earth's magnetic field. The pole of the magnet pointing toward the north direction is the north pole of the magnet, and the pole pointing toward the south direction is the south pole of the magnet. (See Figure 2-11.) A compass is a device that uses this principle to help us determine directions.

When the ends of two magnets are brought together, a push or pull force occurs. The law of magnetic poles states that like poles repel and un-

Figure 2-11. *A freely hanging magnet will align with Earth's magnetic field.*

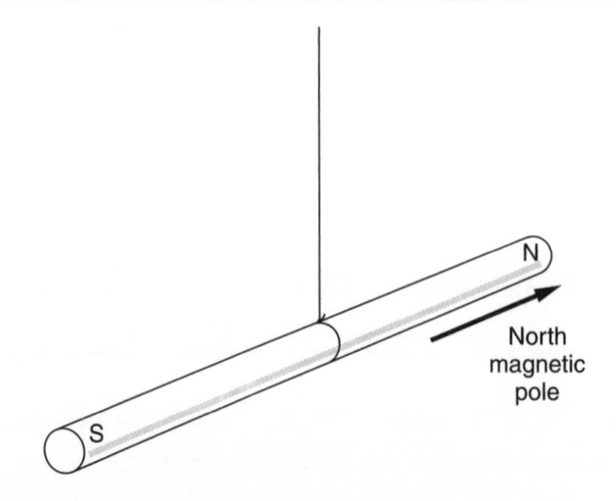

like poles attract. If two N poles or two S poles are brought together, you feel a pushing force as they repel each other; however, if a north and a south pole are brought together, you feel a pulling force as they attract each other. The closer the magnets the greater the force. (See Figure 2-12.)

Figure 2-12. *Like poles of a magnet repel, and unlike poles attract.*

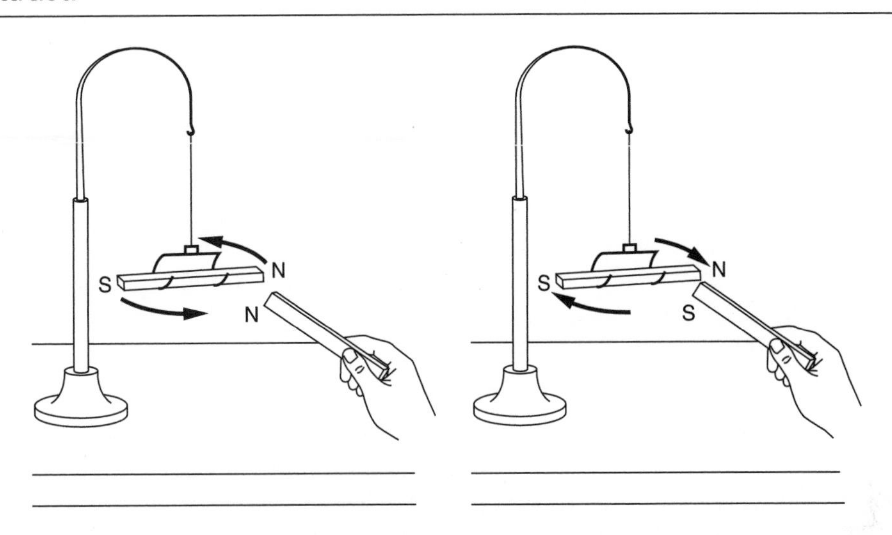

Around each magnet there is a magnetic field. (See Figure 2-13.) Place a sheet of paper over a magnet and sprinkle iron filings on the paper, and you can easily see the shape of the magnetic field. The iron filings are concentrated at the poles, showing that the lines of force are greatest at the poles. However, lines of force can also be seen around the whole magnet. Magnets attract items made of iron, cobalt, or nickel (see Table 2-2 on page 78). Bar magnets have their greatest strength at the ends of the bar called poles.

Figure 2-13. *The magnetic field around a magnet is concentrated around the poles.*

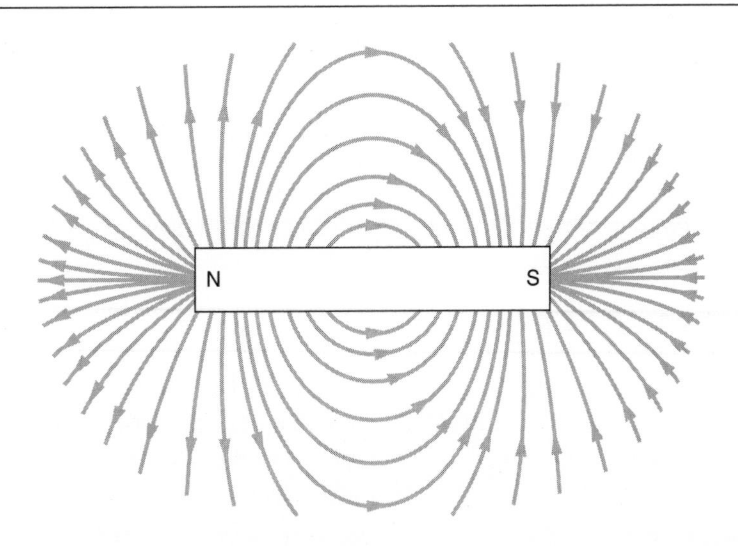

Table 2-2. Some Magnetic and Nonmagnetic Items

Magnetic	Nonmagnetic
Paper clip	Eraser
Staples	Pencils
Iron filings	Pieces of paper
File cabinet	Chalk
Carpenter's nail	Plastic pen

Static Electricity

Atoms normally contain an equal number of electrons (negative [−] particles) and protons (positive [+] particles). Therefore, they are neutral. If electrons are added, the atom becomes negatively charged. If electrons are removed, the atom becomes positively charged. Objects contain many atoms, so they can gain or lose many electrons. An object can become negatively charged if it has gained electrons and positively charged if it has lost electrons.

When a glass rod is rubbed with a silk cloth, electrons are transferred from the glass rod to the silk. The glass rod becomes positively charged. (The silk becomes negatively charged.) When a rubber rod is rubbed with a wool cloth, electrons are transferred from the wool to the rod. The rubber rod becomes negatively charged. (The wool becomes positively charged.) The surface of these objects develops an electric charge called *static electricity*. This is the same static electricity that gives you a shock when you walk across a wool or nylon rug, then touch a metal object. On a much larger scale, lightning is also a form of static electricity.

The law of electric attraction and repulsion states that objects with like charges repel each other while objects with unlike charges attract each other. Hang two balloons from the ceiling so they just touch each other. Rub each with a wool or flannel cloth. Each balloon now has a negative charge, and they repel each other. Rub a glass rod with a silk cloth. Bring the positively charged glass rod toward a negatively charged balloon, and the balloon is attracted to the glass rod. (See Figure 2-14.)

When you run a comb through your hair, you sometimes hear a crackling noise. What you hear is the result of static electricity. The comb gains electrons and becomes negatively charged. Rip a sheet of paper into tiny pieces about 1 cm across and place them on the tabletop. The electrically charged comb will attract and pick up the pieces of paper.

Electric Energy

Electricity is a form of energy produced by the flow of electrons from one place to another. Electricity is usable when the flow of electrons is contained in a metal wire. People have found many uses for electric energy. For instance, the electricity in your home can be used to power lightbulbs, air conditioners, television sets, and many other appliances.

Figure 2-14. *(A) Negatively charged balloons will repel, and (B) a negatively charged balloon is attracted to a positively charged glass rod.*

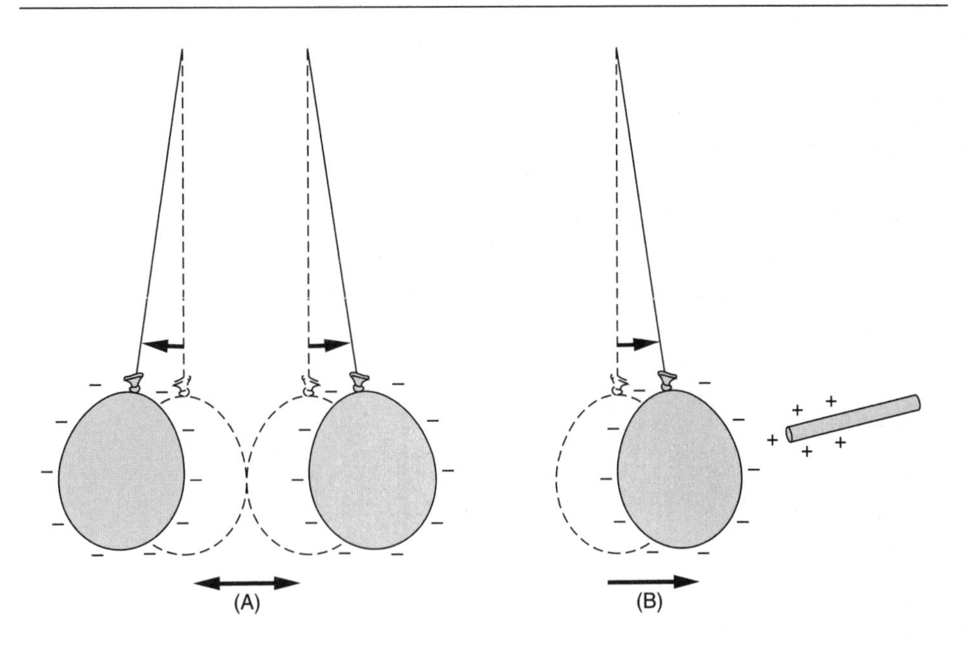

Electric ***conductors*** are materials through which electricity moves easily. Most metals are good conductors. Substances that resist the flow of electric current are ***insulators***. Rubber, glass, and plastic are common insulators.

Electric wires in your home are made of a conductor, like copper or aluminum, wrapped in a protective coating of an insulator, like rubber or plastic. (See Figure 2-15.) These wires provide a safe path for electricity.

Figure 2-15. *An electric wire consists of a conductor (metal) wrapped in an insulator (plastic). The three-pronged plug is also coated with a plastic insulator.*

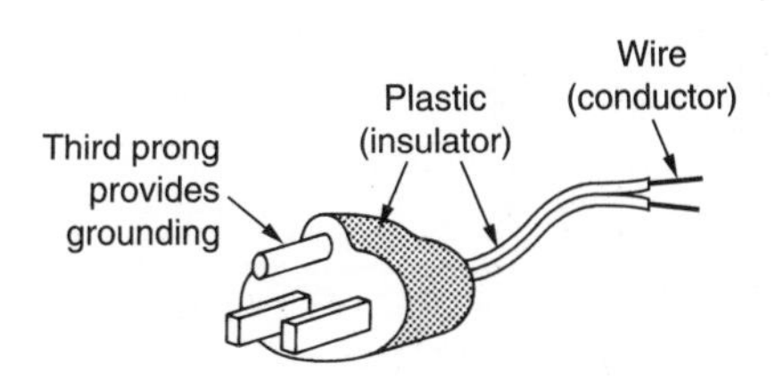

Production of Electricity

An electric current is produced when electrons move through a conductor, such as a copper wire. The two most common methods of producing electricity are electromagnetic induction and chemical action.

Moving a magnet through a coiled copper wire causes, or induces, a flow of electrons through the wire. (See Figure 2-16 on page 80.) Generators (see Figure 2-3 on page 65.) are devices that spin a coiled wire within a magnetic field to produce electricity. In a generator, the mechanical energy of a spinning turbine is transformed into electric energy for human use. The turbine may be powered by running water or wind, or steam produced by burning coal or oil. The electricity in our homes, schools, and office buildings is produced primarily by this method.

Figure 2-16. *Moving a magnet through a coiled wire produces an electric current.*

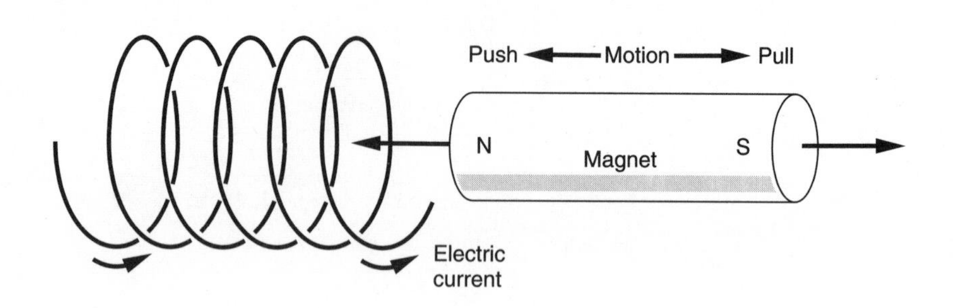

A battery is a device that changes chemical energy into electric energy. Certain chemicals, such as zinc and carbon, produce a flow of electrons when placed in a conducting material. (See Figure 2-17.) We use batteries in cars, portable radios and clocks, laptop computers, and many other devices.

Figure 2-17. *Cross section of a dry cell battery.*

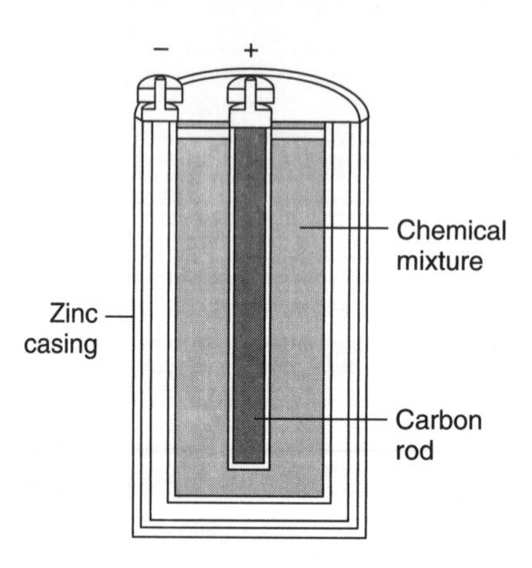

An ***electric circuit*** is a complete path for the flow of electrons. A circuit must contain a source of electric energy, a conducting path for the flow of electrons, and a device that uses the electric current. The source of electricity could be a battery or a generator. Wires usually provide the path for the electricity in a circuit, and the device could be a lamp or a CD player.

At least two wires are needed for a complete circuit, one wire to carry a flow of electrons to the device and another to carry electrons back to the source of electricity. In addition, a circuit often includes a switch that turns the flow of electricity on and off. Figure 2-18 shows a simple electric circuit with an on-off switch.

Figure 2-18. *A simple electric circuit, with an on-off switch.*

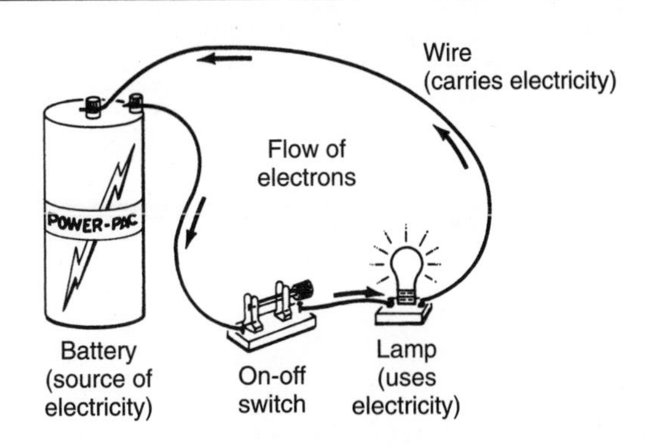

There are two types of electric circuits: series and parallel.

A ***series circuit*** has a single path for electricity as it passes through devices attached to it. (See Figure 2-19.) The same amount of electric current flows through each of the devices. Some colored holiday lights are wired in a series circuit. If any of the lightbulbs are removed or burn out, the electric path is broken, and all the bulbs go out.

Figure 2-19. *In a series circuit, electricity flows through each lightbulb in a single, continuous path.*

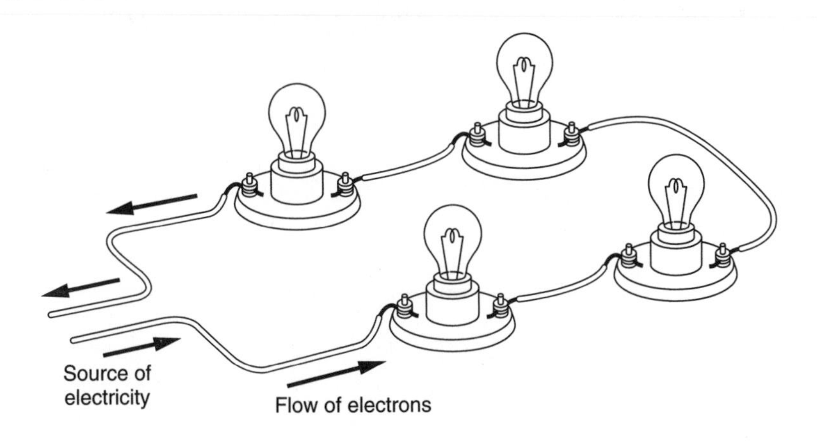

A ***parallel circuit*** has two or more devices wired so that the electricity flows through a separate branching path to each device. (See Figure 2-20 on page 82.) In a string of lightbulbs connected in a parallel circuit, if any of the bulbs are removed or burn out, the electricity continues to flow through the remaining bulbs, so they stay lit.

Figure 2-20. *In a parallel circuit, electricity enters different branches to flow to each lightbulb.*

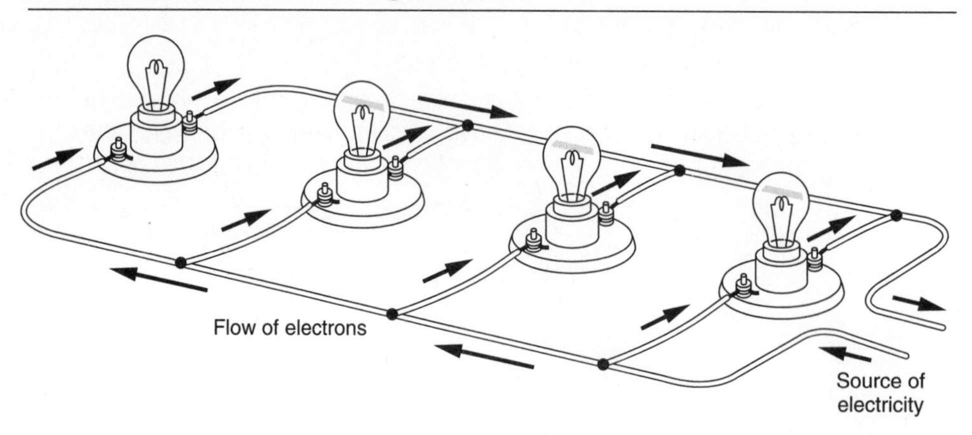

Safe Use of Electricity

Electricity is dangerous and can injure or kill living things. Though it has become a common part of our lives, electricity should always be handled carefully.

When too many appliances are using electricity from one circuit at the same time, the circuit can become overloaded. This overload makes the wires heat up, which can cause an electrical fire. ***Fuses*** and ***circuit breakers*** prevent overloading of circuits by automatically stopping the flow of electricity when it reaches a dangerous level. Figure 2-21 shows how a fuse works.

Figure 2-21. *How a fuse works: When the electric circuit begins to overheat, the metal strip in the fuse melts, breaking the flow of electricity.*

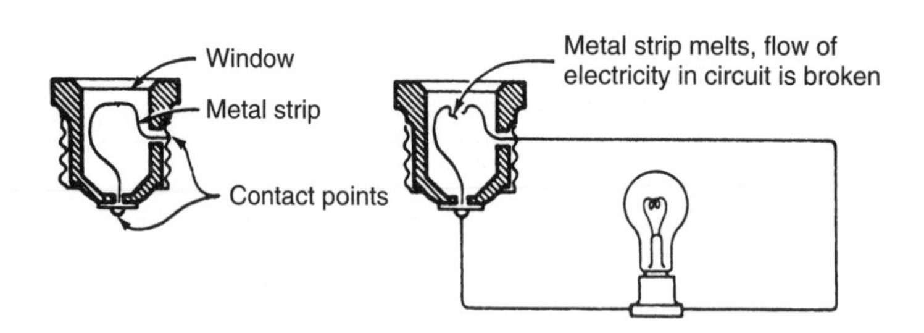

Grounding of electric appliances is another important safety precaution. Sometimes an electric charge can build up in an electric device. A person touching the device may receive a severe shock from this excess charge. Grounding helps prevent such accidents.

To ground an electric device, a wire is run from the device to a conducting material, such as a metal pipe, that is in contact with the earth. This provides a safe outlet for any excess charge in the device by conducting it to the earth, where it is absorbed. Some appliances come equipped with a three-pronged plug. (See Figure 2-15 on page 79.) The third prong provides grounding and should always be used.

You should observe commonsense safety rules whenever using electricity. Always unplug electric devices before cleaning or repairing them. Do not use electric wires if they are broken or frayed, and replace plugs if the wire is broken near the plug. Never handle electric appliances while you are in water or if you are wet, because water can act as a conductor.

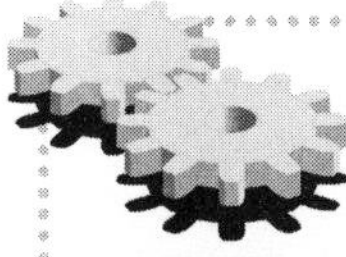

Laboratory Skill

Measuring Electrical Conductivity

A simple circuit can be used to measure the conductivity of solids. The diagram shows an electric circuit containing a source of electricity, a bulb, a switch, and wires with alligator clips. Strips or bars of various test materials are easy to hold with the clips. To test an item, you attach the clips to the item, close the switch, and note whether the bulb lights. If the bulb lights, the item conducts electricity, and if the bulb does not light the item is a nonconductor.

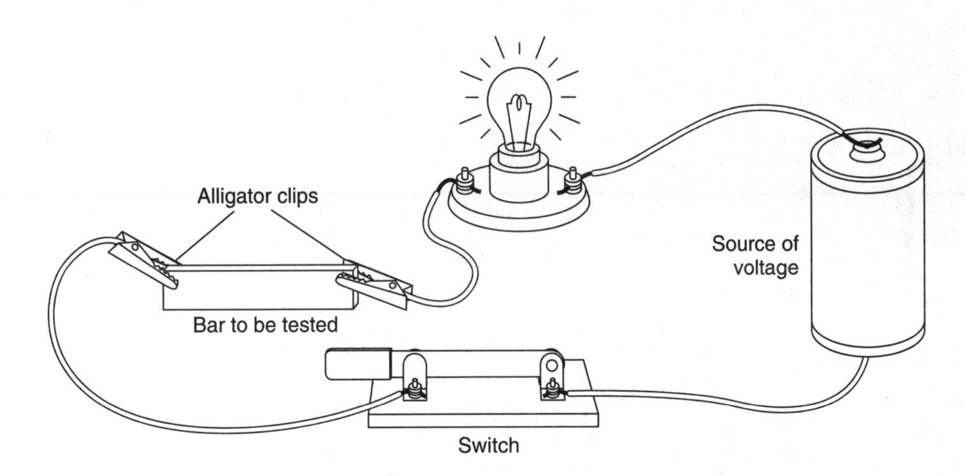

Questions

1. A student used a simple circuit to test the conductivity of the items listed below. Which of these items would most likely conduct electricity?

(1) plastic comb
(2) glass stirrer
(3) plastic stirrer
(4) aluminum foil
(5) brass screw
(6) wooden splint
(7) copper wire
(8) pencil

2. Two unknown wires (wire A and wire B) were tested with the conductivity apparatus. When wire A was used, the lightbulb was bright, and when wire B was used, the lightbulb was dim. Explain what this tells you about the conductivity of the wires.
3. Why are electrical wires covered with plastic or rubber?

Review Questions

Part I

31. Base your answer on the diagram below. Describe what happens when the north pole of magnet A is brought close to the north pole of the suspended magnet B. The magnets

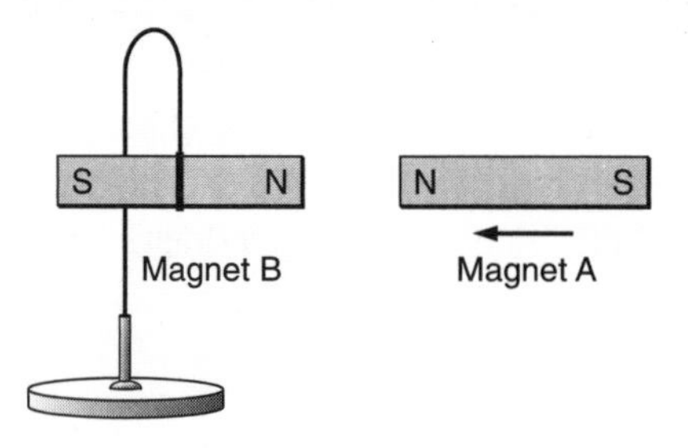

(1) attract each other
(2) have no effect on each other
(3) repel each other
(4) first repel then attract each other

32. If a bar magnet is cut in half, each piece would have

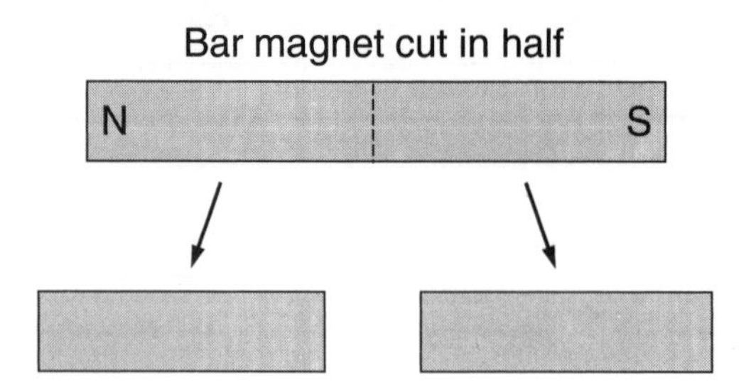

(1) two north poles
(2) two south poles
(3) no north or south pole
(4) one north pole and one south pole

33. Harold removed his sweater in a dark room. As he took his sweater off, he saw static sparks in the material. What caused the static sparks?

(1) dampness in the air

(2) faulty electric wiring

(3) a transfer of heat

(4) movement of electrons

34. Which is a good insulator of electricity?

(1) plastic

(2) copper

(3) aluminum

(4) nickel

35. When the switch is closed in the electric circuit illustrated below, the bulb does not light. What item is missing from the electric circuit?

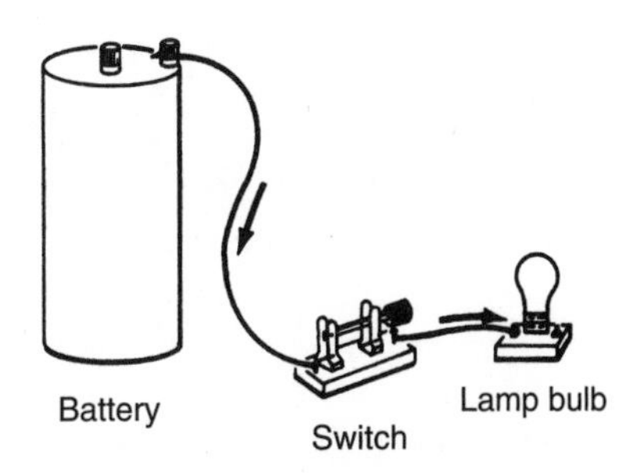

(1) a wire to complete the flow of electricity

(2) a source of electricity

(3) a device to use the electricity

(4) a fuse

36. Electricity is defined as a

(1) flow of electrons

(2) flow of heat

(3) complete circuit

(4) magnetic field

37. Which device is capable of transforming mechanical energy into electrical energy?

(1) motor

(2) generator

(3) battery

(4) fuse

Part II

Base your answers to questions 38 through 40 on the diagram below and your knowledge of science. The diagram shows an electric circuit.

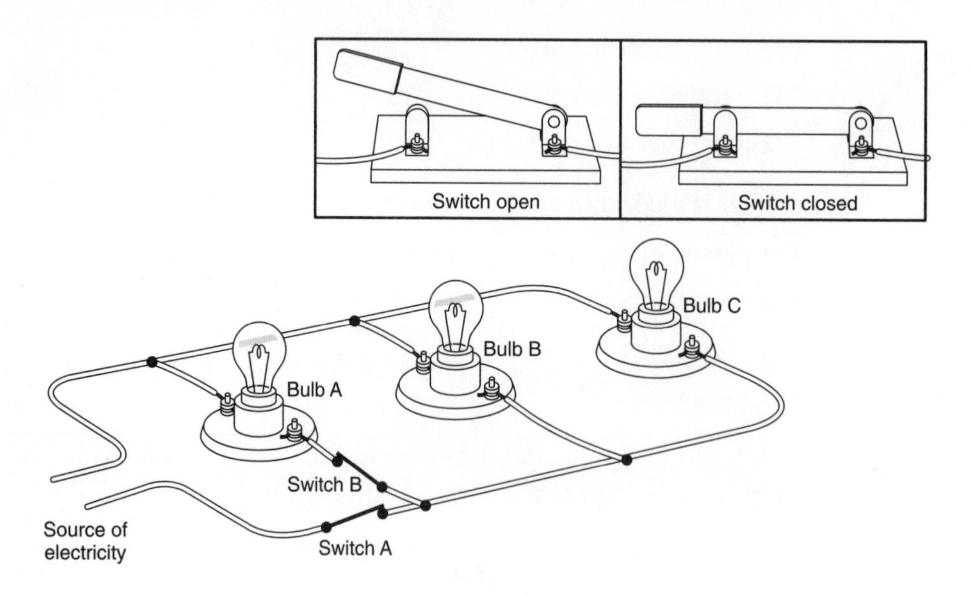

38. What type of electrical circuit is shown in the diagram?
39. If switch A is open and switch B is closed, which lightbulbs will be on and which lightbulbs will be off?
40. Explain what will happen to lightbulbs A and C if lightbulb B is removed.
41. Describe how you can prove your home is a parallel circuit and not a series circuit.

Base your answers to questions 42 through 44 on the diagram below and on your knowledge of science. The diagram shows a balloon with a negative charge and a glass rod with a positive charge.

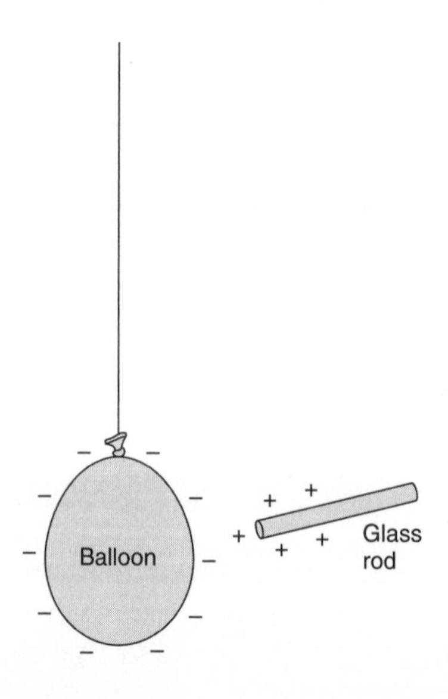

42. Describe what will happen as the glass rod is brought closer to the balloon.
43. How did the glass rod obtain the positive charge?
44. Describe another example of static electricity.

Sound

Sound Energy

Sound is a form of mechanical energy produced by a vibrating object. A sound wave is created when an object moves back and forth rapidly. In air, this back-and-forth motion pushes and pulls the surrounding air molecules, producing alternating layers of molecules that are closer together (compressed) and farther apart (expanded). The compression and expansion of molecules travel in the form of longitudinal waves (see Figure 2.6 on page 67) called ***sound waves***. (See Figure 2-22.) These sound waves spread outward in all directions from their source, somewhat like the circular ripples that are produced when you toss a pebble into a calm pool of water.

Figure 2-22. *A vibrating bell produces sound waves that spread outward in all directions.*

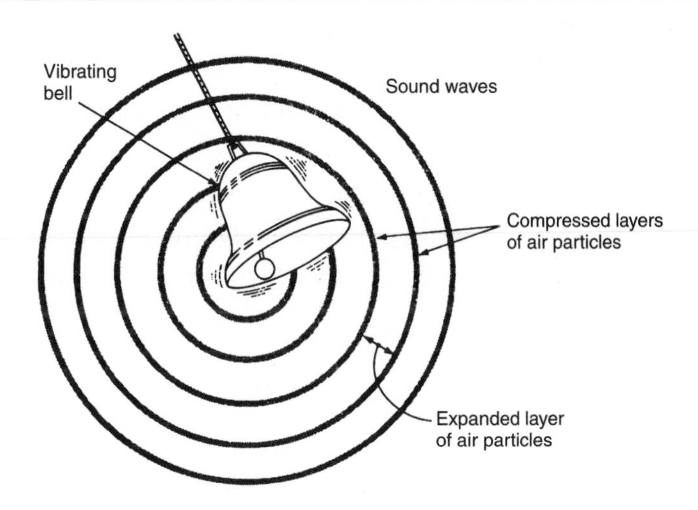

Objects that can produce sound include bells, radio speakers, guitar strings, or anything else that can vibrate. For instance, the sound of your voice is caused by the vibrating vocal cords in your throat. If you place your hand on your throat while you speak, you can feel the vibrations that produce the sound.

To be heard, a sound must be transmitted from an energy source to your ears. The substance that sound travels through is called a *medium*. A medium can be a solid, liquid, or gas. Most sounds we hear travel through air, a gas.

Sound waves cannot travel through a vacuum because there are no particles of matter in a vacuum to vibrate and transmit the sound.

Sound Waves

Although sound waves are longitudinal waves, they can also be represented on a graph that resembles a transverse wave. (See Figure 2-6 on page 67.) The crest of each wave represents the compressed-particle portion of the wave (the compression), and the trough represents the expanded-particle portion (the rarefaction). The larger the amplitude of the wave, the louder the sound.

Wave frequency is commonly expressed as the number of vibrations per second. Humans can hear sounds that range from 20 to 20,000 vibrations per second. A normal speaking voice ranges between 100 to 1000 vibrations per second and has a wavelength of 0.3 to 3.5 meters. The pitch describes how high or low the sound is. A high-frequency sound has a high pitch, and a low-frequency sound has a low pitch. A trumpet can produce high-pitched sounds, and a tuba can produce low-pitched sounds.

To a motionless observer, the frequency of sound waves produced by the horn on a moving train or the siren on an ambulance appears to change. The frequency increases as the source moves toward the listener. The frequency decreases when the source moves away from the listener. This change in sound frequency, called the ***Doppler effect***, is recognized as a change in pitch. When a blaring ambulance siren is moving toward you, the sound waves are crowded together, producing a higher frequency and a higher pitch. When the siren is moving away from you, the sound waves spread out, producing a lower frequency and a lower pitch. (See Figure 2-23.)

Figure 2-23. *The Doppler effect: When the source of sound moves toward the listener, the frequency and pitch increase; when the source moves away form the listener, the frequency and pitch decrease.*

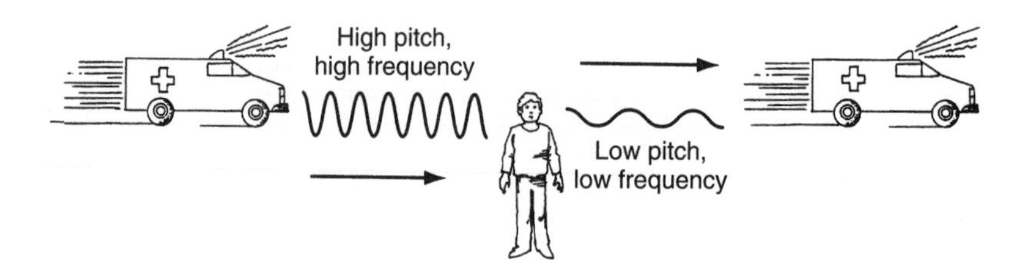

The Speed of Sound

The speed of sound depends primarily on the density of the substance, or medium, through which it is passing. The denser the medium, the faster the sound waves travel through it. Generally, sound travels fastest through solids, which have the greatest density, and slowest through gases, which have the least density. Table 2-3 gives the speed of sound through several substances.

To a lesser extent temperature also affects the speed of sound. In air, the higher the temperature, the faster sound waves travel through it.

Table 2-3. *Speed of Sound Through Different Substances (at 25°C)*

Medium	State	Speed (m/s)
Iron	Solid	5200
Glass	Solid	4540
Water	Liquid	1497
Air	Gas	346

Although the speed of sound can vary, it is always much slower than the speed of light. During a thunderstorm, for instance, a lightning bolt produces a flash of light and a clap of thunder at the same time. The speed of light is so fast that the light reaches us almost instantly. The sound of the thunder travels much more slowly, so we usually hear the thunder several seconds after we see the lightning.

Process Skill

Determining the Speed of Sound From a Graph

Sound travels at different speeds through different substances. The speed of sound is faster in solids, such as stone or metal, and slower in liquids and gases, such as water and air. The speed of sound is also affected by temperature. The graph shows the relationship between air temperature and the speed of sound in air. Study the graph and answer the following questions.

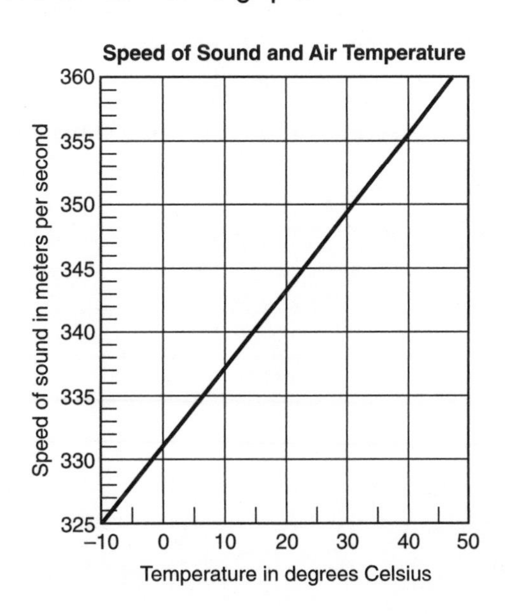

Questions

1. In air, sound travels at a speed of 350 meters per second at about which temperature?
(1) −10°C (2) 31°C (3) 39°C (4) 47°C

2. At a temperature of 15°C, sound travels at about
(1) 330 meters per second
(2) 337 meters per second
(3) 340 meters per second
(4) 344 meters per second

3. What does the graph suggest about the relationship between air temperature and the speed of sound?
(1) As air temperature decreases, the speed of sound increases.
(2) As air temperature increases, the speed of sound remains the same.
(3) As air temperature increases, the speed of sound increases.
(4) As air temperature increases, the speed of sound decreases.

Review Questions

Part I

45. Sound is a form of energy produced by
(1) an expanding object
(2) a contracting object
(3) a heated object
(4) a vibrating object

46. A common characteristic of sound waves is that they
(1) travel much slower than the speed of light
(2) travel much faster than the speed of light
(3) travel fastest through a vacuum
(4) always travel at the same speed

47. The line represents sound in the form of a transverse wave. The diagram shows

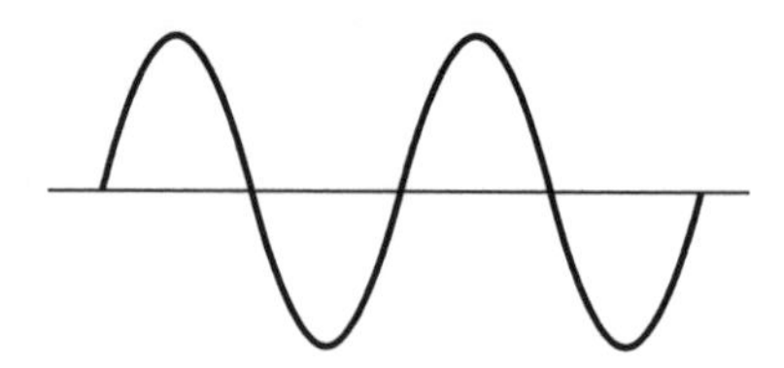

(1) one wave
(2) two waves
(3) four waves
(4) an incomplete wave

48. The diagram below shows a ringing bell in a jar. When the air is pumped out of the jar, you can no longer hear the bell ringing. This demonstrates that sound

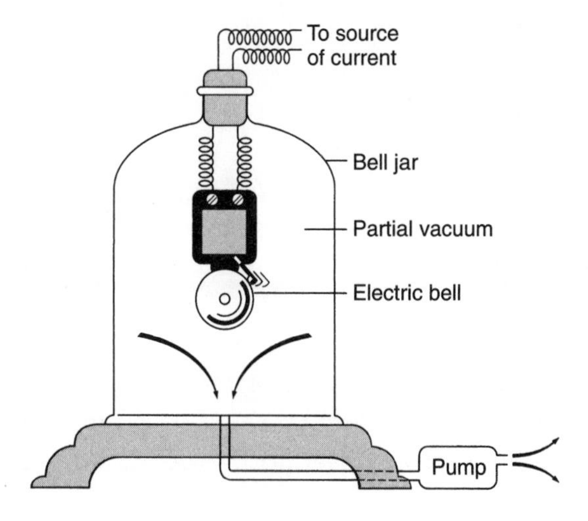

(1) can travel through glass
(2) cannot travel through glass
(3) can travel through a vacuum
(4) cannot travel through a vacuum

49. An ambulance blows a siren at a constant pitch as it approaches, crosses, and leaves a road intersection. A person standing by the road intersection hears the pitch of the siren rise as the ambulance approaches, and then the pitch gets lower as the ambulance moves away. This is caused by the
(1) loudness of the sound
(2) Doppler effect
(3) amplitude of the sound wave
(4) different sirens on the ambulance

Part II

Base your answers to questions 50 through 52 on the passage, diagram, and your knowledge of science. The passage and diagram discuss and show how an echo is produced.

An echo is produced when sound waves reflect from a surface and return to the sender. Sound travels at about 340 m/s in air. Josh yelled "hello" across a canyon. The sound traveled across the canyon and returned as an echo. Josh heard the echo two seconds after he yelled "hello." Use the relationship distance equals velocity multiplied by time ($d = v \times t$) to help you answer the questions.

50. How far did the sound travel from Josh, across the canyon, and back to Josh?
51. What is the distance across the canyon?
52. If the canyon is 680 meters across, how long would it take for Josh to hear the echo?

Base your answers to questions 53 through 55 on the diagram below and your knowledge of science. The diagram shows four sound waves with different properties.

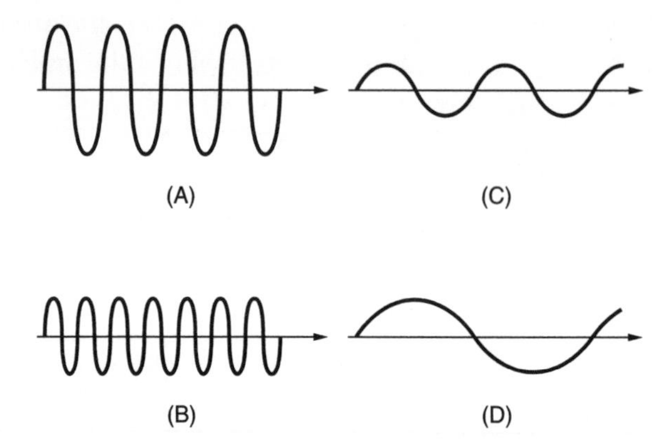

53. Which wave has the highest amplitude?
54. Which wave has the longest wavelength?
55. Assuming all waves are traveling at the same speed, which wave has the highest frequency?

---

56. In air at 20°C, sound travels at 343 m/s. As the temperature of air rises, the speed of sound increases at the rate of 0.6 m/s per degree Celsius. What is the speed of sound at 30°C? (Show all work.)

Light Energy

We can define light as a visible form of radiant energy that travels in waves. We can detect light with our eyes. Some properties of light indicate that it consists of tiny bundles of energy. However, other properties indicate that it travels in waves. The waves travel in straight-line paths called rays, which move away from an energy source in all directions. Light rays cannot bend around solid objects. Objects in their path block the light rays and may form shadows. (See Figure 2-24.)

Figure 2-24. *Light waves travel in straight paths. Shadows are produced when an object blocks their path.*

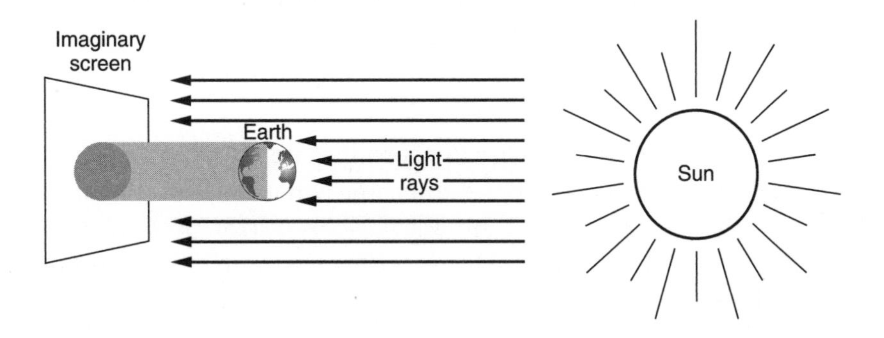

Unlike sound, light can travel through a vacuum. Light from the sun travels through the vacuum of space to reach Earth. At the speed of light, 300,000 km/s (186,000 mi/s), sunlight takes about 8 minutes and 20 seconds to reach Earth.

The sun is our main source of light energy. Fire and lightning are other sources of natural light. Light can also be produced artificially, as it is in a lightbulb.

Light Can Be Reflected, Absorbed, or Transmitted

When light strikes the surface of an object, three things can happen to the light rays, as shown in Figure 2-25 on page 94:

1. Light rays can bounce off the surface of the object. This is *reflection*. A shiny, metal surface reflects most of the light that strikes it.
2. Light rays can be *absorbed* by the object and transformed into heat energy. Much of the light that strikes a blacktop road is absorbed and transformed into heat energy.

3. Light rays can pass through the object. This is *transmission*. Clear glass allows most light that strikes it to pass through it.

Figure 2-25. *When light strikes a surface, it may be reflected, absorbed, and/or transmitted.*

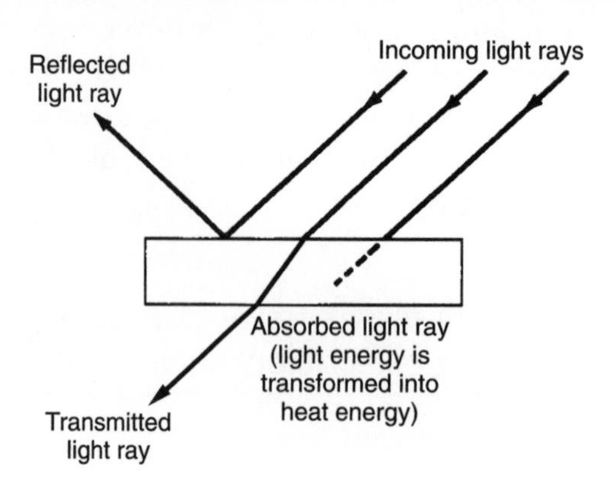

We see objects because they produce light or they reflect light. Light rays enter our eyes, and our brain interprets what we see. Accurate reflections are produced by smooth and shiny surfaces. A mirror gives an accurate reflection because it has a smooth, shiny surface. A wall in your home, because its rough surface, scatters light in all directions.

Colored objects absorb light to varying degrees. Dark-colored objects absorb more light than do light-colored objects. Because absorbed light changes to heat energy, dark-colored objects heat up more than light-colored objects when exposed to sunlight. For this reason, people usually wear light-colored clothing to keep cool during hot, sunny weather.

Materials also differ in their ability to transmit light. *Transparent* materials, such as window glass, permit almost all of the incoming light to pass directly through them. *Translucent* materials, such as wax paper, let some light pass through but scatter the light rays, so the images are not transmitted clearly. *Opaque* materials, like wood and iron, do not allow any light to pass through.

Lenses

Refraction is a change in the path (bending) of light rays that happens when light passes at an angle from one medium into another. For example, when a ray of light from an object passes from water into air, or from air into glass, the light is bent, causing the object to look larger, smaller, or moved out of place. Refraction causes a pencil in a glass of water to look broken or bent where it enters the water. (See Figure 2-26.) Light rays reflecting from the part of the pencil that is underwater are refracted as they pass out of the glass from the water into the air. The light rays reflected from the part of the pencil that is not underwater are not refracted, causing the pencil to look bent. The principles of refraction are used in making lenses for cameras, eyeglasses, and telescopes.

Figure 2-26. *The refraction, or bending, of light rays as they pass at an angle other than 90° from one medium to another makes the pencil look bent or broken.*

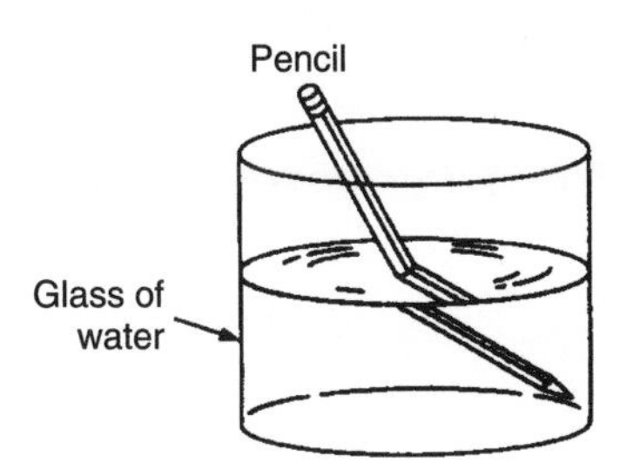

A ***lens*** is a piece of transparent glass or plastic that has curved surfaces. The curved surfaces refract light rays that pass through the lens. The shape of a lens determines how it bends light, as shown in Figure 2-27. Lenses with surfaces that curve outward (*convex*) refract light rays so that they are focused in toward a common point. Lenses with surfaces that curve inward (*concave*) refract light rays so that they spread out.

Figure 2-27. *The shape of a lens determines how it bends light.*

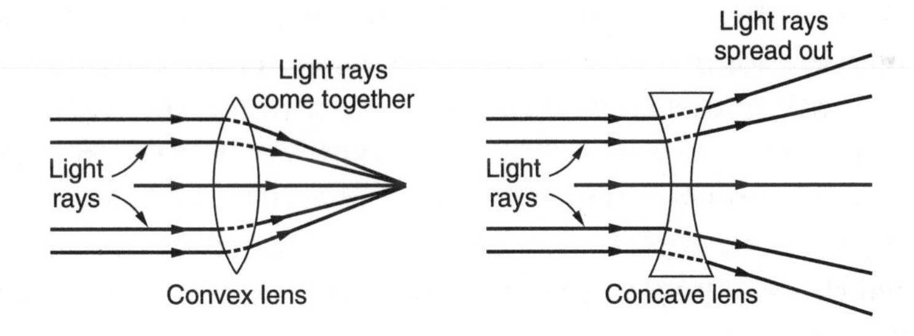

Images of objects seen through lenses may be larger than, the same size as, or smaller than the objects themselves. For instance, the lens of a camera forms smaller images of objects. A photocopy machine has a lens that forms images the same size as the original object. Binoculars contain lenses that magnify objects, making them appear larger.

The Electromagnetic Spectrum

Electromagnetic energy, which can travel through the vacuum of space, moves in waves. They are called electromagnetic waves because they contain both an electrical field and a magnetic field. The waves are created by electrically charged particles that vibrate rapidly. Types of electromagnetic energy include radio waves, microwaves, infrared waves, visible light, ultraviolet light, x-rays, and gamma rays. Together, these energy waves form a continuous band of waves called the ***electromagnetic spectrum***. (See Figure 2-28.) The waves differ in frequency and wavelength, but all travel at 300,000 km/s (186,000 mi/s) through empty space. Light travels at different speeds through different mediums. For example, light travels more slowly through water than it does through air.

Figure 2-28. *The electromagnetic spectrum shows the frequency and wavelength range of electromagnetic waves.*

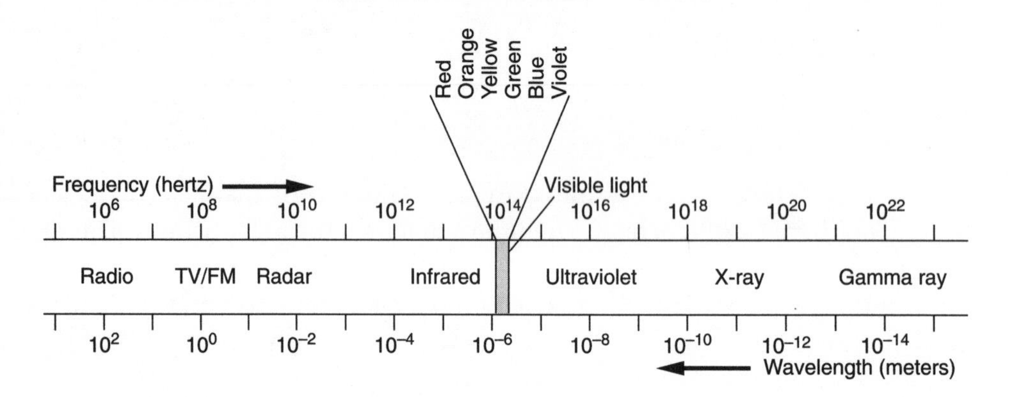

Electromagnetic waves affect our daily lives in many ways. Radio waves are used for radio and television broadcasting. Microwaves are used in communications and microwave ovens. X-rays are used to diagnose illnesses and injuries. Ultraviolet rays cause the skin to produce vitamin D. Visible light is also an electromagnetic wave.

Ordinary white light makes up only a narrow part of the electromagnetic spectrum. White light is divided into the color spectrum, which is based on wavelength. The colors of the spectrum, arranged in order from longest wavelength to shortest wavelength are red, orange, yellow, green, blue, and violet. A rainbow appears in the sky when "white" sunlight passes through raindrops and is scattered into different wavelengths, or colors that make up white light.

The color that you see when you look at an object is the result of the wavelengths of light it reflects and absorbs. For example, a blue car looks blue because it reflects the blue wavelengths of light and absorbs all other wavelengths.

Overexposure to some electromagnetic waves can be harmful to living things. We should be especially careful to limit our exposure to electromagnetic radiation, such as x-rays and ultraviolet rays, which have been linked to genetic mutations and cancer.

Review Questions

Part I

57. A track official fired a gun to start the one-mile race. A person seated far across the stadium saw the flash of the gun and the racers started running. A few seconds later the person heard the sound of the gun. The reason for the delay between the flash of the gun and the sound was that

    (1) sound cannot travel through air

    (2) the gun was fired twice

    (3) light travels much faster than sound

    (4) light travels much slower than sound

58. In the diagram below, the student can see the flame when he looks through the three holes in the cards. What property of light does this demonstrate?

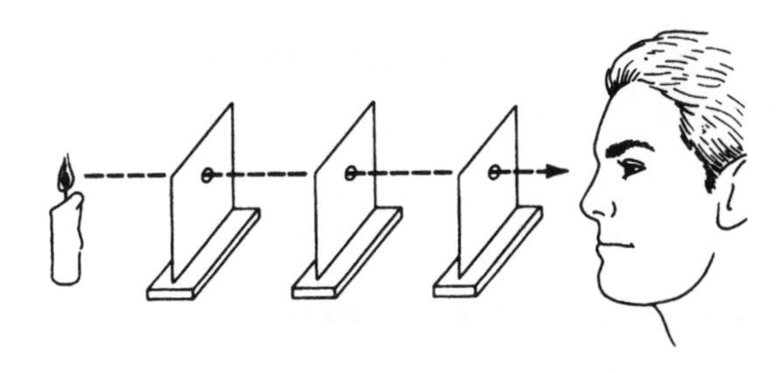

    (1) Light rays travel through transparent objects.

    (2) Light rays are absorbed by some objects.

    (3) Light rays reflect off shiny objects.

    (4) Light rays travel in straight paths.

59. Objects are visible when they

    (1) reflect light

    (2) refract light

    (3) transmit light

    (4) absorb light

60. When white light strikes a red stop sign,

    (1) all of the light is reflected except red light, which is absorbed

    (2) all of the light is absorbed except red light, which is reflected

    (3) all of the light is reflected

    (4) all of the light is absorbed

61. Light is a form of
    (1) heat energy
    (2) electromagnetic energy
    (3) mechanical energy
    (4) chemical energy

62. A glass lens can make an object look larger by
    (1) refracting light
    (2) reflecting light
    (3) transmitting light
    (4) absorbing light

63. Visible light, ultraviolet light, and infrared light are forms of
    (1) mechanical energy
    (2) radiant energy
    (3) heat energy
    (4) potential energy

64. All types of electromagnetic energy
    (1) have the same wavelength
    (2) have the same frequency
    (3) have the same amplitude
    (4) travel at the same speed

65. Which diagram best shows the property of refraction?

(1)

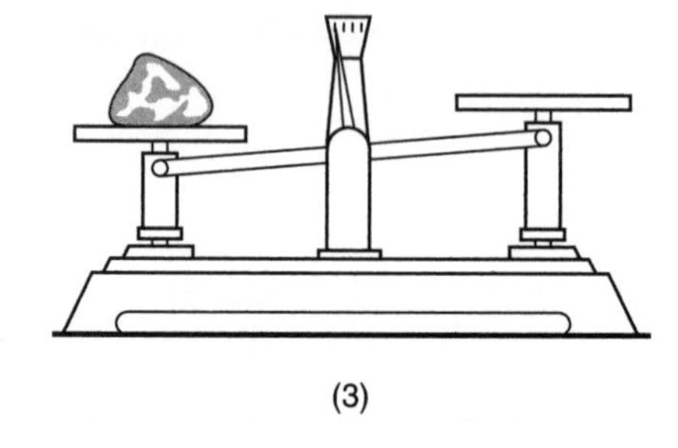
(3)

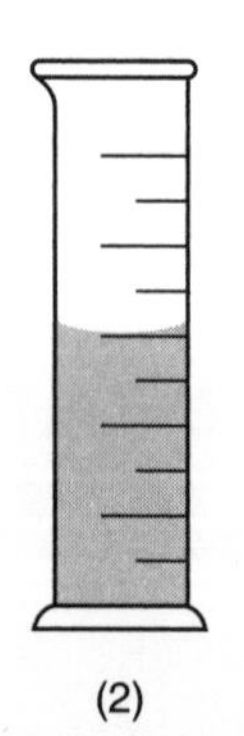
(2)

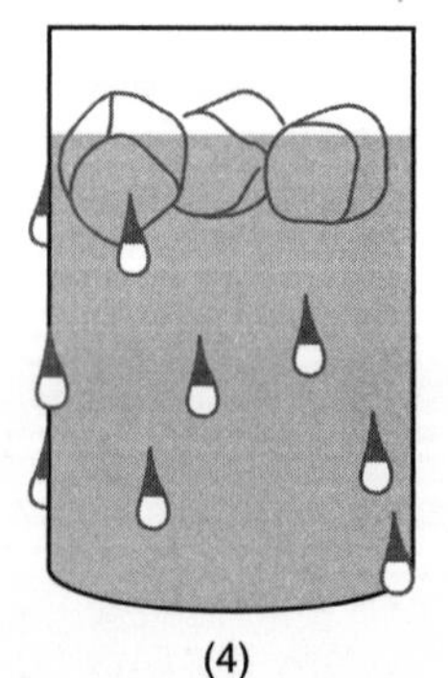
(4)

Part II

Base your answers to questions 66 through 68 on the diagram below and your knowledge of science. The diagram shows a ray of white light hitting an opaque surface.

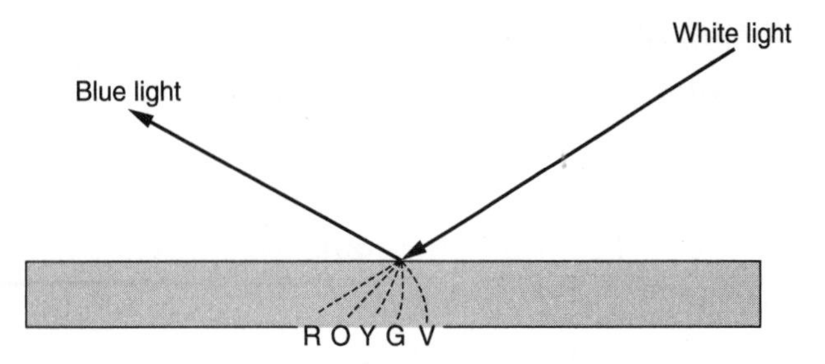

66. What happened to the blue portion of the light?
67. What happened to the red portion of the light?
68. What color will the surface appear?

Base your answers to questions 69 through 71 on the diagram below. The diagram shows some characteristics of different types of electromagnetic waves.

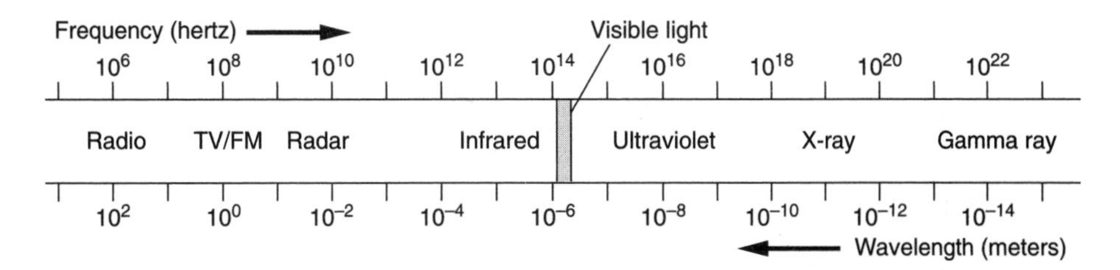

69. Describe two differences between X-rays and radio waves.
70. What is the relationship between wave frequency and wavelength?
71. Which type of electromagnetic wave has a frequency of 10^{10} hertz and a wavelength of 10^{-2} meter?

Base your answers to questions 72 and 73 on the diagram below and your knowledge of science. The diagram shows what happens to light rays when they strike three different surfaces.

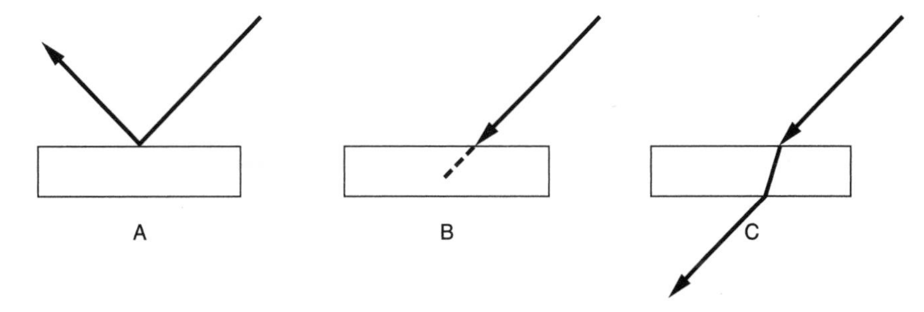

72. Which diagram represents a mirror?
73. Explain the energy transformation that is taking place in diagram B.

Base your answers to questions 74 through 76 on the passage below and your knowledge of science. The passage describes how you can determine your distance from a lightning bolt.

Robert knows that light traveling a relatively short distance arrives almost instantly, and that sound travels at about 340 meters per second in air. Robert estimated that for every 3-second difference in time between a flash of lightning and the clap of thunder, the lightning bolt was about 1 kilometer away.

74. How far away is a lightning bolt if there is a 6-second difference between the flash of lightning and the clap of thunder?

75. What would the time difference be between the flash of lightning and the clap of thunder if the distance were 3.4 kilometers (3400 meters) away?

76. What is the relationship between the speed of sound and air temperature?

Chapter 3 Energy and Resources

Wind farms and nuclear power plants are two different sources of electricity.

Vocabulary at a Glance

calorie
coal
conservation
fossil fuel
hydroelectric energy
insulation
natural gas
nonrenewable resource
nuclear waste
oil
renewable resource
solar energy
thermal pollution
uranium
watt

Major Concepts

- Fossil fuels, our main sources of energy, come from the remains of dead plants and animals. Fossil fuels are nonrenewable resources.
- Burning fossil fuels produces gases such as carbon dioxide and other pollutants, which affect Earth's atmosphere.
- Other sources of energy include hydroelectric, nuclear, solar, and wind. Each energy source has advantages and disadvantages.
- It is important to conserve our energy supply. Methods of conserving include reducing use, recycling materials, and reusing materials.

Forms of Energy

Everything that occurs in the universe involves energy. As you have learned, energy is the ability to do work—to make something move. Heat, light, sound, and electricity are all forms of energy. Humans have learned to describe, explain, and measure energy and to harness it for their use.

Measuring Energy

To compare the amount of energy stored in various substances, we need some unit of measurement. The energy in foods and fuels can be measured and compared using a unit called the ***calorie***. One calorie is the amount of heat energy needed to raise the temperature of 1 gram of water by 1 degree Celsius.

When describing the energy in food, we use the word *Calorie* spelled with a capital "C." This "food Calorie" is equal to 1000 ordinary calories, or 1 kilocalorie (*kilo* means one thousand). Calories indicate how much energy you can get from various foods. Digested food containing potential chemical energy that is not needed by the body is stored, usually as fat. When you go on a diet, you count Calories to make sure you don't eat more food than your body needs for energy.

The rate at which energy is used (the amount of energy used over a certain period of time) can also be measured. The rate at which electric energy is used is measured in a unit called the ***watt***. We can use watts to compare the rates at which different electric devices use energy. For instance, a 100-watt lightbulb uses twice as much electricity each second as a 50-watt bulb. Table 3-1 lists some common electric devices and their wattage.

Table 3-1. *Some Electric Devices and Their Wattage*

Electric Device	Wattage
Hair dryer	1200
Lightbulb	100
Electric shaver	7
Small air conditioner	860
Microwave oven	1200
Stereo	240
Toaster oven	1400

Energy Use

Energy use is constantly increasing because the world's population is increasing and more countries are becoming industrialized. Fuels are sources of energy. As the demand for energy increases, so does the demand for fuel. We need more fuel to cook meals, heat homes, run industries, and power cars, ships, trains, and airplanes. We also use more fuel to produce electricity.

Electricity has become essential to our society. It is used for many purposes, such as heating and cooling buildings, running machines and appliances, and providing lighting. Electric energy can be transmitted easily over conductors such as metal wires. However, electric energy must be produced from other energy sources.

Fossil Fuels

The main energy sources used to produce electricity are ***fossil fuels***. They are called fossil fuels because they were formed from the remains of plants and animals that lived and died long ago. Over time, these organic remains were changed into energy-rich substances. The most commonly used fossil fuels are oil (also called *petroleum*), coal, and natural gas.

Oil is a sticky, black liquid usually found trapped within rock layers deep underground. ***Coal*** is a black rock that occurs in layers, or seams, between other rock layers. ***Natural gas*** is commonly found underground with oil deposits. Each of these fossil fuels can be burned to provide energy for the production of electricity. Figure 3-1 shows the relative amounts of oil, coal, natural gas, and other energy sources used to produce electricity in the United States.

Figure 3-1. *Relative percentages of energy sources used to produce electricity in the United States.*

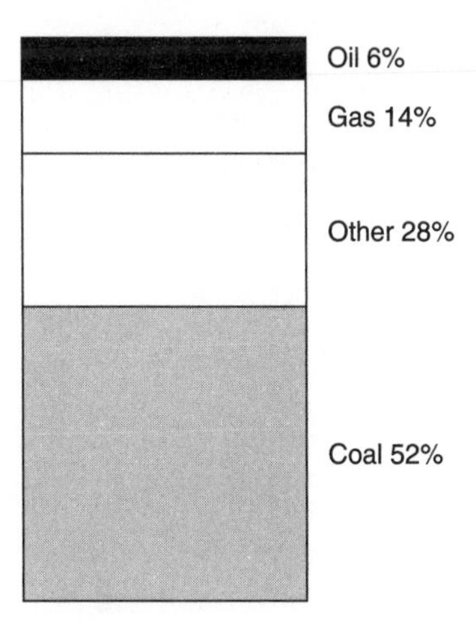

Other Uses of Fossil Fuels

Fossil fuels have many other uses besides producing electricity. Gasoline, used to power cars, and heating oil, used to heat homes, are both fossil fuels made from petroleum. Natural gas is used to heat homes and industries, and for cooking. (Gas stoves use natural gas.) Coal is used in industrial processes, such as the making of steel.

Fossil fuels are used to make many other important substances. Oil, in particular, has many such uses. Plastics, fertilizers, certain drugs, and synthetic fabrics like nylon and polyester are all products made from petroleum.

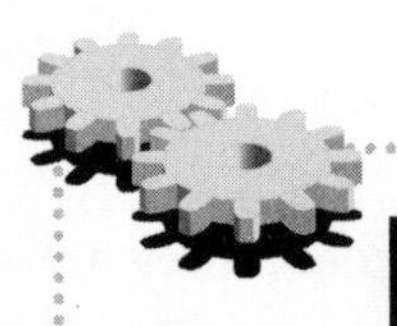

Process Skill

Interpreting a Graph

A pictograph represents numbers, or quantities, by using pictures. In the accompanying diagram, each picture of a barrel represents one million barrels of petroleum. For example, the category "Fuel Oil" is represented by five barrels, which means that five million barrels of petroleum are used each day for these purposes.

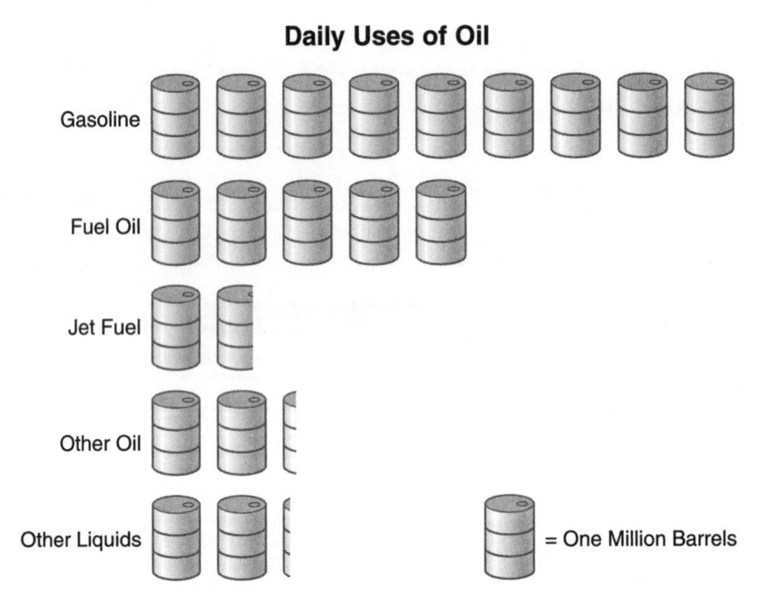

How much petroleum is used daily for "Other Oil"? Looking at the row of barrels for that category, you see there are two complete barrels plus part of a third barrel. This part is about one-quarter of a complete barrel, so it represents one-quarter of a million barrels of petroleum. This makes the total daily use of petroleum for "Other Oil" equal to 2 $\frac{1}{4}$ million barrels. Use the diagram to help you answer the following questions.

Questions

1. Which products use the least amount of petroleum each day?

(1) gasoline (3) other liquids
(2) jet fuel (4) fuel oil

2. How many barrels of petroleum are used for gasoline every day?
(1) fewer than 1 million barrels
(2) 2 ½ million barrels
(3) 5 million barrels
(4) 9 million barrels

3. Oil used for heating homes is included in the category called "Fuel Oil." Explain how the pictograph might differ in summer and winter?

Other Sources of Energy

While most of the energy we use comes from fossil fuels, there are several other energy sources. The most important of these are *hydroelectric energy* and *nuclear energy.*

1. *Hydroelectric energy* is electricity produced by the energy in flowing water. Water in motion has kinetic energy, which can be transformed into electric energy. When water flows swiftly downhill at a waterfall or a dam on a river, it can be used to operate a *generator,* which produces electricity. A generator contains a *turbine,* a device similar to a paddle wheel on an old-fashioned steamboat (see Figure 3-2). The moving water turns the turbine, which spins a coil of wire inside an electromagnet in the generator. This creates an electric current.

Figure 3-2. *A generator in a hydroelectric dam used to produce electric energy.*

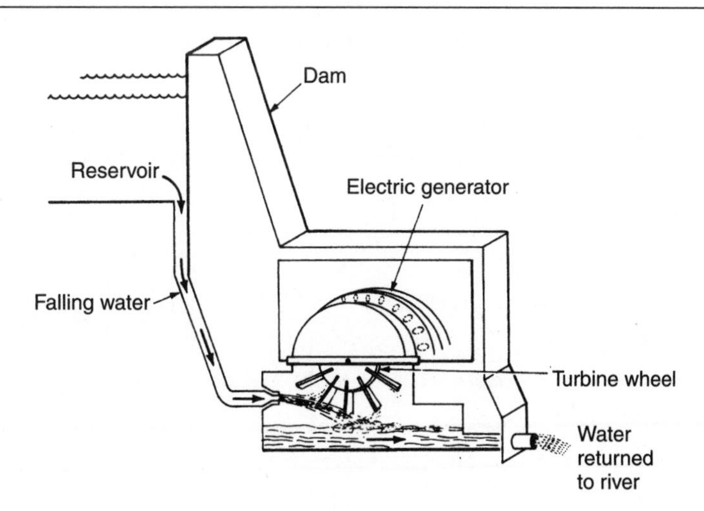

Hydroelectric energy is usually clean and inexpensive. Niagara Falls is a major source of hydroelectric energy in New York State. Figure 3-3 on page 106 compares the energy sources used by electric companies in New York with those used in Connecticut. Notice that New York uses a much greater percentage of hydroelectric energy than does Connecticut, while the percentage of nuclear energy used in New York is much smaller. The choice of energy sources varies from one state to the next, as different states have different resources available to them.

Figure 3-3. *A comparison of energy sources used to produce electricity in New York and Connecticut.*

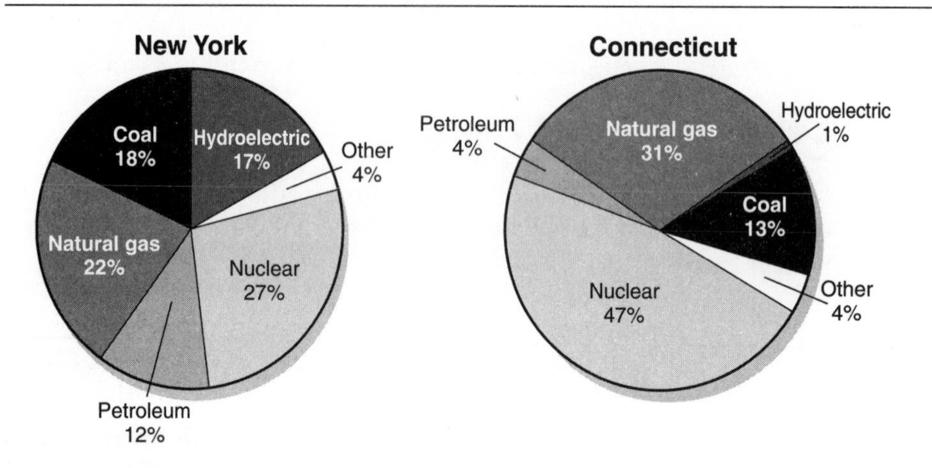

2. *Nuclear energy* is the energy stored in the nucleus of an atom. When this energy is released, some if it is released as heat. The heat can be used to produce electricity. Nuclear power plants use ***uranium*** (a radioactive element found in certain rocks) as their fuel source. Uranium atoms are naturally unstable and can be split apart easily to release heat energy. The heat released by the uranium fuel in a nuclear reactor is used to boil water, thereby producing steam. The steam turns turbines that generate electricity. This is illustrated in Figure 3-4.

Figure 3-4. *A nuclear power plant: Heat from a nuclear reaction changes water into steam, which turns turbines to produce electricity.*

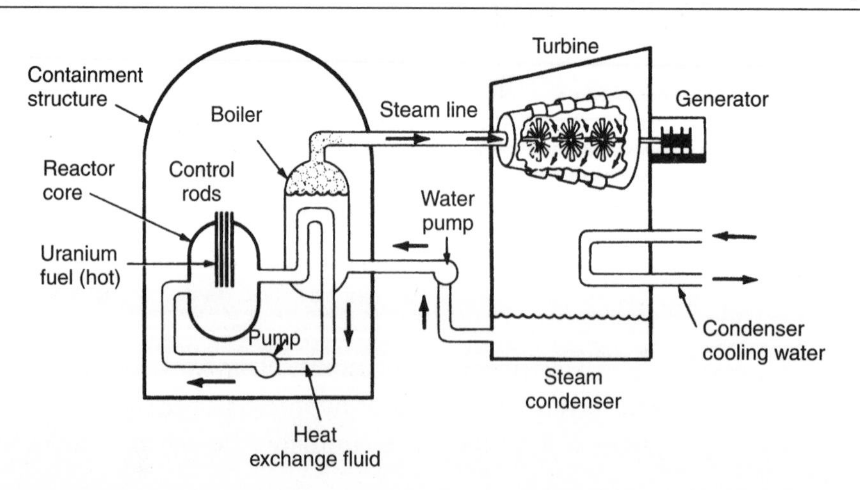

Review Questions

Part I

1. Which is a form of energy?
   (1) heat
   (2) coal
   (3) gasoline
   (4) petroleum

2. The energy contained in food is measured in units called
   (1) degrees
   (2) watts
   (3) barrels
   (4) Calories

3. To reduce the amount of electricity used in his home, Joe has convinced his parents to replace the traditional incandescent lightbulbs in the house with the newer compact fluorescent lightbulbs. A compact fluorescent lightbulb gives off as much light as an incandescent lightbulb but does not give off as much heat. The rate at which electricity is used by these new bulbs is much less than that of incandescent bulbs. The rate at which electricity is used is measured in a unit called a
   (1) volt
   (2) watt
   (3) Calorie
   (4) degree

4. Which energy source is nonpolluting and inexpensive?
   (1) hydroelectric
   (2) coal
   (3) oil
   (4) petroleum

5. Over long periods of time, the remains of dead plants and animals can be converted into
   (1) wind
   (2) uranium
   (3) fossil fuels
   (4) moving water

6. Gasoline and home heating oil are made from
   (1) coal
   (2) petroleum
   (3) natural gas
   (4) uranium

Part II

Base your answers to questions 7 through10 on the graphs below, which compare the sources of the energy used in four different states.

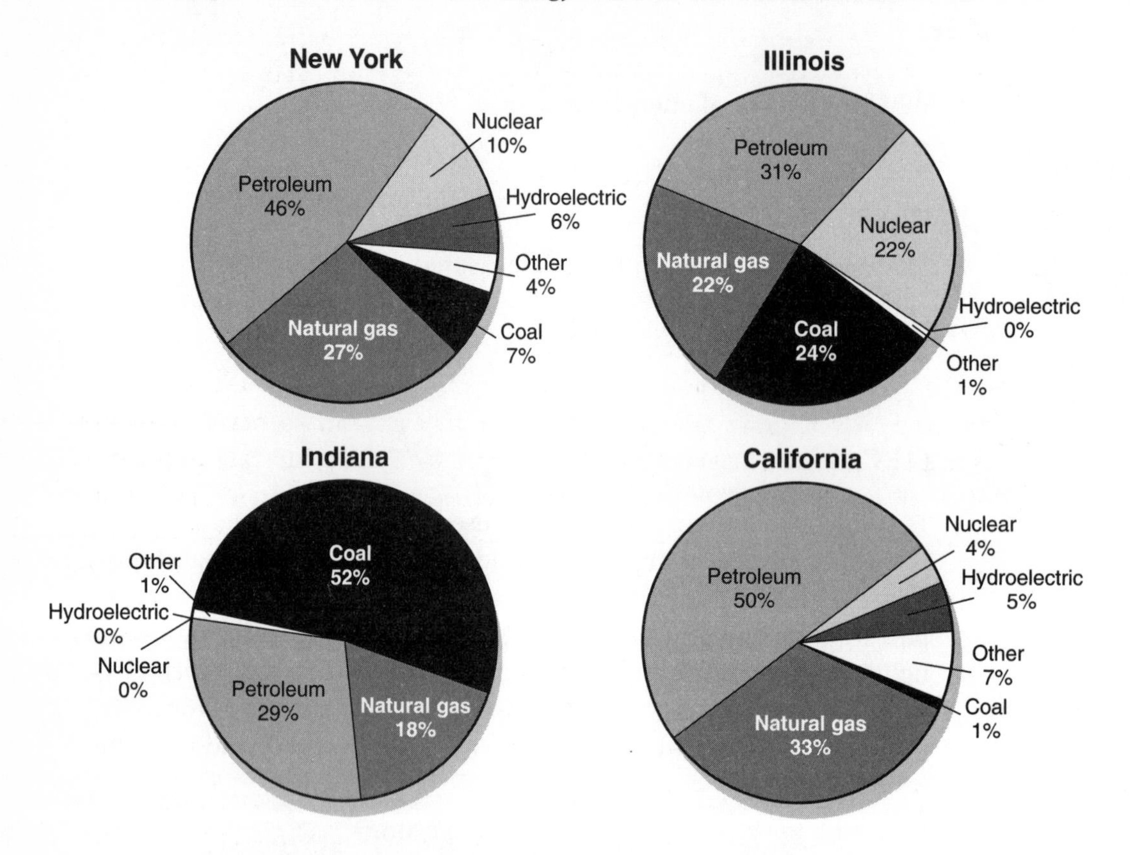

7. Compared with New York, Indiana uses
   (1) a larger percentage of petroleum and a larger percentage of coal
   (2) a smaller percentage of petroleum and a larger percentage of coal
   (3) a smaller percentage of petroleum and a smaller percentage of coal
   (4) a larger percentage of petroleum and a smaller percentage of coal

8. Approximately what fraction of California's energy comes from petroleum?

9. Which two states use running water as a source of energy?

10. Which state uses the highest percentage of fossil fuels to supply its energy needs?

Base your answers to questions 11 and 12 on the information below and your knowledge of science.

Niagara Falls provides New York State with a nonpolluting, inexpensive source of electricity.

11. What is the source of energy that produces electricity at Niagara Falls?

12. Electrical plants that burn fossil fuels are considered sources of energy that generate pollution. What does a fossil-fuel-burning electrical plant produce that Niagara Falls does not?

---

13. In 2007, petroleum (oil) prices reached $100 per barrel for the first time. How does the price of petroleum affect your everyday life?

Problems with Energy Sources

Problems with Fossil Fuels

In the United States, most of the demand for energy is met by the fossil fuels: oil, coal, and natural gas. In 2000 and 2001, California experienced shortages of electricity. The shortages were caused by a sharp increase in the price of fossil fuels needed to generate the electricity. Because state law did not allow an increase in cost to the public, some companies found that they could no longer afford to produce electricity. This led to a decrease in the supply of energy available in California.

The burning of fossil fuels creates air pollution. When fossil fuels burn, chemicals that pose dangers to living things and the environment are released into the air. This is especially true of coal.

1. *Coal.* The supply of coal found in the United States is much greater than that of oil or natural gas. Because there is so much, coal is relatively inexpensive. However, there are serious environmental and health problems involved with its use. Burning coal produces air pollution. Smoke from coal-burning power plants is the main cause of *acid rain,* which is harmful to the ecology of lakes and forests.

   In addition, coal mining is dangerous for people who work in the mines. Breathing air that contains coal dust is unhealthful and can lead to *black lung disease.* Certain coal-mining techniques damage the environment. Sometimes large areas of land are dug up to reach the coal. This practice, called *strip mining,* destroys topsoil and scars the landscape.

   Mining companies are now required by law to restore the land they have damaged. Advances in technology have made coal mining safer and reduced the amount of pollution caused by burning coal. Nevertheless, these measures have only begun to solve the problems with using coal as a fuel.
2. *Oil and Natural Gas.* Although oil and natural gas produce fewer air pollutants than coal, they cause other environmental problems. Offshore drilling for oil and transporting oil by ship can lead to accidental oil spills that kill marine wildlife and cause

severe pollution of land and sea. Pipelines built to transport oil and gas over land may change the ecology of areas they cross.

3. *Global Warming.* Burning any fossil fuel produces carbon dioxide. Carbon dioxide traps heat in Earth's atmosphere much as the glass of a greenhouse traps heat. Most scientists believe that the buildup of carbon dioxide in the atmosphere caused by using fossil fuels is producing a *greenhouse effect,* leading to a warming of Earth's climate. Such global warming could have many harmful consequences for life on Earth.

Problems with Hydroelectric and Nuclear Energy

The production of electricity using moving water or nuclear reactors is generally much "cleaner" (produces less pollution) than energy production with fossil fuels. This is because hydroelectric and nuclear power plants do not involve burning anything. However, even these "clean" energy sources have environmental costs.

1. *Hydroelectric energy.* Building a dam on a river to produce hydroelectric power changes the surrounding area, as you can see in Figure 3-5. The area upriver (behind the dam) is flooded, creating a large lake over land that may have once been a habitat for wildlife or been used for farming. The area downriver from the dam receives a reduced flow of water. These changes greatly affect the ecology of the area around a dam.

Figure 3-5. *The effect of a hydroelectric dam on the environment.*

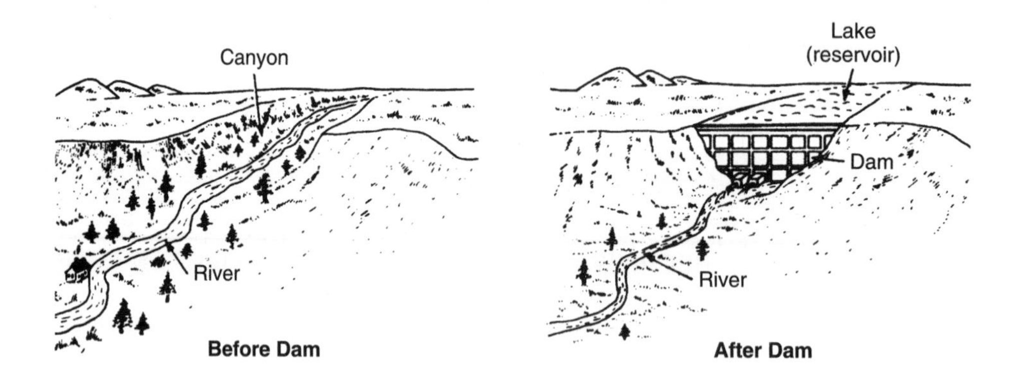

2. *Nuclear energy.* Although nuclear power plants do not cause air pollution, they use water from nearby lakes or rivers to cool their reactors. After cooling the reactor, the water is returned to the environment several degrees warmer than it was before. This increase in the temperature of the water, called ***thermal pollution,*** can be harmful to organisms living in the water.

An even more serious problem with nuclear power is how to safely dispose of the used-up uranium fuel, known as ***nuclear waste***. This poisonous, radioactive material must be stored where it will never leak into the environment. Most people do not want nuclear waste stored in, or even transported through, their communities. Disposal of nuclear waste is a difficult problem, and scientists continue to disagree on whether any of the proposed solutions is acceptable.

Review Questions

Part I

14. Global warming results from increased levels of carbon dioxide in the air. This increase in carbon dioxide levels may be caused by

(1) increased use of solar-powered cars

(2) increased use of fossil fuels

(3) increased use of nuclear energy

(4) increased use of hydroelectric energy

15. Which of the following problems is associated with nuclear energy?

(1) air pollution

(2) cooling the water in the lakes and rivers

(3) safe disposal of waste materials

(4) acid rain

16. Some of the lakes in New York State have become too acidic to support life. This environmental change is most associated with

(1) reducing energy usage

(2) burning coal

(3) using nuclear energy

(4) using hydroelectric energy

17. Which of the following is not considered a fossil fuel?

(1) uranium (3) oil

(2) coal (4) natural gas

18. When the tanker *Exxon Valdez* ran aground in Alaska on March 24, 1989, there was tremendous damage to the environment. Its 10.9-million-gallon cargo was spilled into the waters of a bay. It is estimated that at least half a million birds, fish, and marine mammals were destroyed. What was the cargo spilled by *Exxon Valdez*?

(1) coal (3) natural gas

(2) uranium (4) petroleum

Base your answers to questions 19 and 20 on the reading passage below and your knowledge of science.

Ethanol is an alcohol that can be produced from corn. Like gasoline, ethanol can be burned to produce energy. Although ethanol is not a fossil fuel, it does produce carbon dioxide when it burns. With gasoline prices constantly increasing, engineers are designing automobiles that can run on ethanol, which may eventually become a less expensive fuel than gasoline.

19. According to the reading passage, what advantage might ethanol have over gasoline?

(1) It produces more energy.

(2) It does not pollute the air.

(3) It does not contribute to global warming.

(4) It may become less expensive.

20. Many scientists argue that ethanol is not a good solution for our energy problems, because the use of ethanol

(1) is too expensive

(2) cannot replace gasoline

(3) adds carbon dioxide to the air

(4) cannot produce energy

Part II

21. Identify four sources of energy that are used to produce electricity in the United States. For each source, identify one problem or disadvantage associated with its use.

Use the following reading passage and your knowledge of science to answer questions 22 through 24.

Iceland: One Hot Country!

Most sources of energy involve the production of heat. The heat can be used directly to heat a home or indirectly to produce electricity. Often, the heat is produced by burning fuel. However, in some places, there are sources of heat just below the ground. Iceland is a small island nation that has no oil or natural gas of its own. Iceland does have many waterfalls, volcanoes, geysers, and hot springs. The waterfalls produce hydroelectric energy. The volcanoes, geysers, and hot springs provide geothermal energy (*geo* means Earth and *thermal* means heat). The energy from beneath Earth's surface is used to heat homes and produce electricity. The large amount of inexpensive electricity may allow Iceland to become completely free of fossil fuels. The Icelanders are not giving up their cars and trucks, though. They are using their electricity to produce hydrogen fuel cells, which can power automobiles. Fuel cells produce water as their only waste product.

22. What are two advantages of geothermal energy over fossil fuels?

23. What waste product do fossil fuels produce that hydrogen fuel cells do not?

24. Geothermal energy is a growing resource in the state of Hawaii. Why might geothermal energy be more available in Hawaii than in New York?

Energy for the Future

Energy Conservation

Most energy resources are limited. To guarantee enough energy for the future, we must practice conservation. ***Conservation*** is the saving of natural resources for the future through wise use. This means using resources more efficiently and getting rid of unnecessary waste. The methods used in conserving our resources are often summarized as the three R's of conservation—*reduce, reuse,* and *recycle*.

Reducing Energy Use

1. *High-efficiency appliances.* We can practice conservation by purchasing high-efficiency appliances. These appliances use less energy than do less efficient appliances while doing the same job. For instance, a car that can travel 48 kilometers (30 miles) on a gallon of gasoline is more efficient than a car that gets only 24 kilometers (15 miles) per gallon. An air conditioner with a high "energy efficiency rating" uses less electricity than one with a low rating, but it cools a room just as well. Fluorescent lightbulbs use less electricity than incandescent lightbulbs giving the same amount of light. New *compact fluorscent lightbulbs* (CFL) (see Figure 3-6) can replace traditional incandescent lightbulbs and reduce energy use.

Figure 3-6. *Compact fluorescent lightbulbs can replace traditional incandescent lightbulbs.*

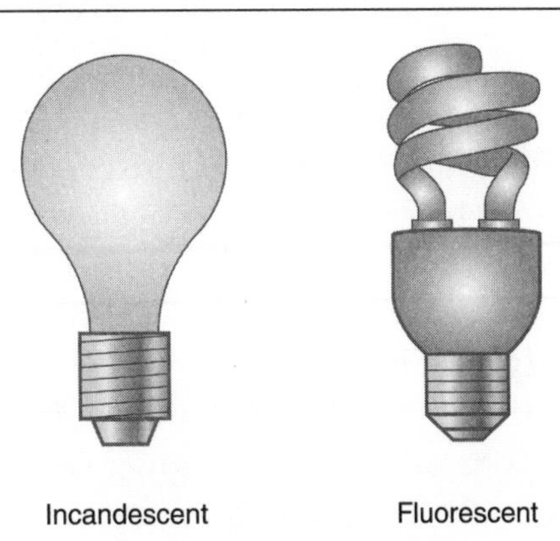

Although a high-efficiency appliance may cost more to buy, it costs less to use. This means that it will save money in the long run, while helping to conserve energy. Many appliances carry an "energy efficiency rating" label that shows how they compare with other models. Figure 3-7 is an example of such a label. For instance, if every home in America replaced just one lightbulb with a CFL bulb, the country would save more than $600 million in energy costs in one year.

Figure 3-7. *Example of an appliance's "energy efficiency rating" label.*

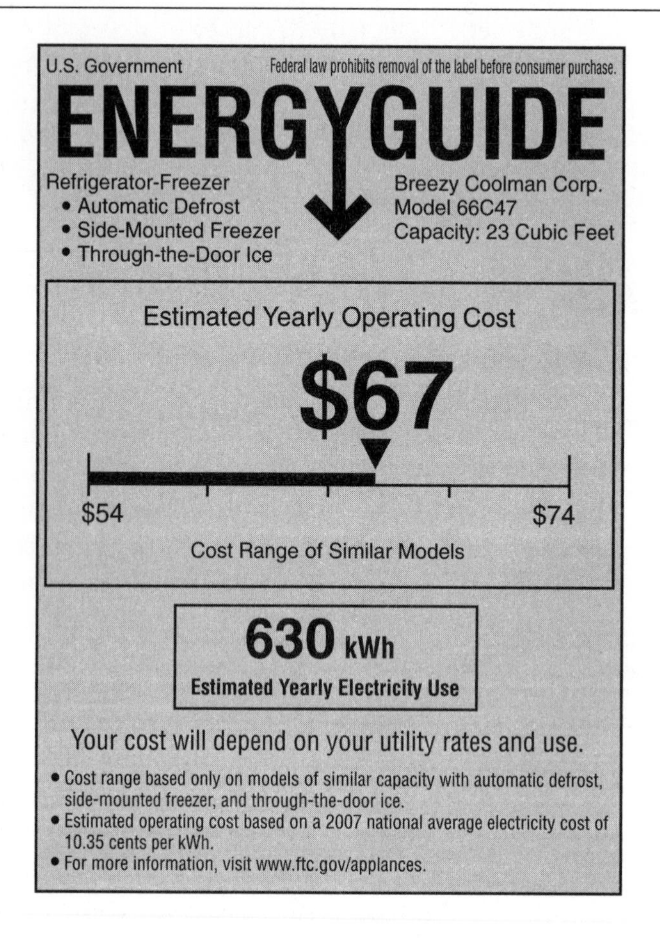

High-efficiency appliances do have some disadvantages. They often require the use of more expensive materials. Many people will not buy more expensive appliances unless they can see that they will save money over time. Scientists and engineers are constantly working to develop machines that combine high efficiency with low cost.

Sometimes, increasing the energy efficiency of a machine affects its performance. Lighter weight automobiles burn less gasoline than do heavier ones but may not be as safe in a collision. Cars with larger engines, which can produce greater acceleration, usually burn more gasoline than do cars with smaller engines. Yet many people prefer big, fast cars despite their greater cost and possible harm to the environment.

New hybrid automobiles combine gasoline and electric engines. These new vehicles use less gasoline. However, it takes more energy and materials to manufacture these cars so they cost more to buy. Car manufacturers are working to create new fuel efficient technologies to reduce gasoline use.

2. *Insulated buildings.* Energy use at home and at work can be reduced through improved ***insulation***. A well-insulated building prevents heat loss in winter and keeps heat out in summer. These benefits can be achieved by constructing walls in two layers, with insulating material in between. Cracks around doors and windows can be sealed with weather stripping for further insulation. With these improvements, less energy is needed to keep the indoor temperature comfortable year-round.

   There are some disadvantages to insulation as well. Some materials that have been used as insulators give off toxic fumes when burned. Safe, efficient insulating materials can be expensive. In some cases, insulation can affect the performance of an appliance. For example, better insulation can greatly increase the efficiency of a refrigerator, but the space occupied by the insulating material may decrease the amount of food the refrigerator can hold.

Reusing Materials

Every time you go to the supermarket you receive one or more bags in which to carry your purchases. Many of these bags could be used more than once, yet we usually throw them away. Today, in an effort to save energy and improve the environment, some supermarkets are offering a slight discount to shoppers who bring their own bags. Reusing resources is efficient in terms of energy and waste management. Other reusable products include glass and plastic bottles, cans, and even worn clothing. Some disadvantages of reusing material are the time and effort required for saving and collecting them.

Recycling

This practice also helps conserve energy resources. The graph in Figure 3-8 on page 116 shows that making bottles and cans from recycled materials uses less energy than does making those products from raw materials.

Figure 3-8. *Less energy is used when products are made from recycled materials than when they are made from raw materials.*

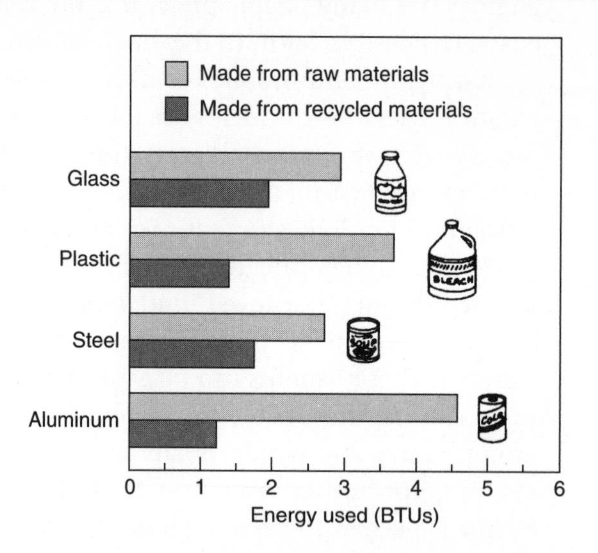

As you can see in the figure, less energy is used to make glass containers from recycled materials than from raw materials. However, the collection of glass for recycling requires the cooperation of large numbers of people, who must separate glass containers from other waste. These containers must be cleaned, collected, and transported to a recycling plant. Sometimes, the cost of collection and transportation is so high that the process no longer saves money. However, recycling is important even when money is not the issue. When materials such as glass and paper are recycled, they no longer become trash. Recycling benefits the environment by reducing the amount of waste that must be dumped in landfills or burned in incinerators.

It's Your Decision

In many places in the United States, recycling is the law. People are required to sort their trash so that recyclable materials can be collected. This requires extra time and effort, but they have decided, through their elected representatives, that the time and effort are worth it. You will be making decisions like these in the future. Let us examine some of the issues you should consider in making your decisions.

Have you ever noticed that a greeting card printed on recycled paper costs more than one printed on new paper? Why should we recycle paper when it saves us neither energy nor money? Paper is made from wood from our forests. By using recycled paper, we are cutting fewer trees, which benefits the environment. Table 3-2 lists some arguments for and against the recycling of paper.

The decision of whether to recycle affects people's jobs and businesses as well as the environment. Environmentalists and businesspeople will try

Table 3-2. *The Cases For and Against Recycling Paper*

For	Against
Preserves forests, which soak up carbon dioxide and provide habitat for wildlife	Trees can be replaced by new trees, minimizing damage to forests; may cause job layoffs in timber industry
Decreases amount of solid waste in landfills	Expensive
Decreases air pollution in areas where paper is burned in incinerators	Separation of used paper from other waste is time-consuming

to convince you of their points of view. Listen carefully to both sides before you make your decisions.

Renewable and Nonrenewable Resources

Earth's supply of fossil fuels is being used up rapidly. We continually remove these resources from the earth, but we cannot replace them. Nature does not create new deposits of oil, coal, and natural gas within the time span of human history. For this reason, fossil fuels are considered ***nonrenewable resources***. Uranium is also a nonrenewable resource.

Renewable resources are those that can be replenished by nature within a relatively short amount of time. Moving water, wind, plants, and sunshine do not run out as we use them. Natural processes are constantly replacing them. Table 3-3 lists some renewable and nonrenewable energy resources.

Table 3-3. *Energy Resources*

Renewable	Nonrenewable
Hydroelectric	Oil
Solar	Coal
Wood	Natural gas
Wind	Nuclear

Using Renewable Resources

Even if we practice conservation, our supply of nonrenewable energy resources may not be enough to meet the energy demands of the future. Renewable resources offer alternatives to fossil fuels and radioactive minerals. Unlike nonrenewable resources such as oil, coal, and uranium, renewable energy resources cannot run out. They also cause fewer environmental problems than fossil fuels and nuclear energy do. For these reasons, scientists and engineers are seeking more and better ways to use renewable resources for our growing energy needs.

As you know, moving water can be used to run generators and produce electricity. The natural water cycle of evaporation, condensation, and precipitation renews the water supply that feeds the rivers used for this purpose. However, not all areas have rivers suitable for producing hydroelectric energy.

The wind can be used to generate electricity by turning the blades of a *wind turbine* (see Figure 3-9). In windy areas, wind turbines can provide safe, clean electricity. But the wind is not as constant and reliable as a flowing river. When there is only a slight wind or the air is calm, little or no electricity is produced.

Figure 3-9. *Wind turbines use the renewable energy of the wind to produce electricity.*

Plant matter and animal wastes can be burned to produce heat, or they can be changed to other fuels. For instance, decaying plant matter and animal wastes produce *methane,* the main component of natural gas. Methane produced in this way is a renewable resource, unlike natural gas found underground. At present, however, converting these materials into fuel on a large scale is too expensive to be practical. Unfortunately, *every* energy resource has both advantages and disadvantages, as outlined in Table 3-4.

Solar Energy

The primary source of energy on Earth is the sun. Energy from the sun is called ***solar energy***. The energy in fossil fuels came originally from sunlight. This light was absorbed by plants during photosynthesis millions of years ago. The moving water used for hydroelectric energy is replenished by the water cycle, which is powered by the sun's energy. Wind, which can be used to make electricity, is caused by the sun's heating of the atmosphere.

People have found ways to use the sun's energy directly to provide heat and hot water for homes, offices, and factories. For example, a device called a *solar collector* absorbs solar energy and converts it into heat energy. The heat is transferred to water circulating through the collector. This hot water can be used to run a home heating system, as shown in Figure 3-10.

Table 3-4. *Advantages and Disadvantages of Energy Resources*

Energy Resource	Advantages	Disadvantages
Petroleum	Efficient; can be converted into different types of fuel	Causes air pollution; risk of spills while drilling or transporting; limited reserves in U.S.; nonrenewable
Natural gas	Available in U.S.; clean	Difficult to store and transport; mostly nonrenewable
Coal	Plentiful in U.S.; inexpensive	Causes air pollution and acid rain; mining practices may be harmful to miners' health and destructive to the environment; nonrenewable
Nuclear (uranium)	Plentiful fuel in U.S.; does not cause air pollution; can meet long-term energy needs	Causes thermal pollution; creates radioactive waste; risk of accidents releasing radioactivity into environment; uranium mining harmful to miners' health; nonrenewable
Hydroelectric (water)	Does not cause air pollution; inexpensive; renewable	Not available in all areas; affects local ecology
Wind	Does not cause pollution; clean, inexpensive, renewable	Not practical for large-scale generation; not always reliable (wind is not constant)
Solar (sunlight)	Does not cause pollution; clean; renewable	Expensive to convert into usable form; not always reliable (depends on the weather)
Plant matter and animal wastes	Renewable	Expensive to convert into usable form; inefficient

Figure 3-10. *Example of capturing and using solar energy to run a home heating system.*

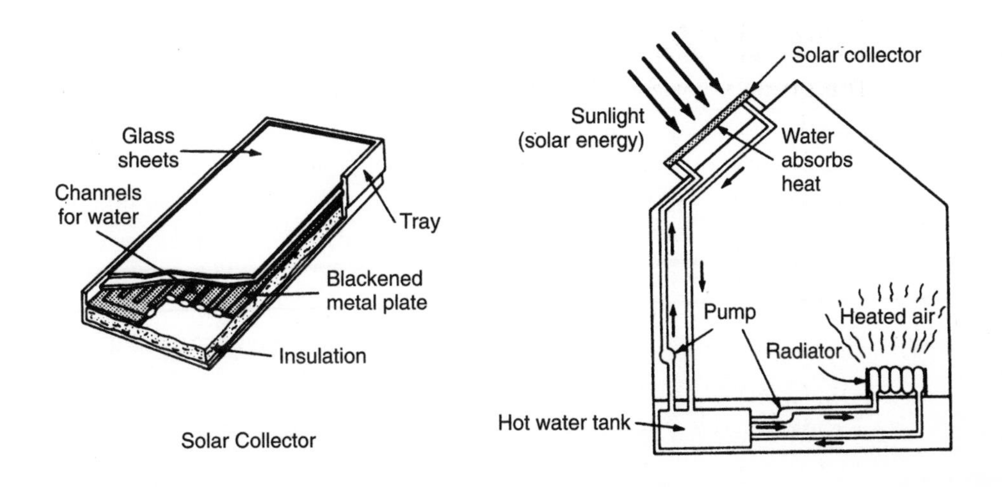

People have developed ways to transform solar energy into electric energy. For instance, a *solar cell* is a device that converts light directly into electricity. Some calculators and light meters in cameras use solar cells instead of batteries for energy. However, generating large amounts of electricity this way requires a huge number of these cells, which is very expensive.

Heating water with solar energy can also produce electricity. Water is heated to a boil by using mirrors to focus and concentrate sunlight (see Figure 3-11). The boiling water changes into steam, which turns turbines to produce electricity. This method is more economical than using solar cells to make electricity, but it is still more expensive than using fossil fuels. However, as fossil fuels become more scarce, this situation may change.

Figure 3-11. *Solar energy changes water into steam, which turns turbines to produce electricity.*

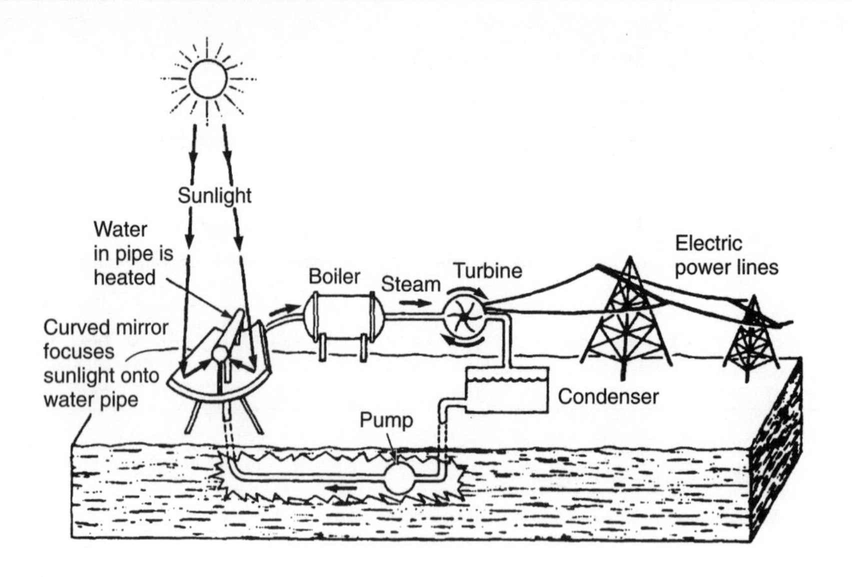

Review Questions

Part I

25. Which source of energy is *renewable*?

    (1) coal (3) oil
    (2) wind (4) fossil fuel

26. Which source of energy is not recycled in the environment?

    (1) coal (3) solar
    (2) wind (4) running water

27. Trees are considered a renewable resource because
    (1) burning them does not pollute the air
    (2) burning them pollutes the air
    (3) more trees can be planted
    (4) they are easy to cut down
28. Recycling newspapers is important because
    (1) it preserves forests
    (2) it is expensive
    (3) it takes time and effort to separate and package the paper
    (4) newspapers are made from a renewable resource
29. Which method will *not* help conserve our resources?
    (1) reduce the use of natural resources
    (2) reuse products more than one time
    (3) recycle materials to form new products
    (4) replace items before they wear out
30. All energy on Earth except nuclear energy comes originally from
    (1) fossil fuels (3) electricity
    (2) the sun (4) the oceans

Part II

Use the picture below and your knowledge of science to answer questions 31 and 32.

31. Identify two renewable energy sources shown in the picture.
32. Identify two uses of nonrenewable energy resources shown in the picture.

33. State one benefit of replacing incandescent lightbulbs with more expensive compact fluorescent lightbulbs.

34. Match each of the energy sources in the left column with the problem it causes in the right column.

1 – Nuclear	A – Building dams can affect the local ecology
2 – Coal	B – May increase the temperature of surrounding rivers or lakes
3 – Petroleum	C – Produces acid rain
4 – Hydroelectric	D – Transporting can lead to spills that harm the environment

You can use the dichotomous key below to identify six different energy sources. Refer to this table to answer questions 35 through 38. Here are two examples: Which couplet is used to distinguish wind from solar? Answer: couplet 6. State two things that wind and solar have in common. Answer: They are both renewable and do not use water.

A Key to Identifying Matter

Couplet	Description	
1a 1b	Nonrenewable Renewable	Go to 2 Go to 5
2a 2b	a fossil fuel not a fossil fuel	Go to 3 Nuclear
3a 3b	Liquid Not a liquid	Petroleum Go to 4
4a 4b	Solid Gas	Coal Natural Gas
5a 5b	Uses water Does not use water	Hydroelectric Go to 6
6a 6b	Uses moving air Does not use moving air	Wind Solar

35. What do hydroelectric, solar, and wind-energy sources have in common?

36. Which couplet is used to distinguish coal from natural gas?

37. What property distinguishes uranium from the other nonrenewable fuels?

38. Which two fuel sources involve liquids?

Chapter 4 Motion and Machines

Even the U.S. Air Force Thunderbirds cannot defy the laws of motion.

Vocabulary at a Glance

acceleration
effort
first law of motion
force
fulcrum
inclined plane
inertia
lever
machine
motion
pulley
resistance
screw
second law of motion
simple machine
speed
third law of motion
velocity
wedge
wheel and axle
work

Major Concepts

- A force is a push or a pull. Balanced forces acting on an object cause it to be at rest. Unbalanced forces acting on an object cause it to be in motion.
- Mass is a measure of the amount of matter in an object. The weight of an object is a measure of the gravitational pull on the object. The amount of gravitational force produced by one object on another depends on the mass of the objects and the distance between the centers of the objects.
- Speed, velocity, and acceleration are ways of describing the motion of an object.
- Newton's first law of motion states that an object at rest will remain at rest and an object in motion will remain in motion, unless an unbalanced force acts on it. Newton's second law of motion states that the acceleration of an object is directly proportional to the force applied and indirectly proportional to its mass. Newton's third law of motion states that every action has an equal and opposite reaction.
- Work is done when an applied force moves an object. Machines make work easier. The six simple machines are the lever, inclined plane, screw, pulley, wheel and axle, and wedge.

Force

A *force* is a push or pull. (See Figure 4-1.) To open a refrigerator door you pull the door. To move a computer mouse across a mouse pad you push or pull the mouse. To lift a book from a table you must pull the book up against the force of gravity. Table 4-1 lists examples of pushing and pulling forces.

Figure 4-1. *Pulling (A) and pushing (B) forces illustrated by opening and closing a refrigerator door.*

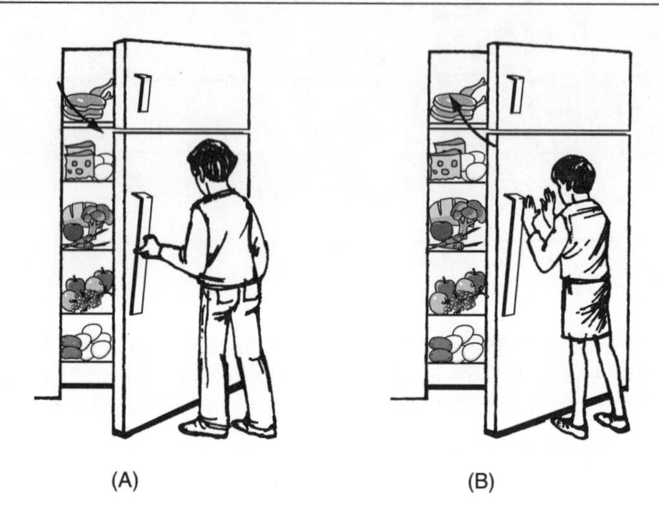

Table 4-1. *Pushing and Pulling Forces*

Pushing Forces	***Pulling Forces***
Hitting a tennis ball	Pulling a rope in a tug-of-war
Closing a refrigerator door	Opening a closet door
Moving a shopping cart	Lifting a shovel full of dirt
Hammering a nail	Climbing a rope
Wind blowing leaves	Apple falling from a tree

A force can also stop an object's motion, change its speed, or change its direction, as the following examples show.

1. Force stops motion: a falling acorn striking the ground; glove catching a baseball.
2. Force slows motion: friction slowing a skateboard; car going from a flat road to an uphill road.
3. Force changes direction: baseball bat striking a ball; wind causing a fly ball to curve.

When the forces acting on an object are balanced, the object is at rest. The teams in a tug-of-war (Figure 4-2a) do not move when the pulling forces are equal. However, balanced forces become unbalanced when one of the forces is changed. Figure 4-2b demonstrates that motion occurs when a force is added to one side of a balanced tug-of-war. Table 4-2 lists examples of balanced and unbalanced forces.

Figure 4-2. *A tug-of-war showing balanced forces (a) and unbalanced forces (b).*

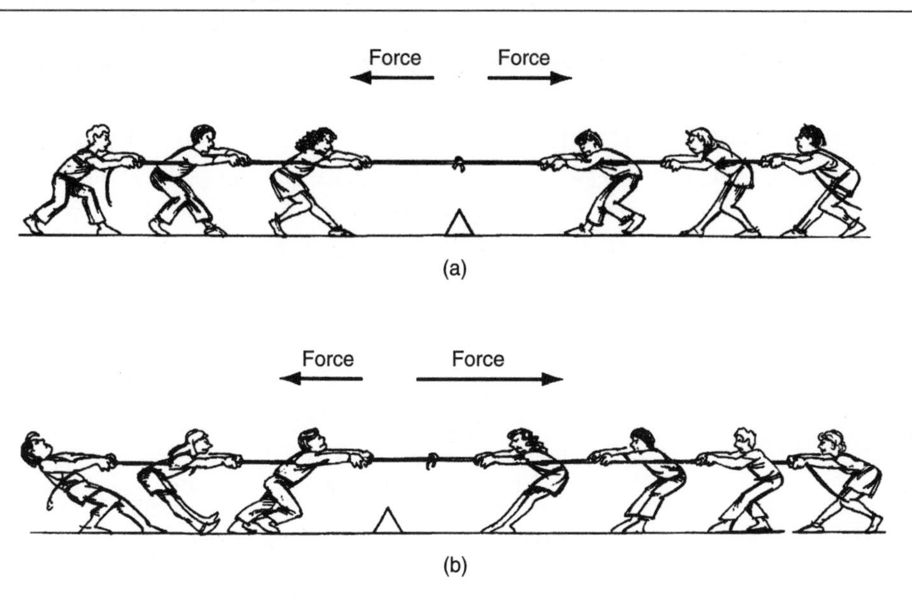

Table 4-2. *Balanced and Unbalanced Forces*

Balanced Forces (At Rest)	***Unbalanced Forces (In Motion)***
Car parked in driveway	Car pulling out of driveway
Apple hanging on a tree	Apple falling to the ground
Roller coaster on top of a hill	Roller coaster coming downhill
Rocket on launching pad	Rocket liftoff

The Force of Gravity

All objects in the universe exert a gravitational pull on every other object. This is most noticeable on Earth's surface when you drop an object and it falls to the ground. The amount of gravitational force between two objects is directly proportional to the masses of the objects and indirectly proportional to the distance between the centers of the objects. That means the greater the masses, the greater the gravitational force; and the farther apart the objects, the less the gravitational force. On Earth, an acorn may fall from a tree and hit the ground. Earth is much more massive than the acorn; thus, the acorn appears to fall to Earth rather than Earth falling toward the acorn. An acorn falling from a branch 10 meters high takes about 1.4 seconds to reach the ground. The gravitational pull on the same acorn 40,000 kilometers above Earth is much less. At this height, it would take the acorn much longer to fall a distance of 10 meters. In other words, as long as the masses of the objects remain the same, the greater the distance between them, the smaller the gravitational force is between them.

The weight of an object is a measure of the gravitational pull exerted on the object. Gravitational force is expressed in pounds (lb), or *newtons* (N) in the metric system. One newton is equal to about 0.22 pound. Figure 4-3 shows an 800-newton box at different distances from Earth's center. As the distance between the box and Earth's center decreases, the gravitational force increases, and the weight of the box also increases. Because Earth bulges at the equator and is flattened at the poles, objects weigh more at the North and South Poles than they do at the equator.

Figure 4-3. *As an object moves toward Earth, its weight increases, but its mass remains the same.*

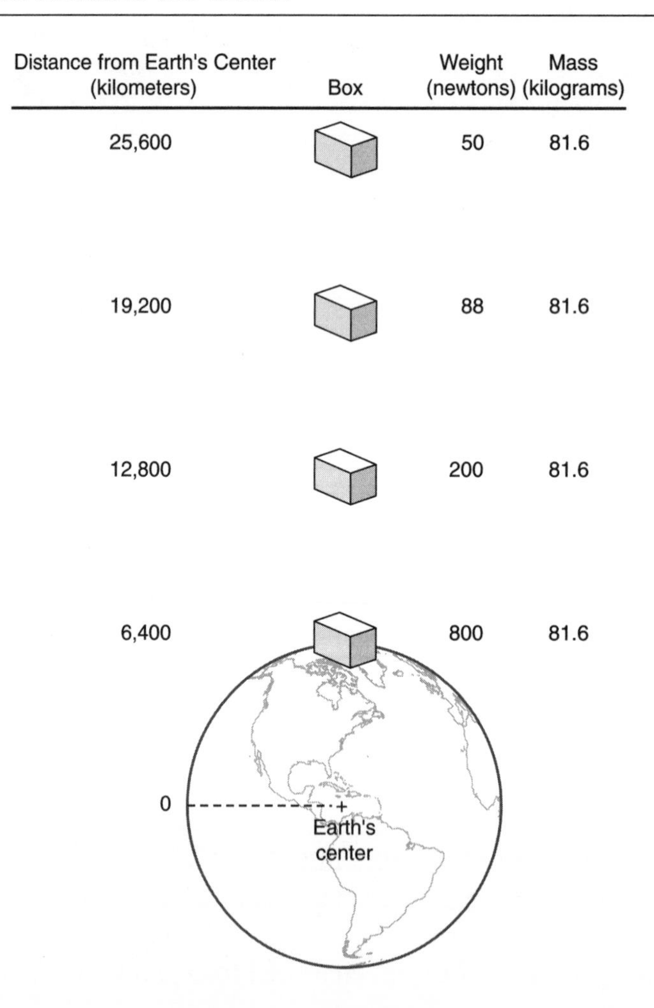

Mass

Mass is a measure of the amount of matter in an object. The mass of an object, unlike its weight, does not change. Figure 4-3 shows that although the weight changes, the mass of the box is unchanged as it is moves toward Earth's center.

A bowling ball and a basketball are about the same size, but their masses are different. A bowling ball contains more matter and therefore has more mass than a basketball. The greater the mass of an object, the greater the force necessary to move the object. Thus, a greater force is needed to throw a bowling ball than to throw a basketball the same distance. In other words, when the same amount of force is applied to a basketball and a bowling ball, the basketball will travel farther than the bowing ball. (See Figure 4.4.)

Figure 4-4. *A greater force is needed to toss a bowling ball (b) than to toss a basketball (a).*

Motion

Motion is a change in the position of an object relative to another object, which is assumed to be at rest. We recognize motion with respect to a nonmoving object or *frame of reference*. Understanding motion is more complex when motion is viewed relative to different frames of reference. For example, a school bus and its driver appear to be in motion to parents watching the bus move down the street. The bus and its driver change position relative to the nearby houses and trees. However, the bus driver does not appear to be moving relative to the windows and seats on the bus. (See Figure 4-5 on page 128.)

Speed, velocity, and acceleration are three ways of describing the motion of an object. ***Speed*** is the distance traveled per unit of time. It can also be described as the rate of change in position of an object. The formula below is used to determine average speed:

$$\text{speed} = \frac{\text{distance}}{\text{time}} \text{ or } s = \frac{d}{t}$$

A car travels from Binghamton to Albany, a distance of about 200 kilometers (125 miles) in 2 hours. The car's average speed is determined by

Figure 4-5. *The bus and driver appear to be moving compared to the houses and trees. However, the driver appears not to be moving compared to the windows and seats on the bus.*

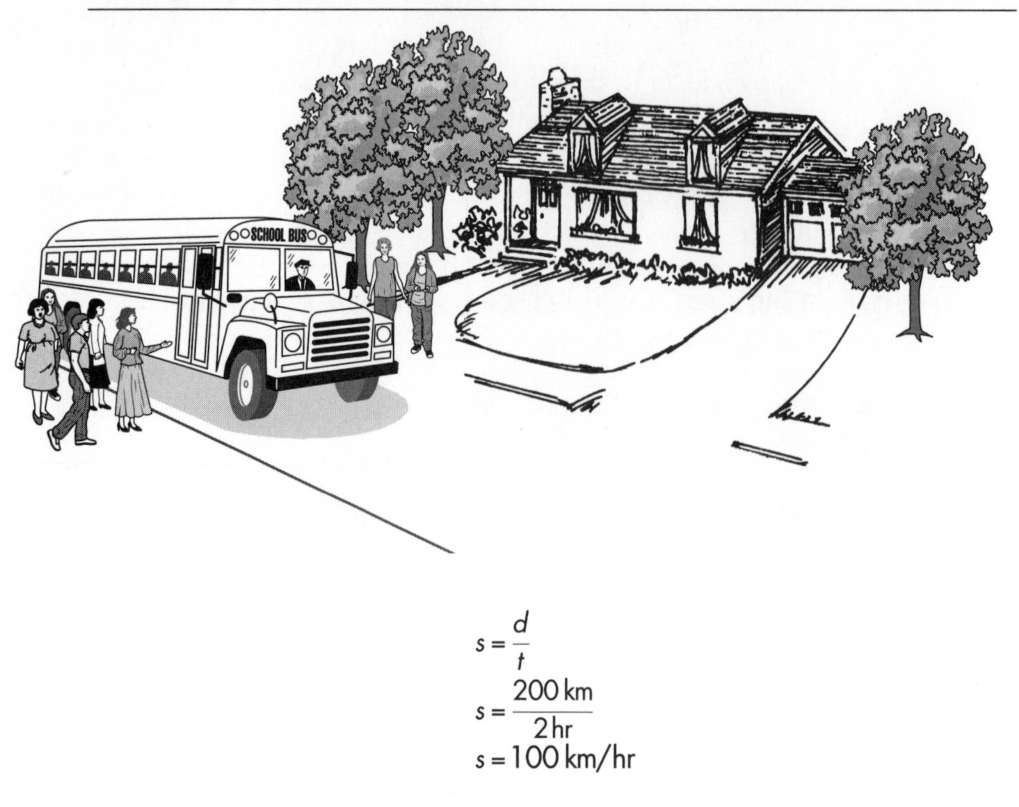

$$s = \frac{d}{t}$$

$$s = \frac{200\text{ km}}{2\text{ hr}}$$

$$s = 100\text{ km/hr}$$

Speed is labeled distance per unit of time. Some units for labeling speed are kilometers/hour (km/hr), meters/minute (m/min), and centimeters/second (cm/s).

Velocity and speed are similar. However, velocity also depends on direction; it has a directional component. ***Velocity*** is the speed of an object in a specific direction. The direction is assumed to be a straight-line direction such as north, east, southwest, etc. The formula below is used to determine velocity. You may notice that this is the same formula as the one used to calculate speed. However, when you solve the problem below, a direction is added to the answer to identify this as a velocity problem.

$$\text{velocity} = \frac{\text{distance}}{\text{time}} \text{ or } v = \frac{d}{t}$$

An airplane travels southward from Milwaukee to New Orleans, a distance of 1400 kilometers (870 miles) in 2 hours. You could determine its velocity by

$$v = \frac{d}{t}$$

$$v = \frac{1400\text{ km}}{2\text{ hr}}$$

$$v = 700\text{ km/hr in a southerly direction}$$

Acceleration is the rate of change in velocity. Since direction is included in velocity, a change in an object's direction is also acceleration. Although acceleration can refer to either an increase or a decrease in velocity, the term *deceleration* is commonly used to describe a decrease in velocity. Acceleration occurs when a car increases its velocity, and deceleration occurs when a car decreases its velocity. The initial velocity is the starting velocity, which for an object at rest is 0. The following formula is used to determine the rate of acceleration:

$$\text{acceleration} = \frac{\text{final velocity} - \text{initial velocity}}{\text{time}} \text{ or } a = \frac{v_f - v_i}{t}$$

A car increases its velocity from 40 km/hr to 60 km/hr in 2 seconds. Its acceleration is determined by:

$$a = \frac{v_f - v_i}{t}$$

$$a = \frac{60\text{ km/hr} - 40\text{ km/hr}}{2\text{ s}}$$

$$a = 10\text{ km/hr/s}$$

Acceleration is labeled distance per unit of time per unit of time. Units for labeling acceleration are kilometers/hour/second (km/hr/s), meters/minute/second (m/min/s), centimeters/second/second (cm/s/s or cm/s^2), etc.

Process Skill 1

Determining the Average Speed of a Car Trip

Mike lives in Glens Falls, New York. Last summer he took four car trips with his family. Mike's journal of each trip indicated the distance and time it took to get to where they were going. The table below summarizes the distance and time for each trip.

From/to	*Distance (km)*	*Time (hr)*	*Average Speed*
Glens Falls to Buffalo	540	6.0	?
Glens Falls to New York City	290	5.2	?
Glens Falls to Cooperstown	170	2.4	?
Glens Falls to Lake Placid	150	1.9	?

Questions

1. Using the formula $s = d/t$, determine the average speed for each trip in km/hr. Round your answer to the nearest tenth.
2. Name two factors that may cause the average speed of the trips to differ.

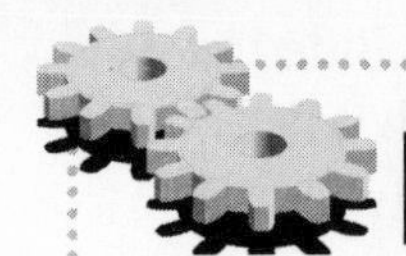

Process Skill 2

Determining the Acceleration of a Moving Car

In a car trip between Binghamton and Middletown, New York, Andrea recorded the car's speed every minute for a period of 12 minutes. The table shows her recorded times.

Time (minutes)	*Velocity (km/hr)*	*Time (minutes)*	*Velocity (km/hr)*
1	80	7	110
2	80	8	95
3	70	9	85
4	80	10	80
5	90	11	75
6	100	12	70

Questions

1. Between the 8-minute mark and the 12-minute mark the car was
(1) accelerating (3) traveling at uniform velocity
(2) decelerating (4) stopped

2. Using the formula $a = \frac{v_f - v_i}{t}$, determine the acceleration of the car between the 3- and 6-minute marks. (Label your answer with the correct units.)

3. In terms of acceleration and deceleration, explain what the car was doing between the 2- and 4-minute marks.

Review Questions

Part I

1. Of the four different balls listed, the greatest force would be needed to move the
   (1) golf ball (3) basketball
   (2) baseball (4) bowling ball
2. You would weigh the least
   (1) on the top of the Empire State Building
   (2) on the 86th floor of the Empire State Building
   (3) on the ground floor of the Empire State Building
   (4) in the subway under the Empire State Building

3. The mass of the Statue of Liberty would be
   (1) greatest in Washington, D.C.
   (2) greatest at the North Pole
   (3) greatest on the moon
   (4) the same everywhere

4. The front door of a house can be closed by use of
   (1) a pushing force
   (2) a pulling force
   (3) either a pushing or a pulling force
   (4) neither a pushing nor a pulling force

Base your answer to question 5 on the graph below, which shows the distance and time traveled by four cars.

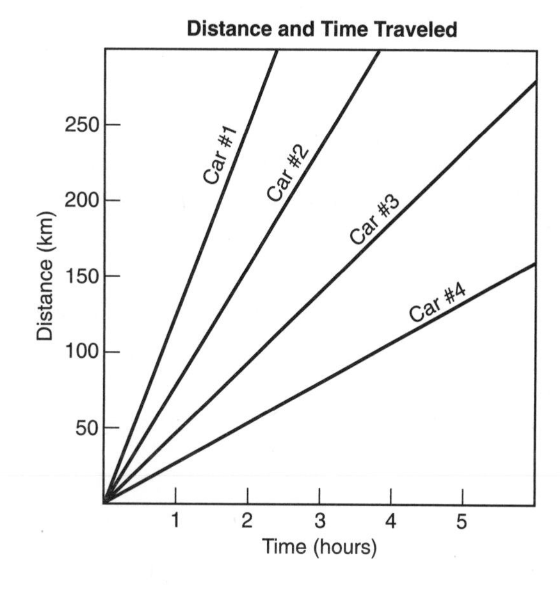

5. Which car traveled the fastest?
   (1) Car #1
   (2) Car #2
   (3) Car #3
   (4) Car #4

Base your answers to questions 6 and 7 on the formula for determining speed if distance and time are known. Speed equals distance divided by time ($s = d/t$).

6. A spider walked across a 2-meter-wide table in a half minute. What is the speed of the spider?
   (1) 1 meter/second
   (2) 4 meters/minute
   (3) 1 meter/minute
   (4) 1 kilometer/hour

7. If it took twice the time to travel the same distance, how would the speed change?
   (1) The speed would be halved.
   (2) The speed would double.
   (3) The speed would triple.
   (4) The speed would quadruple.
8. To determine the velocity of a moving object, you need to know
   (1) time and direction
   (2) time, distance, and direction
   (3) distance and direction
   (4) time and distance
9. We can best recognize the motion of an object relative to
   (1) an object moving in any direction
   (2) an object moving in the same direction
   (3) an object moving in the opposite direction
   (4) a frame of reference
10. A boy on a bicycle rides 15 kilometers in 2 hours. At what speed did the boy travel?
   (1) 30 kilometers per hour
   (2) 15 kilometers per hour
   (3) 3000 meters per hour
   (4) 7.5 kilometers per hour

Part II

Base your answers to questions 11 through 13 on the passage below and your knowledge of science. The passage describes the forces applied in a game of volleyball.

Volleyball

A force is a push or a pull. A force can start motion, stop motion, change the speed of motion, or change the direction of motion. In volleyball, the person serving the ball tosses it up to start the ball moving. The server then strikes the ball, changing its direction and increasing its speed toward the opposing team. When the ball reaches the other team, a player applies a new force, causing the ball to slow and go up. Then a player applies a stronger force to change direction and send the ball back to the serving team. This continues until the point is won and the motion of the ball is stopped.

11. What type of force is applied to the ball when it is served?
12. What is the purpose of the force applied by the person returning the served volleyball?
13. In baseball, relative to the motion of the baseball, what is the purpose of the baseball glove?

Base your answers to questions 14 through 16 on the diagram below and your knowledge of science. Roberto threw a baseball up into the air. It reached a height of 9 meters and then started to come down. He then threw the ball to a height of 12 meters.

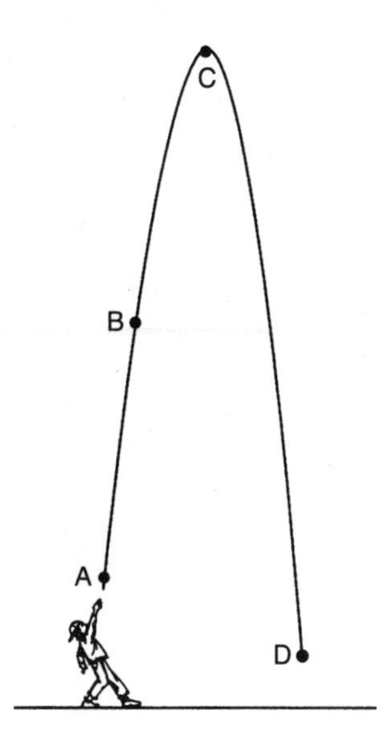

14. (a) What force caused the ball to go from A to C?

    (b) What force caused the ball to go from C to D?

15. At what point in the arc of the baseball's flight were the two forces balanced?

16. Roberto threw the ball a second time and it reached a height of 12 meters. What is the most likely reason that the ball went higher on Roberto's second throw?

Laws of Motion

Newton's Laws of Motion

In the mid-1660s, Sir Isaac Newton, an English mathematician, studied force and motion. He described three laws that describe the movement of objects. The laws explain:

- The properties of objects at rest and in motion
- The mathematical relationship among force, mass, and acceleration
- How forces act in pairs

Newton's laws of motion remain the basis for understanding the motion of objects on Earth.

First Law of Motion

The ***first law of motion*** states that an object at rest will remain at rest and an object in motion will remain in motion, unless an outside force acts on the object. There are two parts to this law. First, any object at rest will not move unless some force acts on it. An empty garbage can will remain at the curb or a dead leaf will remain on the front lawn until some force moves it. The force that moves the garbage can may be a person, and the wind may be the force that blows the leaf away. Second, any moving object will continue to move in the same direction, at the same speed, until a force acts on the object to change its speed or direction. A thrown baseball will move in a straight line at a constant speed until another force affects it. The force of air friction will decrease the speed of the ball, and the force of gravity will change the direction of the ball by pulling it down toward Earth. (See Figure 4-6.)

Figure 4-6. *Air friction is the force that slows the ball and gravity is the force that pulls the ball back to Earth.*

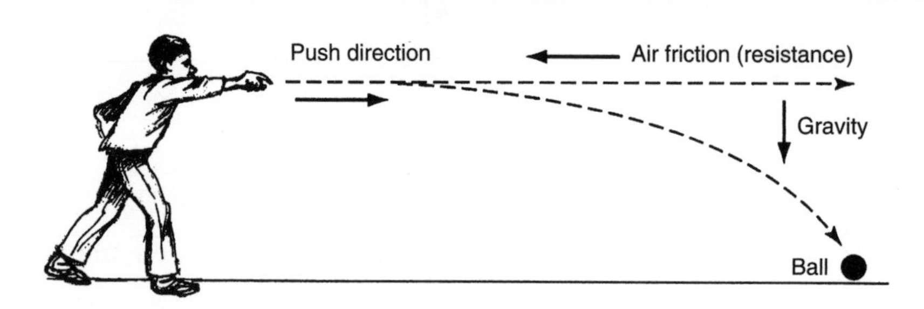

The tendency of an object at rest to remain at rest or an object in motion to remain in motion is called ***inertia***. In other words, inertia is the tendency of an object to resist any change in its motion. The more massive an object is, the greater its inertia, or the greater it will resist a change in motion. When you are riding in a moving car that stops suddenly, your body continues to move forward as it resists the stopping action. Also, when you are seated in a stopped car that suddenly accelerates, you feel your body move backward as it resists being put in motion. (See Figure 4-7.)

Figure 4-7. *The first law of motion: Objects resist a change in motion—an effect you can feel in a starting or stopping car.*

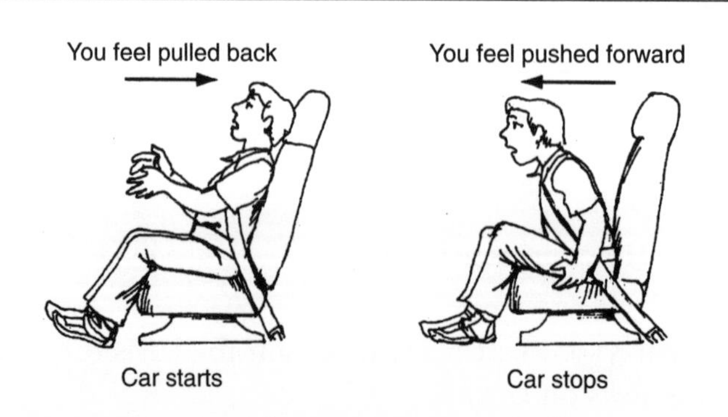

Second Law of Motion

The ***second law of motion*** states the relationship among force, mass, and acceleration. The law is commonly expressed by the formula:

$$\text{force} = \text{mass} \times \text{acceleration} \text{ or } f = m \times a$$

The formula could also be written as $a = \frac{f}{m}$ and as $m = \frac{f}{a}$.

A large force acting on a given mass will cause a greater acceleration than a small force acting on the same mass. For example, an adult (large force) pushing another adult (large mass) on a swing will cause a greater acceleration than a child (small force) pushing an adult (large mass) on the swing. (See Figure 4-8A.)

Also, a small mass acted on by a given force will have a greater acceleration than a large mass acted on by the same force. For example, an adult (large force) pushing a child (small mass) on a swing will cause a greater acceleration than an adult (large force) pushing another adult (large mass) on the swing. (See Figure 4-8B.)

Figure 4-8. *In Diagram (A), if the mass is the same, the larger force will cause a larger acceleration. In Diagram (B) if the force is the same, the smaller mass will have the larger acceleration.*

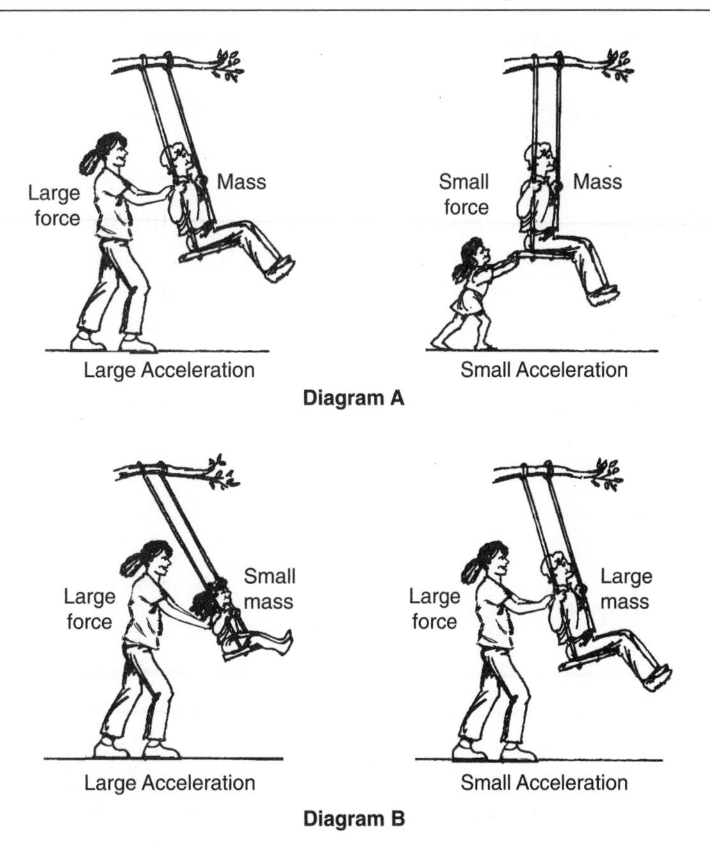

Laboratory Skill

Using Math to Analyze Data

Newton's second law of motion states the relationship among force, mass, and acceleration. This relationship is expressed by the following formula:

$$\text{force} = \text{mass} \times \text{acceleration or } f = m \times a$$

The following table shows five mathematical examples of how the formula is applied. Base your answers to the questions on the data in the table.

	Force (newtons)	*Mass (kilograms)*	*Acceleration (meters/second/second)*
Example 1	1	1	1
Example 2	20	20	1
Example 3	40	20	2
Example 4	20	40	0.5
Example 5	*X*	60	3

Questions

1. According to the table, how much acceleration will 1 newton of force give to a 1 kg mass?

2. If the mass of an object remains the same and the force moving it is doubled, then the object's acceleration is doubled. Which two examples in the table above demonstrate this?

3. If the force remains the same but the first object is replaced with another object whose mass is twice that of the first object, then the acceleration
(1) is doubled
(2) is halved
(3) remains the same
(4) is equal to mass

4. In Example 5 in the table, how many newtons is the unknown force *X*?

Third Law of Motion

The ***third law of motion*** states that for every action there is an equal and opposite reaction. This law is sometimes referred to as the law of action-reaction because it shows how forces work in pairs. Consider what happens when you push a heavy table. If you push the table and the table pushes back on you with an equal force, there is no movement. When your force becomes greater than the table's force pushing back on you, the table moves. A simple demonstration of blowing up a balloon and letting it go also shows how this law works. When the air is released from the balloon, the balloon moves in the opposite direction. (See Figure 4-9.) Kicking a soccer ball, hot gases shooting out of a rocket engine, and walking are all actions that produce an equal and opposite reactive force.

Figure 4-9. *The third law of motion: Every action has an equal and opposite reaction.*

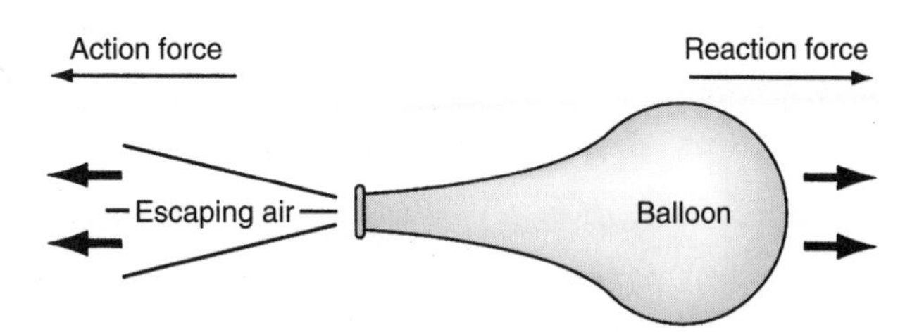

Review Questions

Part I

17. A rocket traveling in space increases its velocity from 30,000 km/hr to 42,000 km/hr. To do this, the rocket must
    (1) increase its mass
    (2) decrease its mass
    (3) increase the force acting on it
    (4) decrease the force acting on it

18. To determine the mass of a moving object using Newton's second law, you need to know
    (1) velocity and distance
    (2) velocity and time
    (3) velocity and force
    (4) force and acceleration

19. When you push on a wall, the wall pushes back against you. This process demonstrates Newton's law that states that
    (1) a body at rest remains at rest unless a force affects it
    (2) a body in motion remains in motion unless a force affects it
    (3) a large mass requires a large force to move it
    (4) every action has an equal and opposite reaction

20. What is the relationship between mass and inertia of an object?
    (1) The greater the mass, the greater the inertia.
    (2) The greater the mass, the less the inertia.
    (3) The less the mass, the greater the inertia.
    (4) Mass and inertia are not related.

21. When a golf ball is placed on a tee, it will remain there until the golfer hits the ball with a club. This demonstrates Newton's law that states that

(1) a body at rest remains at rest unless a force affects it

(2) a body in motion remains in motion unless a force affects it

(3) a large mass requires a large force to move it

(4) every action has an equal and opposite reaction

22. Martin took a full garbage can to the curb. He pushed as hard as he could, but the garbage can moved very slowly. This example best demonstrates the principle that

(1) a body at rest remains at rest unless a force affects it

(2) a body in motion remains in motion unless a force affects it

(3) a large mass requires a large force to move it

(4) every action has an equal but opposite reaction

23. When a rifle is fired, the force of the expanding gases produced by the exploding gunpowder pushes the bullet out the gun barrel. The rifle suddenly moves backward. This is an example of the

(1) law of inertia　　(3) law of action-reaction

(2) law of acceleration　　(4) law of machines

24. The boys in the picture are playing baseball. If the batter swings and misses the ball, the force that eventually changes the motion of the ball is most probably

(1) air friction
(2) gravity
(3) the catcher's glove
(4) the pitcher

25. Alice was standing inside a train when the train suddenly started moving. She immediately lost her balance and stumbled toward the back of the train. This happened because
    (1) Alice was standing in the back of the train
    (2) Alice was at rest and her body resisted a change in its motion
    (3) Alice started walking through the train
    (4) Alice experienced a greater force of gravity on the moving train

Base your answers to questions 26 and 27 on the diagram and description below and your knowledge of science. The passage describes an investigation done by a student.

A CO_2 cartridge was mounted on the top of a toy car with the nozzle pointing toward the rear of the car. When the compressed gas was released from the cartridge, the toy car moved 6 meters across the room and stopped. This was repeated using a car that had a mass 10 times greater than that of the first car.

26. The second car most likely moved
    (1) more than 6 meters
    (2) about 6 meters
    (3) less than 6 meters
    (4) backward

27. The reaction of the cars to the release of gas from the cartridge
    (1) only supports Newton's first law—a body in motion remains in motion and a body at rest remains at rest unless a force affects it
    (2) only supports Newton's third law—every action has an equal and opposite reaction
    (3) supports both Newton's first and third laws
    (4) does not support any of Newton's laws

28. Equal forces are applied to two objects, A and B, causing them to move. The mass of the second object, B, is twice the mass of the first object, A. Which statement correctly compares the motion of the two objects?

(1) Object A will have a greater acceleration than object B.

(2) Object B will have a greater acceleration than object A.

(3) Objects A and B will have the same acceleration.

(4) Object A will accelerate and object B will decelerate.

Part II

Base your answers to questions 29 through 31 on the passage below, the diagram, and your knowledge of science. The passage and diagram explain an investigation performed by two eighth-grade students.

Gail and Jared did an investigation that they read about in their science textbook. The investigation required a glass, several coins (penny, dime, and quarter), and an index card. They placed the index card on top of the glass and they placed a coin on top of the index card. The students flicked the card from under each of the coins to see what would happen.

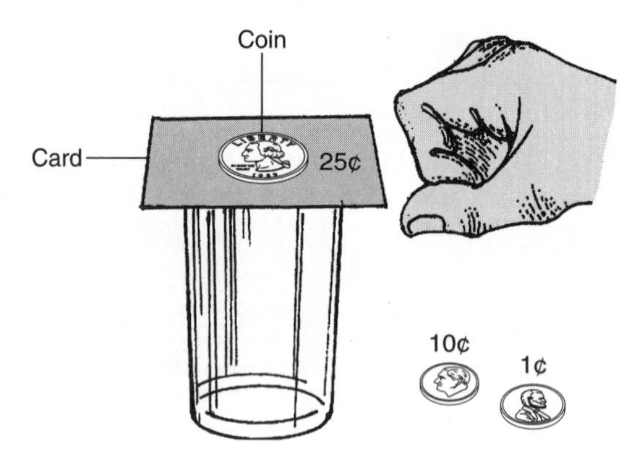

29. Predict what happened to the coin when the card was flicked.

30. Which coin had the greatest resistance to motion? Why?

31. Which of Newton's laws of motion did this investigation demonstrate?

Describe in terms of Newton's first law of motion how each of the following three events affected the participant.

32. A student was riding a skateboard. The front wheels hit a curb, and the skateboard suddenly stopped.

33. A stunt-car driver drove a car into a wall at 80 km/hr. He was wearing a seat belt.

34. A student was riding a roller coaster at an amusement park. The roller coaster climbed upward and then suddenly plunged down, going from 0 to 80 km/hr in 6 seconds.

Base your answers to questions 35 through 37 on the diagram below and your knowledge of science. The diagram shows water coming out of a sprinkler.

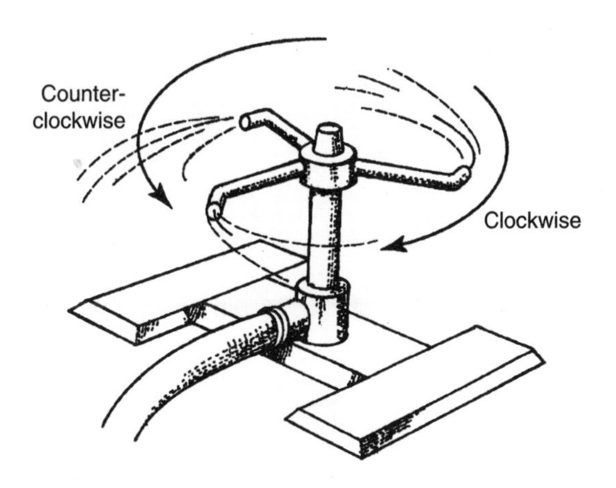

35. When water comes out of the nozzle, in what direction will the sprinkler turn—clockwise or counterclockwise?

36. Which of Newton's laws explains the direction in which the sprinkler moves?

37. If the force of the water coming out of the nozzle is decreased, what will happen to the sprinkler's speed of rotation?

Machines

Work

Scientists use the word "work" in a very specific way. They say that ***work*** is done when a force causes an object to move some distance. The amount of work done depends on the amount of force applied and the distance the object is moved. The relationship between work, force, and distance is given by the formula

work = force × distance or $w = f \times d$

The unit of force is the newton and the unit of distance is the meter; therefore the unit of work is the newton-meter (N-m). One newton-meter of work is done when an object weighing 1 newton is lifted 1 meter.

When a force is applied to an object, the force may or may not cause the object to move. If the force does not produce motion, no work is done. A force produces work only if it produces motion. (See Figure 4-10 on page 142.)

Figure 4-10. *Work is done when a force acts over a distance.*

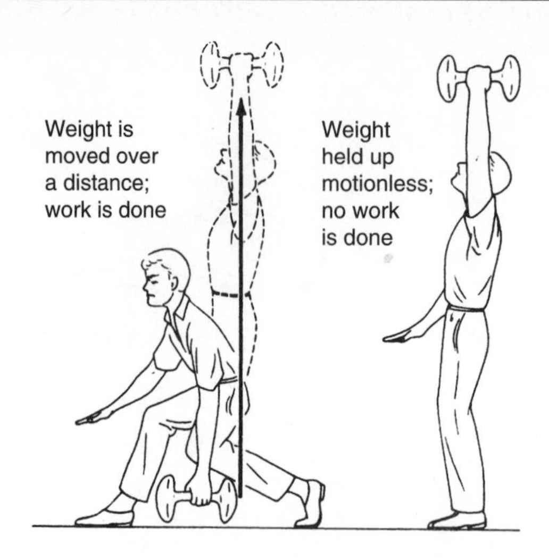

Machines and Work

A ***machine*** is a device that transfers mechanical energy from one object to another object. Machines make work easier. They do this by changing

1. the amount of force or
2. the direction of a force or
3. the distance over which a force is applied

The force a machine has to overcome is called the ***resistance***, and the force applied is called the ***effort***. A machine can reduce the effort needed to overcome a large resistance because the effort is applied over a longer distance, in effect multiplying the force. However, using a machine does not decrease the amount of work.

A wrench multiplies the force (effort) applied to it when removing a tight bolt. The tight bolt is the resistance force. However, the distance the effort is applied is much greater than the distance the bolt turns. (See Figure 4-11.)

A ***simple machine*** is a device that makes work easier by decreasing the effort needed while increasing the distance the effort is applied. A seesaw is a simple machine. A small downward force on one side of the seesaw can lift a large force on the other side. (See Figure 4-12.) The small boy is the effort and the large man is the resistance.

Types of Simple Machines

Simple machines help us do work faster and with less effort. There are six simple machines: the lever, pulley, wheel and axle, inclined plane, wedge, and screw.

Figure 4-11. *A small effort at the end of a wrench is capable of overcoming a great reistance of a tight bolt. However, the effort distance must be applied a greater distance than the bolt turns.*

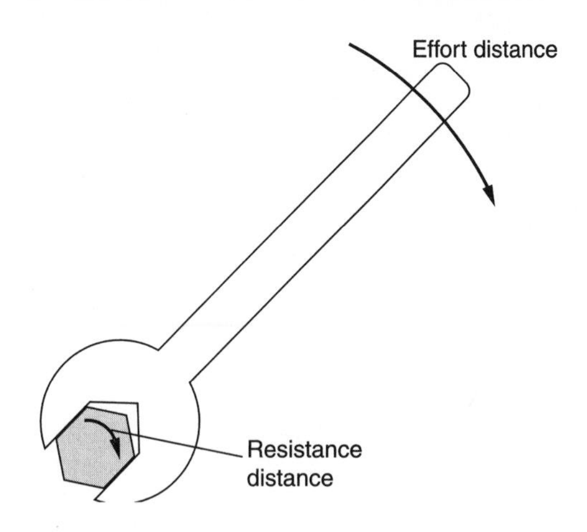

Figure 4-12. *A seesaw is a simple machine. A small effort force down (boy) can lift a larger resistance force (man).*

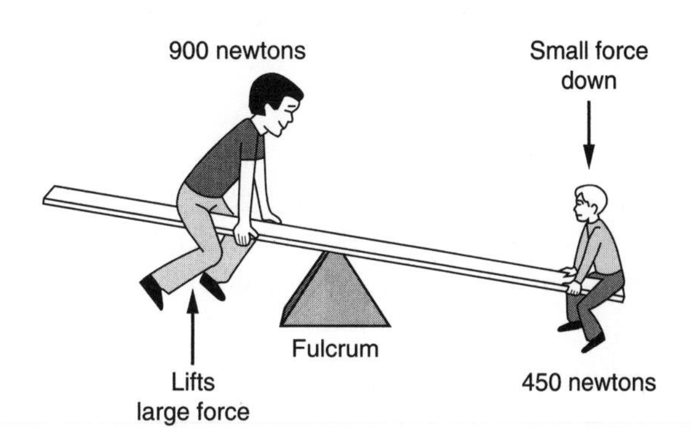

A ***lever*** is a rigid (stiff) bar that can turn around a point called a ***fulcrum***. (See Figure 4-13.) Levers make work easier by multiplying the applied force, or effort. Scissors and crowbars are examples of levers.

Figure 4-13. *A lever multiplies effort, making it easier to uproot a tree stump.*

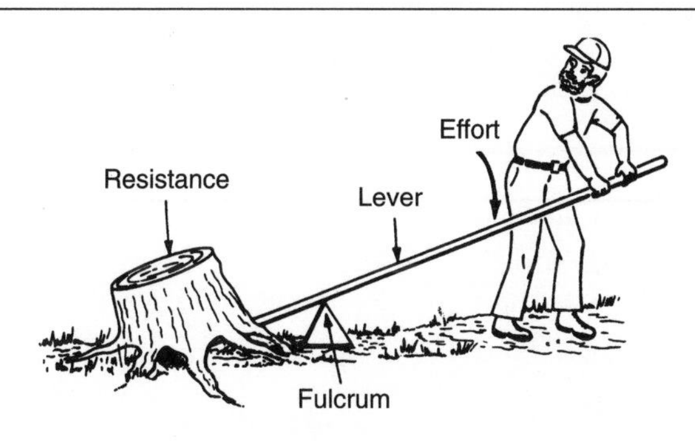

A ***pulley*** is a simple machine consisting of a rope and one or more wheels. The resistance is a weight attached to one end of the rope or to one of the wheels. The effort is applied to the other end of the rope. A single-wheel pulley is used to change the direction of a force. It can be used to raise a flag up a flagpole. Multiple-wheel pulleys can multiply the effort. A heavy car engine can be lifted using a multiple-pulley system. Figure 4-14 shows several types of pulleys.

Figure 4-14. *Three types of pulleys. (R is resistance, E is effort.)*

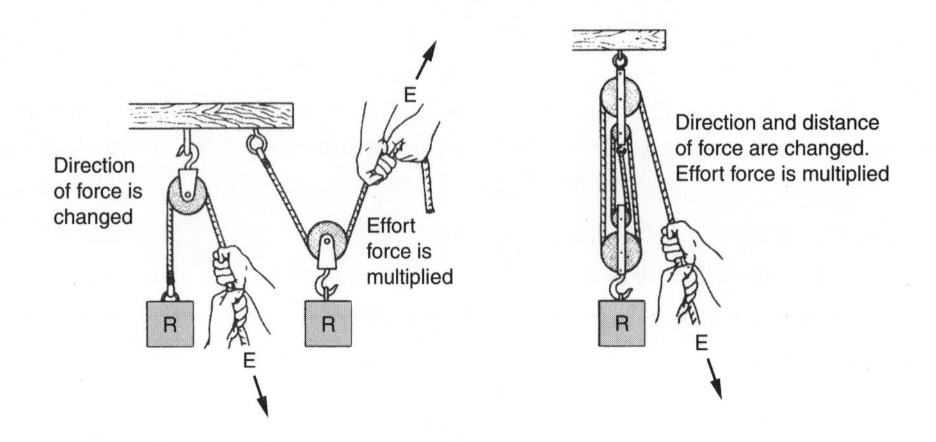

A ***wheel and axle*** is a large wheel with a smaller wheel, or axle, in its center. The wheel is attached to the axle so they turn together. When the outer wheel turns, the axle turns, too (see Figure 4-15). The outer wheel turns a greater distance, multiplying the force applied to the axle. Bicycle handlebars, steering wheels in cars, and doorknobs are examples of a wheel and axle.

Figure 4-15. *A steering wheel is an example of a wheel and axle.*

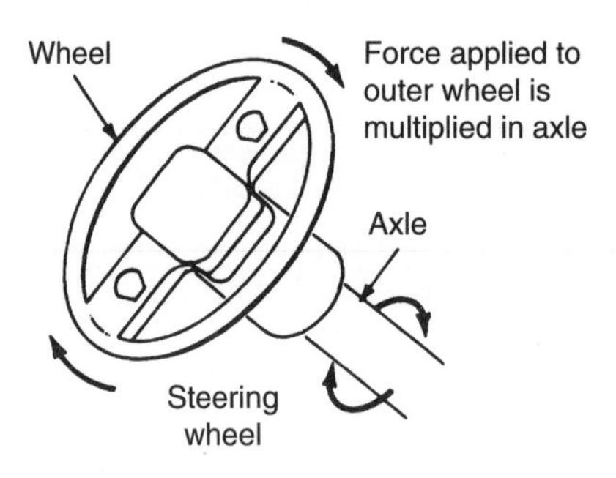

An ***inclined plane*** is a flat surface with one end higher than the other. A wheelchair ramp and truck ramp are inclined planes. A staircase is another example of an inclined plane. Figure 4-16 shows how an inclined plane makes work easier by decreasing the force needed and increasing the distance the force is applied.

Figure 4-16. *A loading ramp is an inclined plane.*

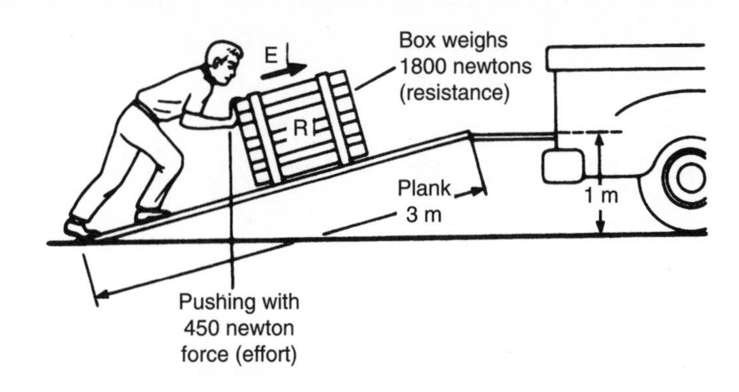

The ***wedge*** is a double-sided inclined plane. The effort is applied by driving the wedge into something. For example, an ax is a wedge driven into a log. Other examples of wedges are knives, wood nails, and chisels. (See Figure 4-17.)

Figure 4-17. *Three examples of wedges: (a) knife, (b) ax, and (c) wood nail.*

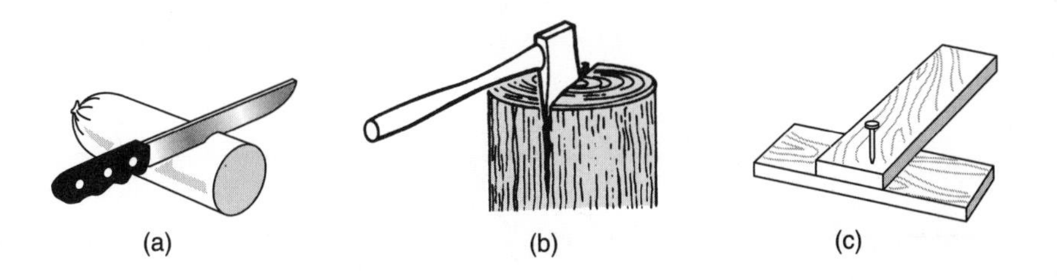

The ***screw*** is an inclined plane wrapped around a wedge or cylinder. Examples of screws are wood screws, bolts, and car jacks. When you use a wood screw, a circular force is applied to the screw's head to overcome the resistance of the wood. As the screw turns, it penetrates the wood. (See Figure 4-18.)

Figure 4-18. *The screw is a spiral inclined plane.*

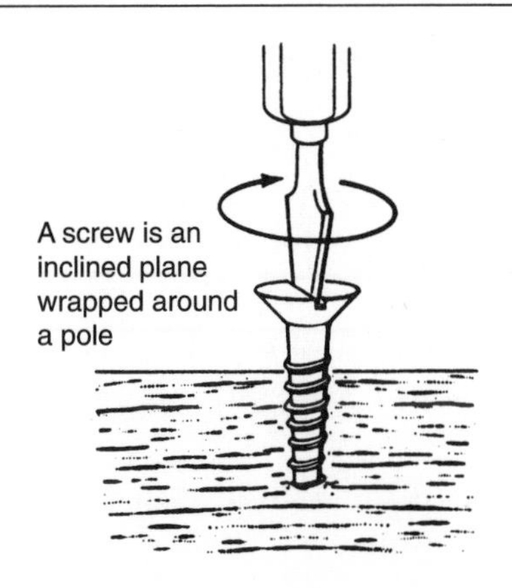

Many machines are called complex machines because they contain two or more simple machines. A bicycle is a complex machine. On most bicycles you should be able to locate a lever, wheel and axle, a pulley system, and a screw. (See Figure 4-19.)

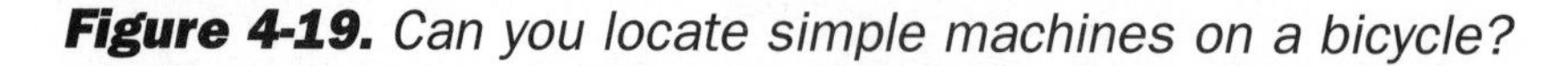

Figure 4-19. *Can you locate simple machines on a bicycle?*

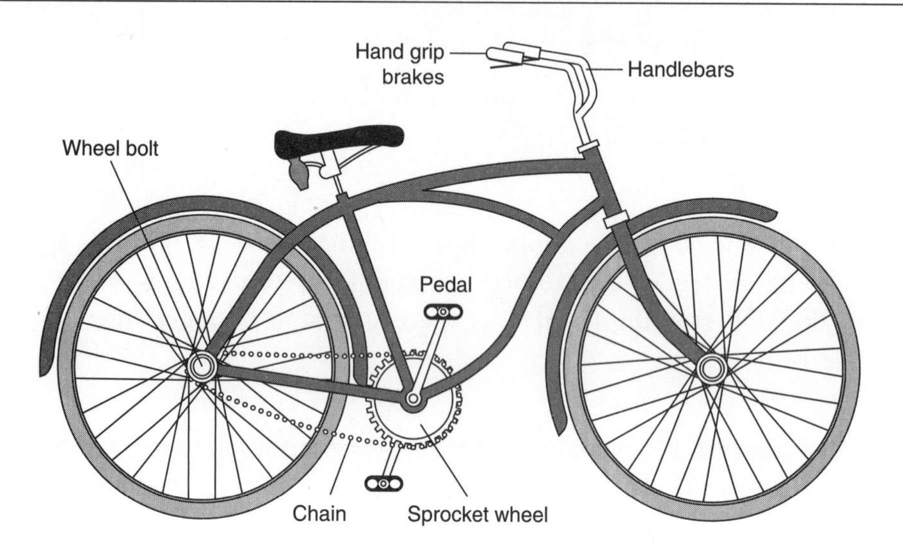

Table 4-3. *Some Simple Machines Found on a Bicycle*

Bicycle Part	Simple Machine
Hand brakes	Lever
Handlebars	Wheel and axle
Chain	Pulley
Wheel bolts	Screw
Sprocket wheel	Wheel and axle
Pedal	Lever

Efficiency of Machines

Ideally, a machine's work output should equal the amount of work put into the machine. However, in reality, machines are never 100 percent efficient. The amount of work done by any machine is always less than the amount of work put into it. This is because some of the work put into a machine is converted into heat energy and therefore wasted. The heat is produced by friction, which occurs where the machine's moving parts come in contact with one another.

Suppose you lift a 450-newton box up 2 meters, using pulleys. Although 900 newton-meters (450 newtons $\times$ 2 meters) of work output were accomplished, you actually had to do more than 900 newton-meters of work. Some of your work input is wasted because friction between the wheel and axle and the rope of each pulley produces heat.

A machine can be made more efficient by reducing friction. A common way to reduce friction is to put grease or oil on the contact surfaces of a machine's moving parts. Other methods include waxing the contact surfaces, sanding the surfaces to make them smoother, or using ball bearings between the surfaces. (See Figure 4-20.)

Figure 4-20. *Ball bearings in the wheel of a Rollerblade reduce friction as the wheel turns.*

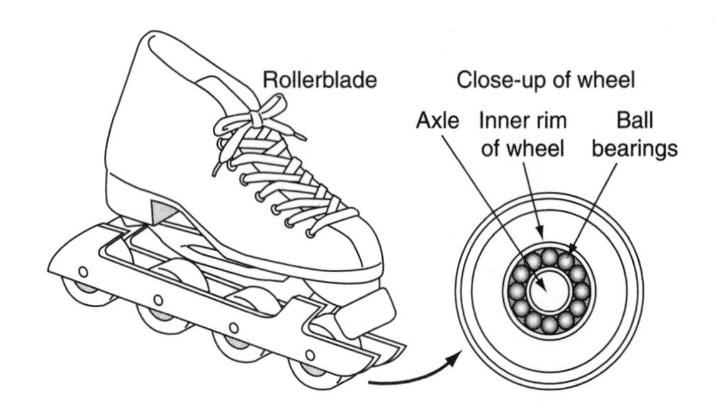

Process Skill 3

Classifying Levers

Many common tools are levers of some kind. For example, scissors, shovels, and salad or ice tongs are levers. Levers can be grouped into three basic classes, depending on where the lever encounters the resistance, where the effort is applied, and where the fulcrum is located. Diagram 1 illustrates the three classes of levers.

Diagram 1. *Three classes of levers.*

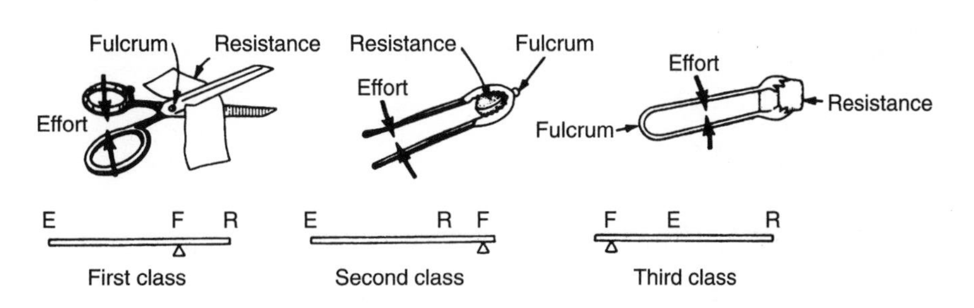

A first-class lever, such as a pair of scissors, has the effort (E) applied on one end, the resistance (R) on the other end, and the fulcrum (F) in between. A second-class lever, like a nutcracker, has the fulcrum and effort at opposite ends, and the resistance in the middle. A third-class lever, such as a pair of ice tongs, has the resistance and the fulcrum at opposite ends, and the effort applied in the middle.

QUESTIONS

1. Diagram 2 shows some examples of levers. Classify each lever in the diagram as a first-, second-, or third-class lever.

Diagram 2. *Examples of levers.*

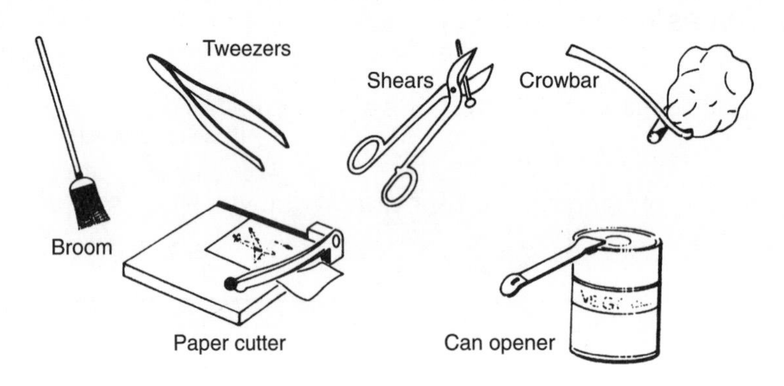

2. What class of lever is a wheelbarrow?

Review Questions

Part I

38. The applied force that is used in a simple machine is called
    (1) effort
    (2) resistance
    (3) friction
    (4) energy

39. Oil is placed in a car engine because it
    (1) cools the engine
    (2) reduces air drag
    (3) reduces friction
    (4) increases resistance

40. The chain on a bicycle is a simple machine. What type of simple machine is the chain?
    (1) wheel and axle
    (2) lever

(3) pulley

(4) inclined plane

41. Allan used a pulley system to remove the engine from his car. If his work output was 1000 newton-meters, his work input was

(1) greater than 1000 newton meters

(2) less than 1000 newton meters

(3) exactly 1000 newton meters

(4) no way to tell

42. The pulley on the flagpole in the illustration is a simple machine. The purpose of the pulley is to

(1) decrease the amount of work required

(2) change the direction of the force applied

(3) increase the amount of work required

(4) decrease the resistance

43. The man in the diagram is splitting wood using a wedge. A wedge is a type of

(1) wheel and axle

(2) pulley

(3) lever

(4) inclined plane

44. In science lab, Jennifer tested the efficiency of four machines and recorded the results in a table. For which machine must she have made an error?

Machine	Efficiency
Lever I	52%
Lever II	76%
Lever III	19%
Pulley	100%

(1) Lever I
(2) Lever II
(3) Lever III
(4) Pulley

Base your answers to questions 45 and 46 on the diagram below.

45. The screwdriver being used to remove the lid from a can of paint is acting as

(1) a wedge
(2) a lever
(3) an inclined plane
(4) a screw

46. Compared with the amount of work put into this simple machine, the amount of work put out by the machine is

(1) greater
(2) less
(3) the same
(4) sometimes greater, sometimes less

Part II

47. What three simple machines are being used in the diagram to help move bricks from position A to position B?

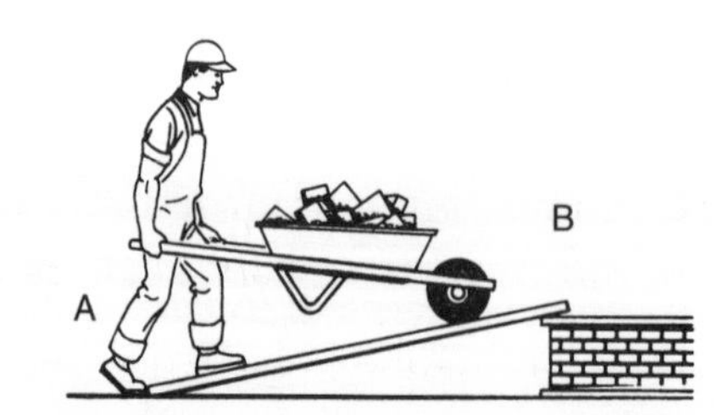

48. Fred tries to lift a heavy box off the ground, but cannot make it move. Even though he exerts great effort, no actual work is done. Explain.

Base your answers to questions 49 through 51 on the diagram and description below and your knowledge of science.

John and Dorothy set up a ramp similar to the one in the diagram below to do an experiment. They attached a 50-newton metal block to a cord and a spring scale, as shown in the first diagram. During the experiment, they made a number of changes as they pulled the block up the ramp, as shown in the two other diagrams.

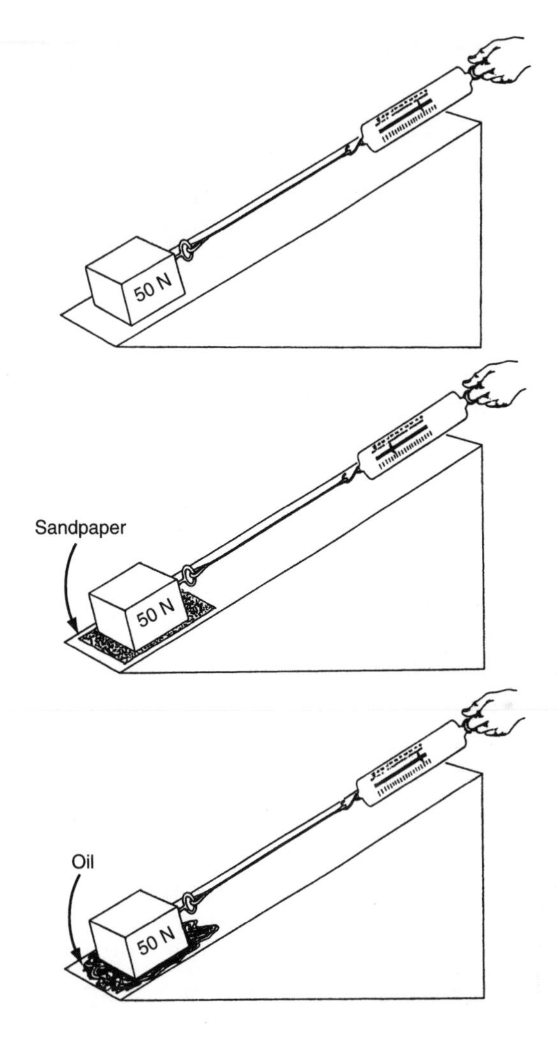

49. What type of simple machine is illustrated by using the ramp to pull the weight?

50. When the students pulled the weight up the wooden ramp, the spring balance indicated an effort of less than 50 N. When they attached a sheet of sandpaper to the ramp and pulled the weight up across the sandpaper, how was the amount of effort affected?

51. What did John and Dorothy notice about the effort needed to move the block when oil was placed on the ramp?

Chapter 5 Organisms

Vocabulary at a Glance

asexual reproduction
bacteria of decay
cancer
cell
cell membrane
chlorophyll
chloroplast
coarse-adjustment knob
compound microscope
cytoplasm
decomposer
digestion
elimination
environment
excretion
fertilization
field of view
fine-adjustment knob
growth
infectious disease
ingestion
life cycle
metamorphosis
microorganism
microscope
noninfectious disease
nucleus
nutrient
nutrition
organism
oxidation
photosynthesis
regulation
reproduction
respiration
sexual reproduction
species
specimen
stain
transport
virus

These tropical fish are interacting with the coral in their environment.

Major Concepts

- All living things are composed of basic units called cells. Plant cells and animal cells have common structures. These include the nucleus, cytoplasm, and cell membrane.
- Cells carry out life processes.
- All living things carry out life processes. These include nutrition, respiration, transport, excretion, regulation, growth, and reproduction.
- Plant cells have cell walls and chloroplasts, which are not found in animal cells. Plant cells manufacture their own food by a process called photosynthesis.
- Animals take in nutrients for energy and growth.
- Some microorganisms are harmful and can cause infectious diseases.

Although living things have their differences, they also have many things in common. They all carry out the same basic life processes.

Living Things Carry Out Life Processes

Living things, or ***organisms***, share certain activities that set them apart from nonliving things. In particular, organisms carry out *life processes,* some of which are listed below in Table 5-1.

Table 5-1. *Life Processes and Their Functions*

Process	Function
Nutrition	Taking food into the body (*ingestion*), breaking it down into a form usable by cells (*digestion*), and eliminating undigested material (*elimination*)
Transport	Moving materials throughout the organism
Respiration	Releasing energy stored in food
Excretion	Removing waste materials produced by the organism
Regulation	Responding to changes in the organism's surroundings
Reproduction	Making more organisms of the same kind
Growth	Increasing in size

The Cell

The basic unit of all living things is the ***cell***. Each cell of an organism carries out life processes. Most cells contain similar structures. The ***nucleus*** controls cell activities. The ***cytoplasm*** is a thick fluid that surrounds the nucleus. Most life processes take place in the cytoplasm. The cytoplasm is contained within the ***cell membrane***, the "skin" of the cell, which regulates the flow of materials into and out of the cell. Figure 5-1 shows a typical animal cell.

Figure 5-1. *Typical animal cell.*

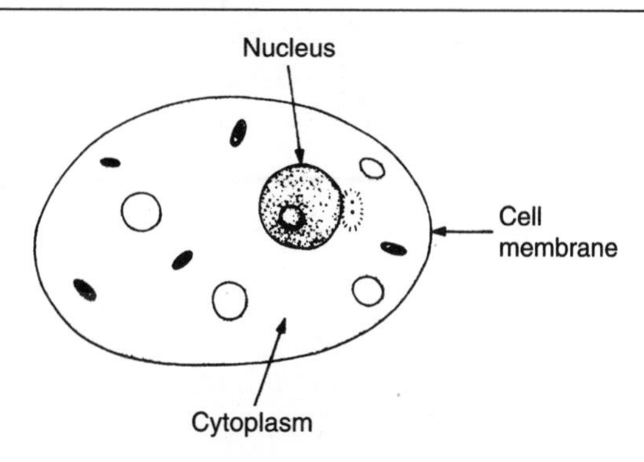

Cells and Life Processes

Living things are made up of one or more cells. For instance, an ameba is made up of a single cell, while a human is made up of trillions of cells. An ameba is therefore unicellular, while a human is multicellular. Each cell carries out basic life processes. Smaller structures within the cell perform these life processes. Table 5-2 lists some of these structures.

Table 5-2. *Some Cell Structures and Their Functions*

Structure	***Function***
Mitochondria	Respiration—where food is "burned" (combined with oxygen) to produce energy. Called the "powerhouse of the cell"
Ribosomes	Synthesis—where proteins are made
Lysosomes	Digestion—where digestive enzymes are stored
Nucleus	Reproduction—where genetic material is stored
Vacuole	Digestion and excretion—where digestion occurs or where excess fluid is stored
Chloroplasts	Photosynthesis—where glucose (sugar) is produced in green plants (present in plant cells only)

Plant cells and animal cells have many structures in common, but they also have differences (see Figure 5-2).

Figure 5-2. *Comparison of plant and animal cells.*

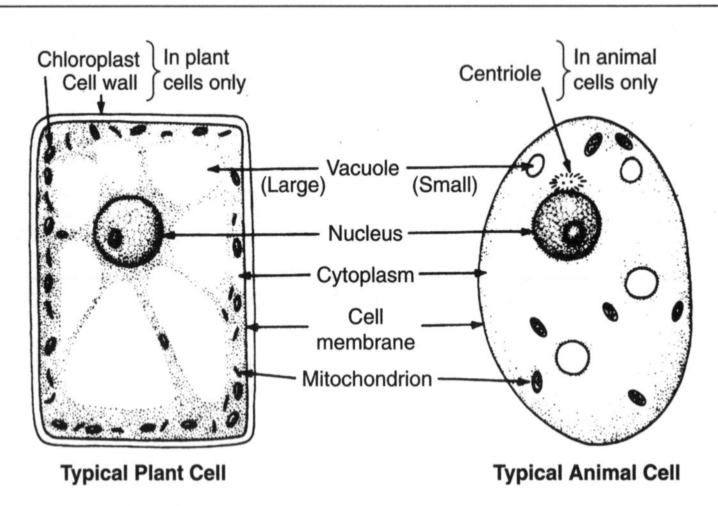

1. Plant cells have a cell wall that encloses the entire cell, including the cell membrane. The tough cell wall gives support to the

plant's structure. Animal cells do not have cell walls but are enclosed only by the cell membrane.

2. Only green plant cells have chloroplasts. Chloroplasts, which contain chlorophyll, are the site of photosynthesis, the food-making process of plants.
3. The centrioles, which participate in cell division, are found only in animal cells.

Organisms and Their Environment

Living things are constantly interacting with their surroundings. These surroundings are called the ***environment***, which includes both living and non-living things. Organisms get food, water, and oxygen from the environment. In turn, they release wastes, such as carbon dioxide, into the environment. There is a continual (constant) exchange of materials between an organism and its environment.

Nutrition

Every organism needs food to stay alive. Food provides an organism with ***nutrients***, which are used for producing energy as well as for growth and repair. Some important nutrients are listed in Table 5-3. The process of ***nutrition*** includes ingestion, digestion, and elimination. ***Ingestion*** is the taking in of food. ***Digestion*** is the breaking down of nutrients into a usable form. ***Elimination*** is the removal of undigested materials from the body.

Table 5-3. *Nutrients and Their Uses*

Nutrient	Use
Proteins	Supply materials for growth and repair
Carbohydrates	Provide quick energy (sugars and starches)
Fats and oils	Provide stored energy
Vitamins	Assist life processes; prevent disease
Minerals	Supply materials for growth and repair; help carry out life processes

Nutrition in Plants

Green plants make their own food by a process called ***photosynthesis*** (Figure 5-3). Plants use energy from sunlight to change carbon dioxide and water from the environment into sugar. The sun's energy is therefore stored in the sugar. Photosynthesis also produces oxygen, which is released into the environment. The green pigment ***chlorophyll***, found in the leaves and green stems of plants, is necessary for photosynthesis to take place.

Cells in the leaf of a plant contain large numbers of ***chloroplasts***. These structures contain chlorophyll, which is necessary for photosynthesis to take place. The roots of a plant, however, do not contain any chloroplasts. Instead, the roots are specialized for growing into the soil and absorbing water. The stem, with its thick cell walls, supports the plant and permits the flow of water and nutrients from the roots to the leaves and back again.

Figure 5-3. *One leaf has been enlarged to show what happens during photosynthesis.*

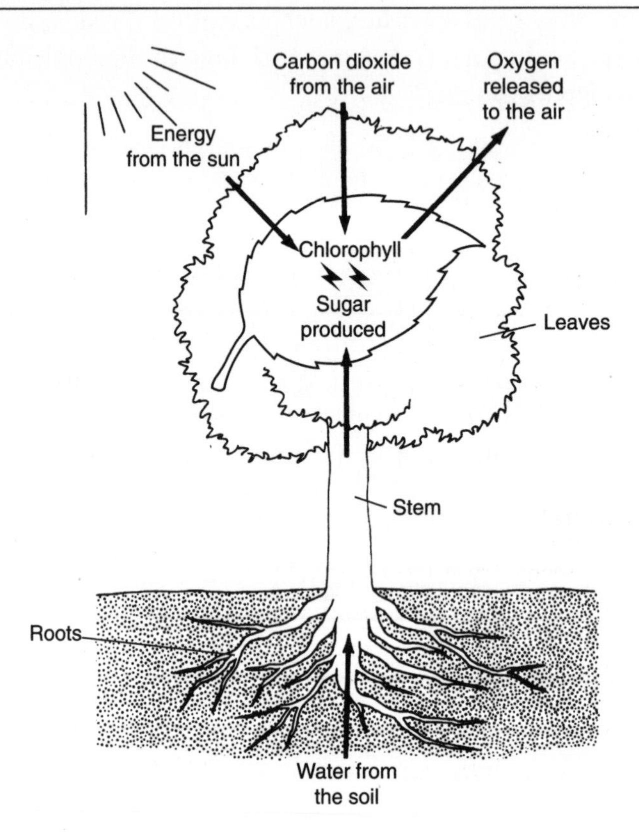

Nutrition in Animals

Animals get nutrients by eating plants or by eating other animals that feed on plants. The sugar in plants is used by animals to produce energy. The original source of energy in food is the sun. Figure 5-4 shows one way that animals get food energy from the sun.

Figure 5-4. *The energy in meat comes originally from the sun.*

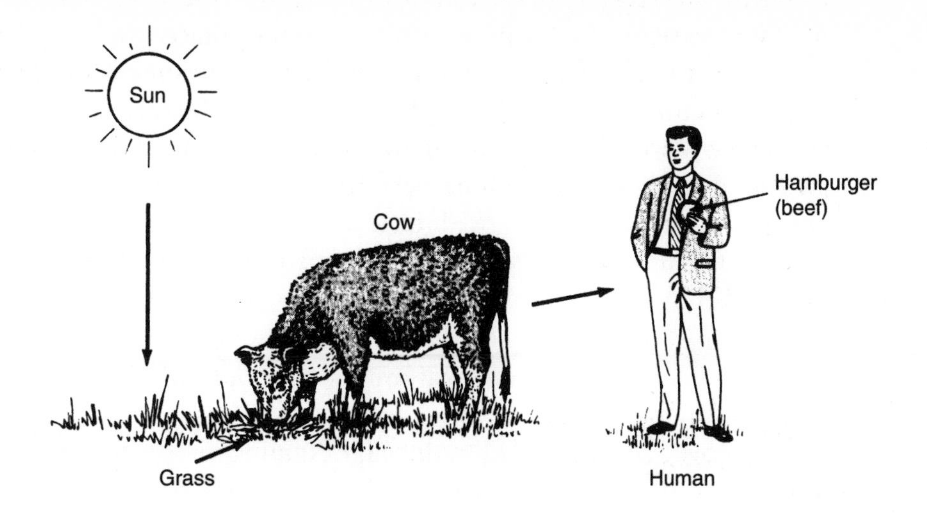

To maintain a balanced state, all organisms must have a minimum daily intake of each type of nutrient. The amount of each nutrient varies based on the size, species, activity, age, and sex of the organism. An imbalance in any of these nutrients may result in weight gain, weight loss, or an unhealthy condition.

Water is an important nutrient, too. Water is constantly exchanged between an organism and its environment. For example, we get water from the environment by eating and drinking, and we give water back by exhaling, perspiring, and urinating.

Transport

Water is necessary for ***transport***, the moving of materials throughout an organism. For instance, blood, which is mostly water, carries nutrients to the cells of your body. Blood takes away wastes from the cells, too. Most of the chemical processes in living things can take place only in a watery environment. In addition, water is necessary for green plants to make food. For all of these reasons, life is not possible without water.

Respiration

Organisms release the energy stored in food through a process called ***respiration***. Respiration occurs in all cells. During respiration, carbohydrates combine with oxygen to release energy and form carbon dioxide and water as waste products. Scientists use the term ***oxidation***, or "burning," to describe changes in which a substance combines with oxygen and releases energy. Therefore, respiration is a form of burning without flames.

Respiration is the reverse of photosynthesis, as shown below:

Photosynthesis: energy + carbon dioxide + water → sugar + oxygen
Respiration: sugar + oxygen → energy + carbon dioxide + water

All living things, including green plants, obtain energy from food through some form of respiration.

Excretion

Carbon dioxide is a waste material produced by cells during respiration and must be removed. In humans, carbon dioxide and other waste materials are carried away by the blood. These wastes are filtered out of the blood and then removed from our bodies through exhaling, perspiring, and urinating. The process of removing wastes from the organism is called ***excretion***.

Regulation

Why do you perspire when it is hot? Your body perspires to cool off. Why do you drink more when it is hot? Your body knows that it must replace water used to keep you cool. Being thirsty is a response to the loss of water. Organisms respond to changes in their internal and external environments. This process, called ***regulation***, helps an organism carry out *homeostasis*, the maintenance of a constant *internal* environment.

Reproduction

Living things come from other living things. ***Reproduction*** is the process by which an organism produces new individuals called *offspring*. Each particular kind of organism is called a ***species***. Lions are a species. Tigers are a different species. Since every individual organism eventually dies, reproduction ensures the continuation of its species.

There are two types of reproduction: asexual and sexual. ***Asexual reproduction*** involves only one parent. The offspring created are identical to the parent. Figure 5-5 shows examples of asexual reproduction.

Sexual reproduction involves two parents and produces offspring that are not identical to either parent. The female parent produces an egg cell, and the male parent produces a sperm cell. The joining of these cells is called ***fertilization***. The fertilized egg grows into a new individual.

Growth

A puppy resembles an adult dog. A young elephant looks like a small version of its parents. As these young animals mature, they will increase in size. This increase in size is called ***growth***. However, a frog or butterfly develops quite differently. Some organisms, such as frogs and most insects, change so dramatically

Figure 5-5. *Asexual reproduction: (A) binary fission; (B) budding; (C) regeneration.*

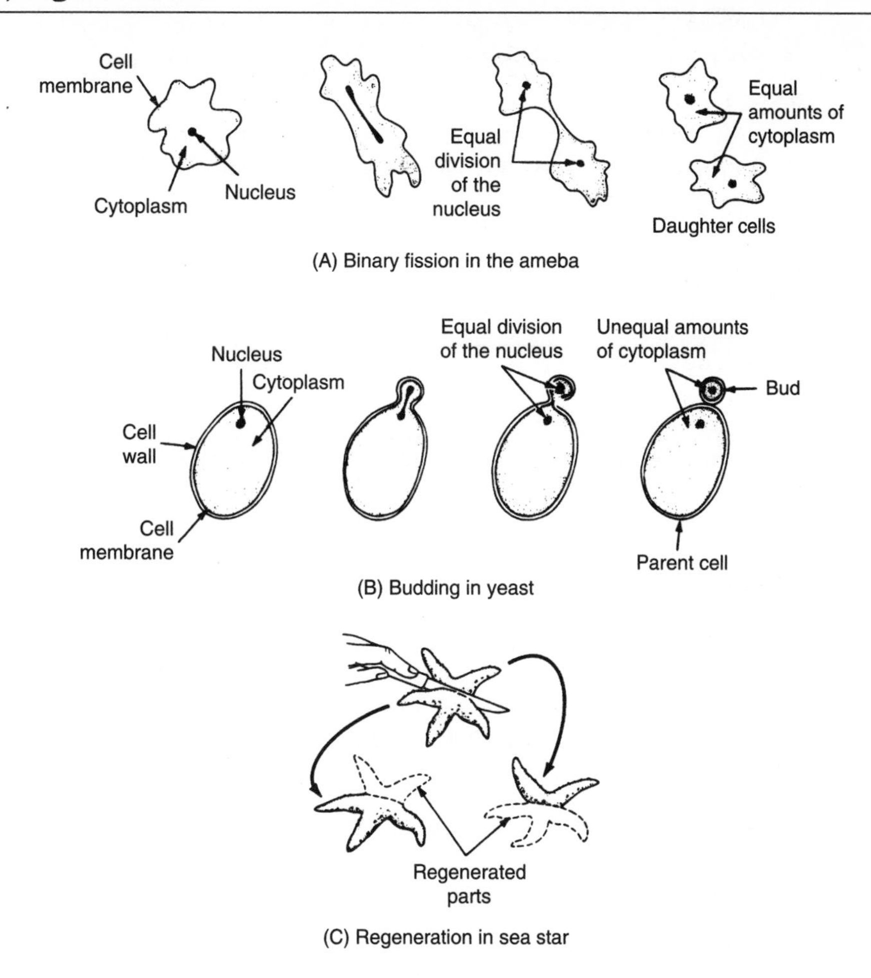

during their lives that the young may not resemble the adults at all. A dramatic change such as this is called ***metamorphosis***. The changes an organism undergoes as it develops and then produces offspring make up its ***life cycle***. Figure 5-6 illustrates the life cycles of a frog, a butterfly, and a human.

Figure 5-6. *Life cycles: (A) frog; (B) butterfly; (C) human.*

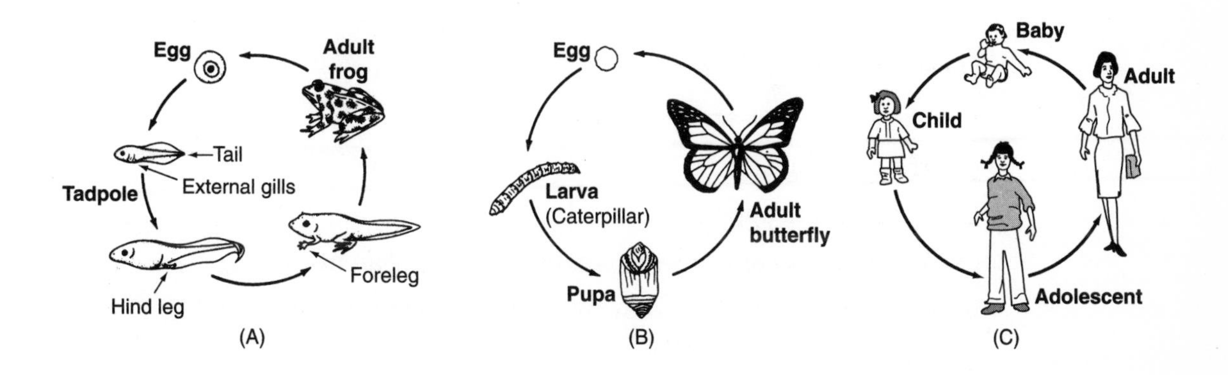

Review Questions

Part I

1. Cells use oxygen to release energy from food in the process known as
   (1) excretion
   (2) reproduction
   (3) digestion
   (4) respiration
2. Which part of the cell controls cell activities?
   (1) nucleus
   (2) cell membrane
   (3) chloroplasts
   (4) centriole
3. Which life process is necessary for the survival of a species but not necessary for the survival of the individual organism?
   (1) excretion
   (2) nutrition
   (3) respiration
   (4) reproduction
4. The diagram shows an example of a cell splitting. This is called

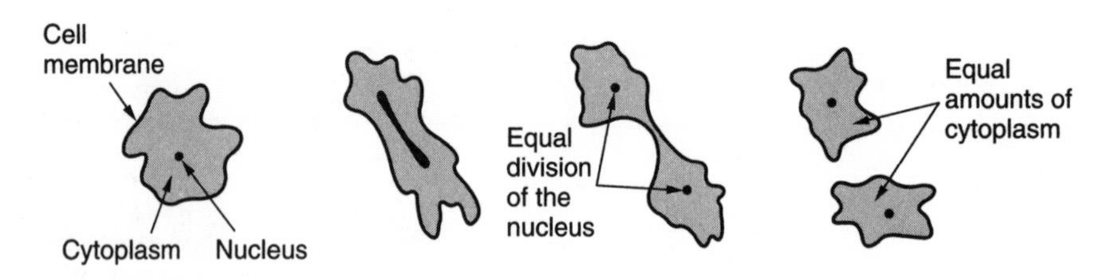

   (1) asexual reproduction
   (2) sexual reproduction
   (3) photosynthesis
   (4) respiration
5. Which part of the cell is found only in plant cells?
   (1) cell membrane
   (2) cytoplasm
   (3) chloroplast
   (4) nucleus
6. The energy for photosynthesis comes from
   (1) oxygen
   (2) sunlight
   (3) wind
   (4) sugar
7. Which statement describes the offspring of asexual reproduction?
   (1) The offpring appears different from either of its parents.
   (2) The offpring appears identical to both its parents.
   (3) The offspring appears identical to its single parent.
   (4) The offspring appears different from its single parent.

8. Which process is illustrated in the diagram below?

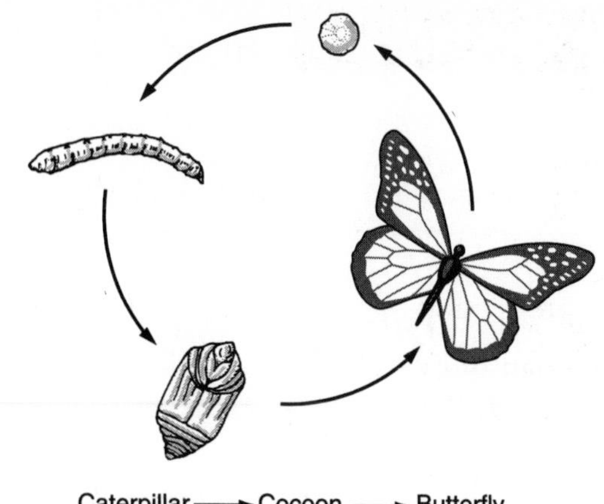

(1) regulation
(2) nutrition
(3) reproduction
(4) metamorphosis

Part II

Base your answers to questions 9 through 11 on the diagrams of Cell A and Cell B below.

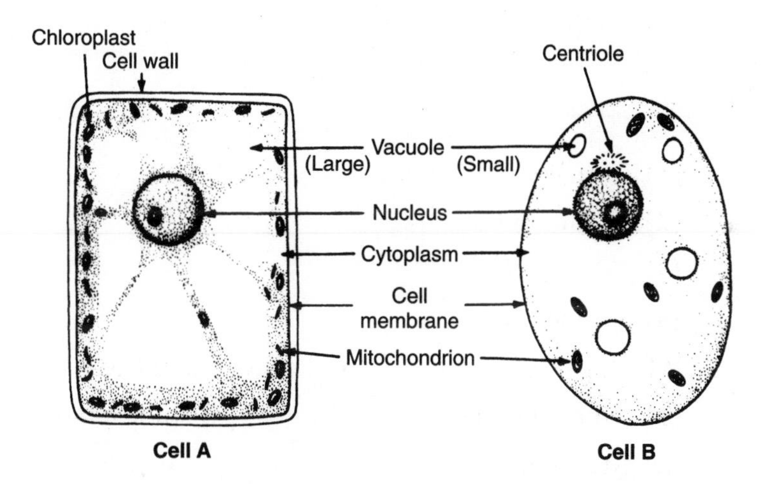

9. Which cell is a plant cell?

10. Identify two structures that are found in plant cells that are not found in animal cells.

11. Plant cells undergo photosynthesis to make their own food. What are the starting materials for photosynthesis?

Base your answers to questions 12 through 14 on the statement below.

Humans produce waste products such as carbon dioxide, water, and urine.

12. What life process produces the waste products carbon dioxide and water?

13. What life process eliminates these waste products from the body?

14. Urination is one method of removing wastes. Name one other process in humans that removes wastes.

Process Skill 1

Interpreting an Experiment

About 300 years ago, the Italian scientist Francesco Redi wondered where maggots—small, wormlike organisms—came from. The popular belief at the time was that rotting meat turned into maggots. This idea, that living things could come from nonliving material, was called *spontaneous generation*. Redi designed an experiment to test this belief. He placed meat into eight jars. Four jars were left open; four were tightly sealed. Diagram 1 shows what Redi observed.

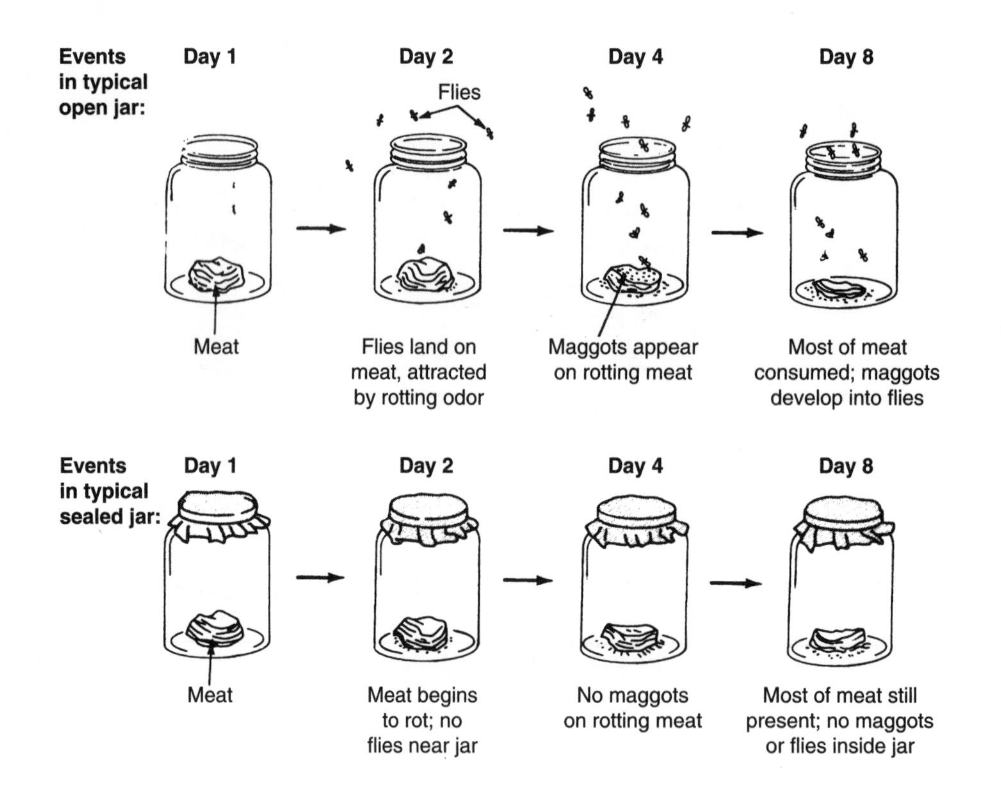

Diagram 1. Redi's first experiment: drawings show events in (*top*) a typical open jar; (*bottom*) a typical sealed jar.

As you can see, no maggots appeared on the rotting meat in the sealed jars. However, not everyone was convinced that Redi's experiment had disproved spontaneous generation. Some people claimed that fresh air was needed for spontaneous generation to occur. Therefore, Redi performed a second experiment. This time the jars were covered by fine netting, which allowed fresh air into the jars but prevented flies from entering and landing on the meat. Diagram 2 shows what Redi observed in his second experiment. Study both diagrams and then answer the following questions.

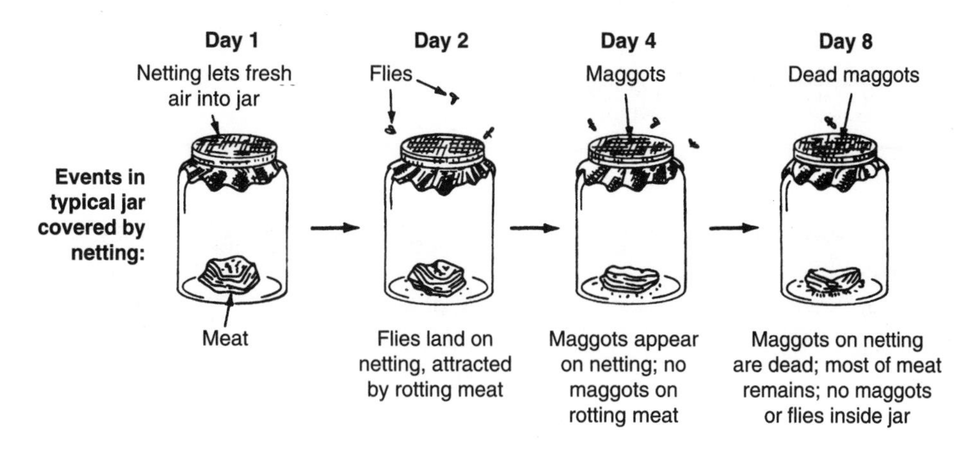

Diagram 2. Redi's second experiment.

Questions

1. Based on Redi's experiments, where do the maggots come from?
2. What conclusion can you draw from these experiments about spontaneous generation?

Microorganisms

A ***microorganism*** is a very small living thing that usually cannot be seen without a *microscope.* Figure 5-7 on page 164 shows several kinds of microorganisms as they would appear when seen through a microscope. Microorganisms include animal-like *protozoa,* plant-like *algae,* and *bacteria.*

The Compound Microscope

A ***microscope*** is a tool used by scientists to magnify tiny objects such as cells. The ***compound microscope,*** used in most classrooms and laboratories, uses two lenses. They are called the *eyepiece* and the *objective.* (See Figure 5-8 on page 164.) To see cells at different magnifications, most compound microscopes have more than one objective lens. The microscopes in Figure

Figure 5-7. *Different types of microorganisms.*

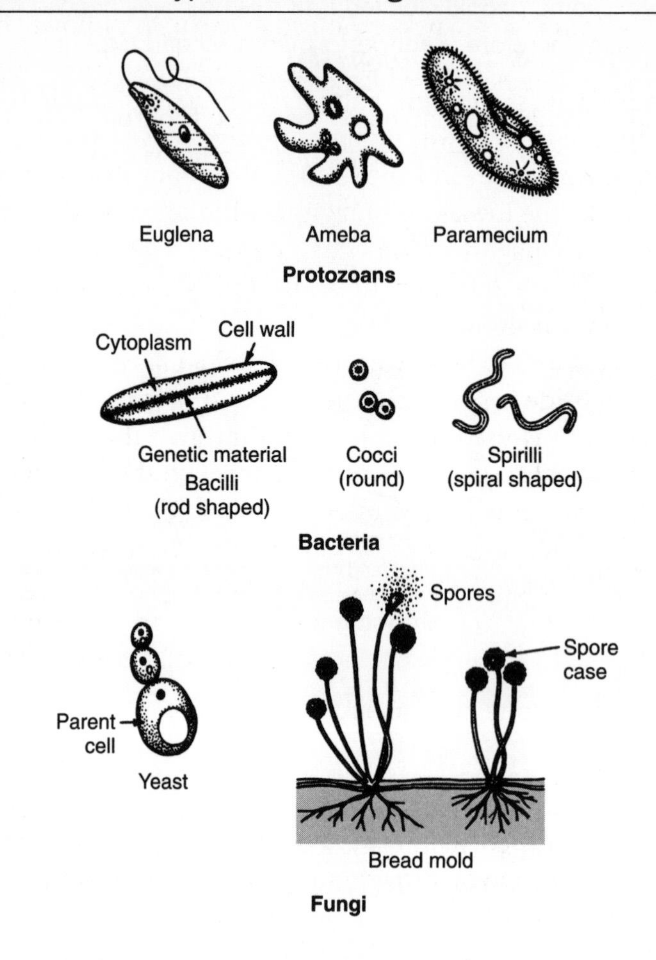

Figure 5-8. *Compound microscopes.*

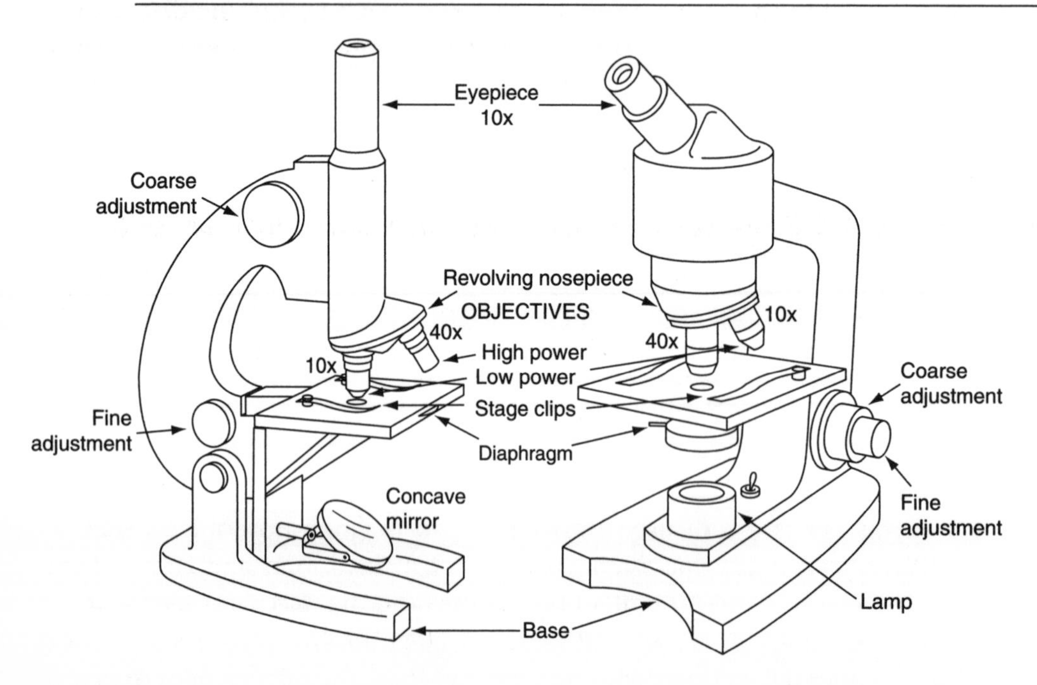

5-8 have 10x and 40x objective lenses. By rotating the *nosepiece* you can switch from the low-power objective to the high-power objective. The low-power objective is the shorter of the two and has a lower magnification number. You can calculate the *magnification* of a microscope by multiplying the eyepiece power by the objective power. For example, the eyepiece in the microscopes in Figure 5-8 is 10x. Since the low-power objective lens is also 10x, the low-power magnification must be 100x. What would be the high-power magnification? The eyepiece is 10x and the high-power objective is 40x. We multiply 10×40 and get 400. The magnification under high power is 400x.

The object to be viewed, called the ***specimen***, is usually placed on a glass *slide*. The slide is placed on the *stage* and kept in place by the *clips*. A mirror or lightbulb is used to project light through the opening in the stage. The light passes through the specimen, through the objective lens, through the eyepiece, and finally to your eye.

The microscope has two knobs that are used to focus the image. The ***coarse-adjustment knob*** is always used first under low power. Once the image is in focus under low power, you can switch to the high-power objective. All additional focusing adjustments must be done with the ***fine-adjustment knob***. This prevents the longer high-power objective lens from hitting the slide and possibly breaking it.

When viewing the image, you will notice that it appears upside down and moves in the opposite direction from the movement of the slide. This happens because the lenses produce the magnification by bending light. The area you see is called the ***field of view***. The field of view is greater under low power than it is under high power. For example, if you were viewing a group of cells (see Figure 5-9), you would see a larger number of cells under low power than you would under high power. If the magnification were four times greater under high power, the diameter of the field would be one-fourth of what it is under low power. However, most microorganisms are so small that you must view them under high power.

Figure 5-9. *More cells are seen under low power than under high power.*

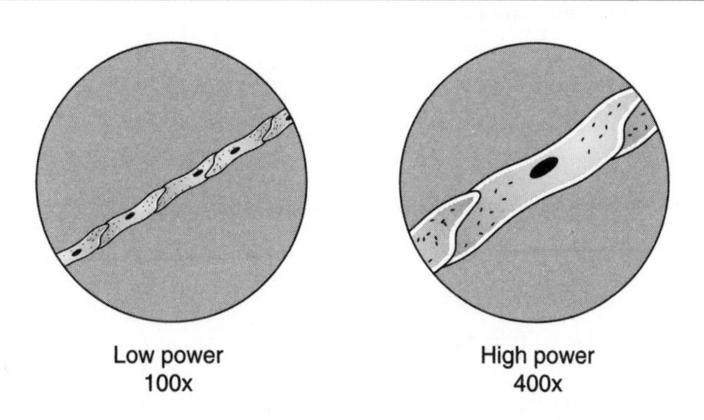

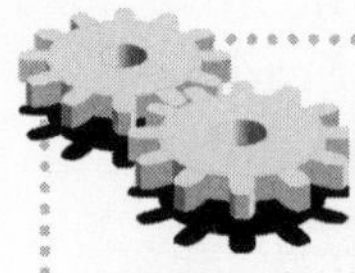

Laboratory Skill

Estimating the Size of a Cell

You use a microscope to view things that are very small. Microscopes are used to look at cells. To clearly see the parts of the cell, a material called a ***stain*** is used to color certain structures. In this experiment, Juan was asked to stain an onionskin and estimate the size of the cells.

The problem for this experiment was: What is the size of an onion skin cell?

His materials were a compound microscope, tincture of iodine solution, glass slide, coverslip, dropper, clear plastic ruler, paper towel.

Procedure 1—Estimating the diameter of the field

- Juan placed a clear plastic ruler with each small line representing 1 millimeter on the microscope stage. See diagram 1.

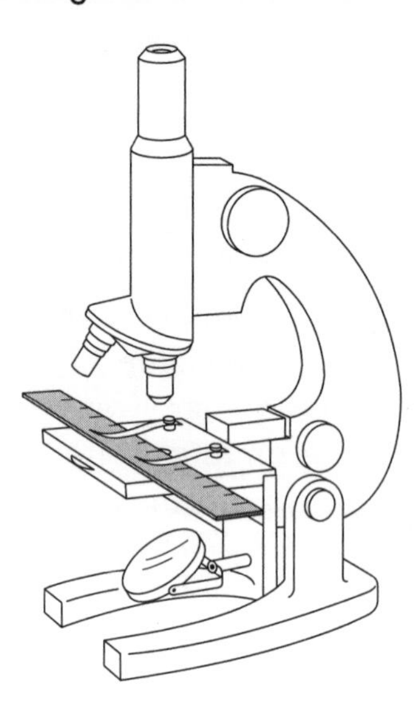

Diagram 1. Placing the ruler on the stage.

- He observed the ruler under low power.

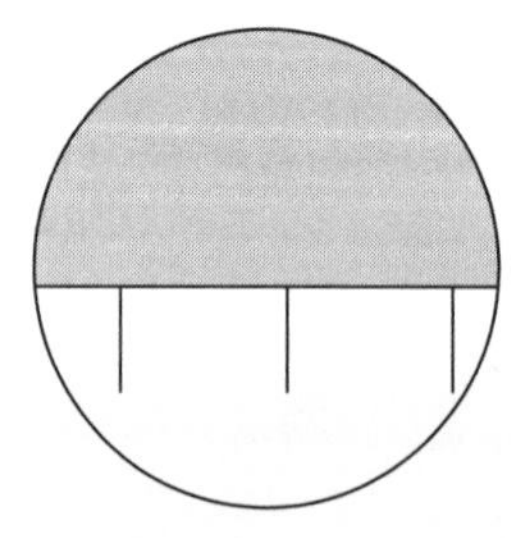

Diagram 2. Measuring the diameter of a field under low power.

Questions

1. What is the approximate diameter of the field under low power?
2. The microscope used provides a magnification of 100x under low power and 400x under high power. What is the approximate diameter of the field under high power?

Procedure 2—Observing cells under low power

- Juan placed a piece of onionskin in the center of a glass slide. He placed a drop of water on it and covered it with a coverslip, as shown in diagram 3.

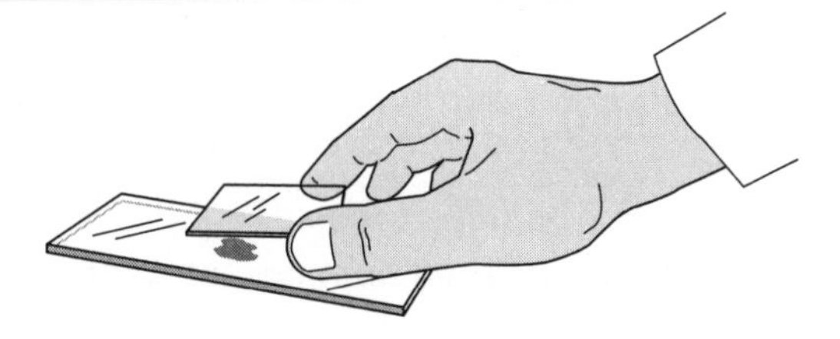

Diagram 3. Covering the specimen with a coverslip.

- Juan stained the onionskin with tincture of iodine, as shown in diagram 4.

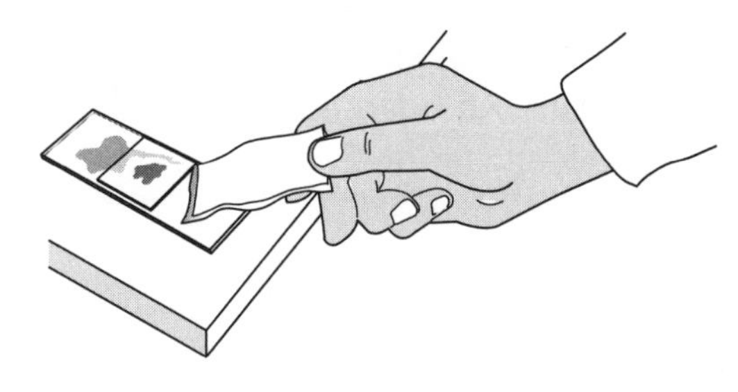

Diagram 4. "Drawing" a stain across a slide.

- When Juan viewed the cells under low power, they appeared as shown in diagram 5. Count the number of cells that fit across the diameter of the field of view.

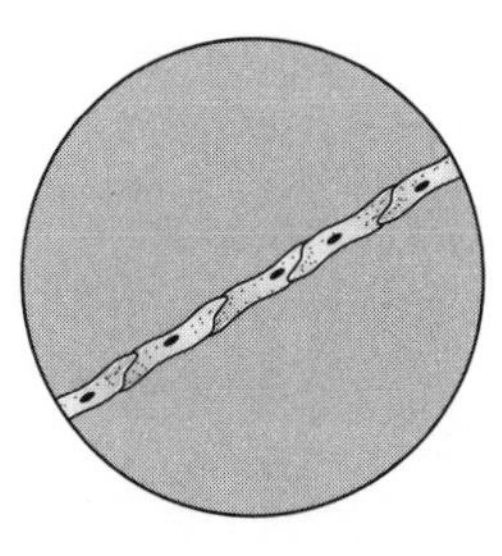

Diagram 5. Viewing onion cells under low power.

- By dividing the diameter of the field (found in question 1) by the number of cells that fit across the field of view, Juan determined the average size of each cell.

Questions

3. What was the average size of each cell?
4. What was the purpose of staining the cells with tincture of iodine?

Procedure 3–Observing under high power

- Juan was asked to predict how many cells he would see under high power and then check his prediction. Recall that the microscope magnifies 100x at low power and 400x at high power.

Questions

5. How many cells could Juan see under high power?
6. Which part of the microscope did Juan use to focus the cells under high power?
7. If Juan measured the average size of the cells under high power, how would his result compare with the value he obtained at low power?

Harmful Microorganisms

Some microorganisms are harmful to humans and other living things. These are often called germs. An ***infectious disease*** is an illness caused by microorganisms that can be transmitted, or passed on, from one person to another. Table 5-4 lists some infectious diseases and the types of microorganisms that cause them.

Table 5-4. *Some Infectious Diseases and Their Causes*

Disease	Cause
Pneumonia	Bacteria and virus
Strep throat	Bacteria
Botulism	Bacteria
Common cold	Viruses
Flu (influenza)	Viruses
AIDS	Virus
Athlete's foot	Fungus
Ringworm	Fungus
Amebic dysentery	Protozoan

Viruses

Have you ever had a cold or the flu? These diseases are caused by viruses. A ***virus*** contains genetic material surrounded by a protein coat. A virus does not carry out any life functions by itself and therefore cannot be considered a living thing. However, viruses do contain the genetic information needed for reproduction. (See Figure 5-10.) A virus can inject its genetic material into a cell, called a host cell, and use the host cell to produce more viruses. Viruses can be transmitted from one person to another. Table 5-4 lists some of the many infectious diseases that are caused by viruses.

Figure 5-10. *A virus contains genetic material in a protein coat.*

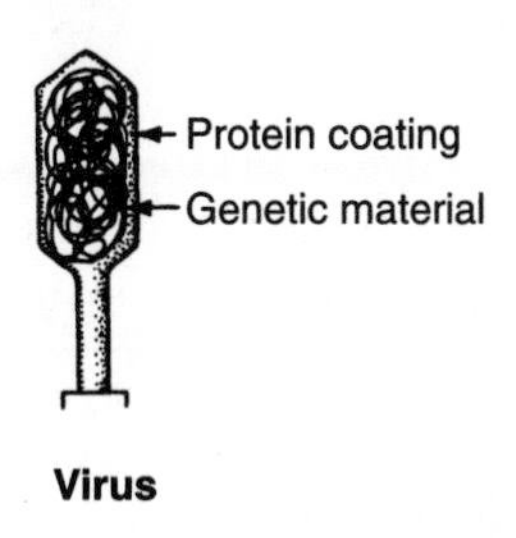

Noninfectious Diseases

Not all diseases are caused by microorganisms. Arthritis, high blood pressure, asthma, and cancer are examples of ***noninfectious diseases***, which cannot be transmitted from one person to another. Causes of noninfectious diseases include poor diet, malfunctioning glands, damaged organs, allergies to substances, and reactions to chemicals or radiation in the environment. Some diseases are hereditary; that is, you inherit them from your parents. These diseases can be transmitted only from parent to child. Table 5-5 lists some noninfectious diseases and their causes.

Table 5-5. *Some Noninfectious Diseases and Their Causes*

Disease	Cause
Scurvy	Lack of vitamin C
Anemia	Lack or iron
Hay fever	Allergy to pollen
Hemophilia	Heredity
Sickle-cell anemia	Heredity
Diabetes	Malfunctioning pancreas
Emphysema	Damaged lungs

Cancer is the result of abnormal cell division. Scientists are not sure exactly what causes this growth. They do know certain environmental factors increase the occurrence of cancerous cells. For example, cigarette smoking is known to increase the chances of getting lung cancer. Excessive exposure to sunlight may cause skin cancer.

Helpful Microorganisms

Only about 5 percent of the known microorganisms are harmful. In fact, many microorganisms are helpful to us and to other living things, and some are even necessary for our well-being. Helpful microorganisms include the ***bacteria of decay***, or other ***decomposers***, which break down dead organisms and return nutrients to the environment.

Some microorganisms are used to help produce certain foods. Yogurt, for example, is produced by the action of bacteria on milk. Some *molds,* a category of fungus, help in making cheese. *Yeasts* are fungi that cause bread to rise and produce alcohol in beer and wine. Humans even need certain kinds of bacteria inside their bodies to aid in digestion.

Process Skill 2

Experimental Design

In the late nineteenth century, the French scientist Louis Pasteur investigated *anthrax,* a disease that kills sheep and cattle. He suspected that the disease was caused by a particular kind of bacteria. To test this *hypothesis,* or educated guess, he performed an experiment.

Pasteur heated a sample of the bacteria just enough to weaken, but not kill, them. Using a group of 50 sheep, he injected 25 sheep with the weakened bacteria and left the other 25 sheep alone. The injected sheep became slightly ill, but soon recovered. Several weeks later, Pasteur injected all 50 sheep with a large dose of active anthrax bacteria, strong enough to kill a normal sheep. After a few days, all 25 sheep that had been injected with weakened bacteria were still alive, while the other 25 were all dead of anthrax.

Through this experiment, Pasteur demonstrated that a type of bacteria does cause anthrax. He also showed that by giving sheep a mild case of the disease, he could protect them from more serious infection in the future. This procedure is called *immunization* and is used today to protect humans from many infectious diseases.

Pasteur's demonstration was effective because he followed the scientific method in designing his experiment. Why was it necessary to use two groups of sheep? The group that was made immune by the first injection was the *experimental group.* To be sure it was the injection that made them immune, he needed another group for comparison—a group that did not receive the injection. This was the *control group.* Both groups had to be treated exactly the same, except for the condition that was being tested. The condition that was different (the immunization) was the *independent variable.*

In an experiment, the condition that is manipulated, or changed, by the scientist is called the independent variable. The condition that responds to the change is called the *dependent variable*. In Pasteur's experiment, the dependent variable was the sheep's immunity to anthrax.

A teacher asked her students to perform an experiment on factors that affect the souring of milk. Betty bought three containers of Surefresh milk. She kept one at room temperature, 20°C. She refrigerated another at 5°C, and the third she kept at 1°C. She checked the milk every day. Examine her results in the table below and answer the questions that follow.

Temperature	*Time Until Milk Turned Sour*
20°C	1 day
5°C	3 days
1°C	7 days

Questions

1. What was the independent variable in this experiment?
2. Betty might reasonably conclude that as the temperature increases, the length of time that milk takes to turn sour
   (1) increases (2) decreases (3) remains the same
3. Based on her experiment, Betty should predict that at 10°C, the milk would turn sour in about ________ days.

Marshall decided to compare the souring times of three different brands of milk. He left a container of Surefresh milk at 20°C, one of Sunshine milk at 5°C, and one of Dairytime milk at 1°C. All three containers had the same expiration date stamped on them, and all were fresh when the experiment began. Here are his results:

Brand	*Temperature*	*Time to Sour*
Surefresh	20°C	1 day
Sunshine	5°C	2 days
Dairytime	1°C	8 days

Questions

4. Which conclusion could Marshall reasonably draw from his experiment?
   (1) "Surefresh" milk turns sour faster than "Sunshine."
   (2) "Sunshine" milk turns sour faster than "Dairytime."
   (3) All brands of milk are the same.
   (4) He could not conclude anything from the experiment.
5. What was the major mistake that Marshall made in his experiment?

Marshall's teacher told him to check all three brands at the same temperature. When he did this, at 5°C, he got the following results:

Brand	*Time to Sour*
Surefresh	3 days
Sunshine	2 days
Dairytime	2 days

All three containers had the same expiration date and were fresh before the experiment began. Marshall concluded that he should always buy "Surefresh." Other students, however, repeated his experiment and found that, on the average, there was no difference between the three brands.

Questions

6. What was wrong with Marshall's conclusion for his second experiment?
(1) He did not measure the time accurately enough.
(2) 20°C is too warm to store milk.
(3) He should have tried several different temperatures.
(4) He should have repeated the experiment several times to see if he always got the same results.

Review Questions

Part I

15. Infectious diseases might be caused by any of the following *except*
(1) viruses
(2) bacteria
(3) chemicals in the air
(4) protozoans

16. Which statement is true about bacteria?
(1) All bacteria are harmful.
(2) All bacteria are harmless.
(3) Most bacteria are harmful.
(4) Most bacteria are harmless.

17. Which organism is best viewed under a microscope?
(1) paramecium
(2) elephant
(3) earthworm
(4) grasshopper

18. Infectious diseases are caused by
(1) damaged organs
(2) poor diet
(3) heredity
(4) microorganisms

19. One example of a noninfectious disease is

(1) flu (3) AIDS

(2) hay fever (4) strep throat

20. Abnormal cell division can result in a disease called

(1) cancer (3) emphysema

(2) ringworm (4) pneumonia

Base your answer to question 21 on the passage below and your knowledge of science.

Many cities have banned smoking in restaurants and other public places. Scientists have determined that cigarette smoke can cause diseases such as lung cancer, emphysema, and asthma. Cigarette smoke can also contribute to heart disease. Cigarette smoke is not only dangerous to the smoker. Second hand smoke, the smoke exhaled by the smoker, can harm others as well when they inhale it.

21. Emphysema is an example of a disease caused by

(1) chemicals in the environment

(2) microorganisms

(3) deficiencies in diet

(4) bacteria

Part II

Use the diagram below to answer questions 22 and 23. The diagram shows a common laboratory microscope with several parts labeled. The microscope is a tool used by scientists to study microorganisms.

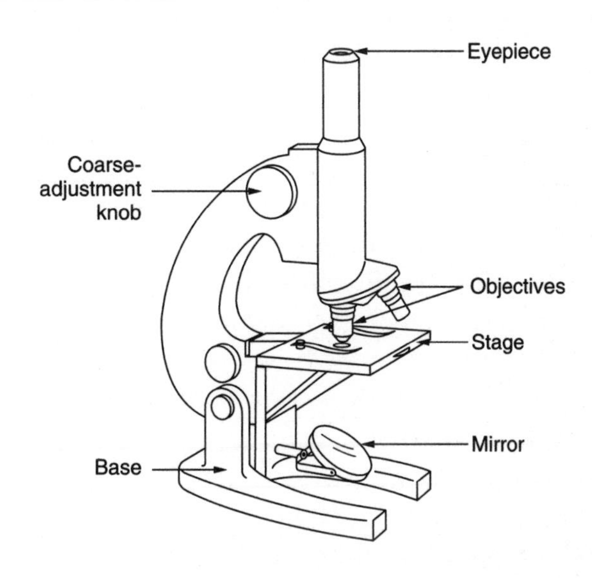

22. Which part of the microscope is used to focus the image?

23. Which part of the microscope is used to change from low power to high power?

---

24. Identify two different diseases that are caused by microorgansims.

Base your answers to questions 25 through 27 on the passage below and your knowledge of science.

One day in science lab, Omar decided to test five substances to see if they were living organisms. He put a small sample of each into five different test tubes, along with 3 mL of bromthymol blue. Test tube 6 is his control; it contains only the bromthymol blue. Omar knew that bromthymol blue turns yellow in the presence of carbon dioxide gas. He corked all the tubes and waited 24 hours. Here is a table of his results.

Test Tube	*Substance*	*Color of Bromthymol Blue*
1	Yogurt	Dark yellow
2	Salt	Dark blue
3	Yeast	Yellow-green
4	Bean seeds	Dark yellow
5	Sand	Dark blue
6	—	Dark blue

25. According to this table, which substances should Omar conclude are living?

26. To come to this conclusion, what inference must Omar make about living things?

    (1) They give off carbon dioxide from respiration.

    (2) They have the element carbon in their cells.

    (3) They are able to reproduce in bromthymol blue.

    (4) They absorb the blue nutrients from the liquid, leaving the yellow.

27. Yogurt is formed from milk that has been treated with a microorganism called *Lactobacillus acidophilus*. Some brands of yogurt contain live bacteria culture, while others do not. How would Omar's results have been different if he had chosen a brand of yogurt that did not contain live bacteria?

Chapter 6 Biological Diversity and Heredity

All living things reproduce. Some live together in families, like this pride of lions.

Vocabulary at a Glance

cell division
evolution
gene
genetic engineering
germination
mutation
natural selection
Punnett square
selective breeding

Major Concepts

- Biologists classify organisms based on characteristics they share.
- Genetic information passes from one generation to the next through chromosomes during reproduction.
- Mitosis produces two new cells with the same number of chromosomes as regular body cells.
- Meiosis produces sex cells that contain half as many chromosomes as regular body cells.
- A Punnett square can be used to predict the probability of inheriting a trait.
- A change in a gene is a mutation. Mutations may cause a species to change. This process of change, called evolution, is usually very slow.

Placing Organisms in Groups

Biologists separate living things into groups by deciding what characteristics they share. A classification system groups things by properties they have in common. Classification is an important technique used by scientists in all areas of science.

If you were asked to classify the items in Figure 6-1 into two groups, you might separate them as living things and nonliving things. If you look closely at the group of living things, you might separate it further into two smaller groups, plants and animals.

Figure 6-1. *All things are classified as either living or nonliving. Identify the living and nonliving things in this illustration.*

When you look at Figure 6-2, you might separate these animals into three groups. This classification may seem easy, but sometimes it can be difficult. You may not always be sure how to classify something.

Figure 6-2. *Scientists classify animals into different groups, such as mammals, birds, and fish. Where would the dolphin be placed?*

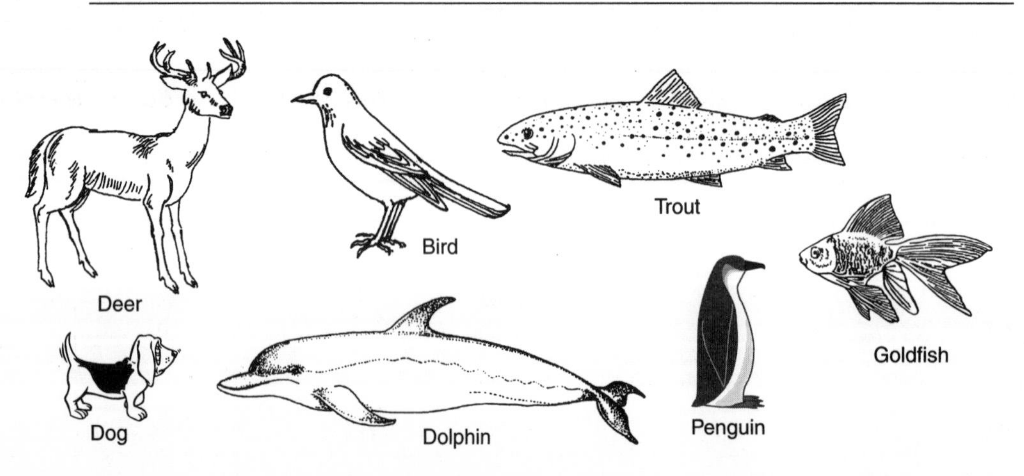

You may have chosen mammals (deer and dog), birds (robin and penguin), and fish (trout and goldfish) as your three groups. Where did you put the dolphin? The dolphin is a mammal. Scientists must carefully define the characteristics they use so they agree on how to classify something. If the three groups you chose were animals that walk, animals that swim, and animals that fly, how would these groups be different? In which group would you place the penguin?

In 1737, a Swedish scientist named Carolus Linnaeus developed a classification system for living things. He grouped them based on internal and external structures and other shared characteristics. Linnaeus called the largest group a kingdom. Scientists now recognize five kingdoms of living things. The two kingdoms you are probably most familiar with are plants and animals. Each kingdom is broken down into subgroups called *phyla* (singular, *phylum*). Each phylum is broken down into classes, each class is broken down into orders, and each order is broken down into families. Each family is broken down into *genera* (singular, *genus*), and each genus is broken down into species. This may seem confusing at first, and you might wonder why we need so many different groups.

Think about how you address a letter. You include the country, state, city, zip code, street, and number. The country contains the largest number of locations. As you go from state, to city, and eventually to street number, the areas get smaller and smaller. The system used by biologists works the same way. A kingdom contains a huge number of different living things. As you move down to species, the number of different kinds of living things in each group gets smaller and smaller.

Kingdom, phylum, class, order, family, genus, and species are assigned to every living thing (see examples in Table 6-1). We usually use only the last two names, the genus and species, to identify a living thing. The first letter of the genus is capitalized; the first letter of the species is not. Both are generally written in italics or are underlined. Together, the genus and species make up the scientific name.

The scientific name for a lion is *Panthera leo,* and for a house cat it is *Felis catus.* A lion and a house cat are both in the family Felidae, but they are in a different genus and species. We say that they are both felines.

Table 6-1. *Classification of Living Things*

Group	House Cat	Lion	Human	Red Maple	Sugar Maple
Kingdom	Animal	Animal	Animal	Plant	Plant
Phylum	Chordate	Chordate	Chordate	Tracheophyte	Tracheophyte
Class	Mammal	Mammal	Mammal	Angiosperm	Angiosperm
Order	Carnivore	Carnivore	Primate	Dicotyledonus	Dicotyledonus
Family	Felidae	Felidae	Hominid	Aceraceae	Aceraceae
Genus	*Felis*	*Panthera*	*Homo*	*Acer*	*Acer*
Species	*catus*	*leo*	*sapiens*	*rubrum*	*saccharum*

Review Questions

Part I

1. According to the system of biological classification (see Table 6-1), human beings, cats, and lions are similar enough to be classified in the same

   (1) genus (2) species (3) family (4) class

2. Japanese beetles and june bugs are classified in the same order. Therefore, they must also be in the same

   (1) kingdom (2) family (3) genus (4) species

3. Trees and humans are classified in

   (1) the same kingdom but different species

   (2) the same kingdom and the same species

   (3) different kingdoms and different species

   (4) different kingdoms and the same species

4. Lions and tigers are most likely classified in

   (1) the same kingdom and the same species

   (2) the same kingdom and different species

   (3) different kingdoms and different species

   (4) different kingdoms and the same species

Part II

5. A dog and a wolf are so similar that they are classified in the same genus. What can you conclude about which family each of these two species is placed in?

Base your answers to questions 6 through 11 on the art and dichotomous key below.

Deep lobes

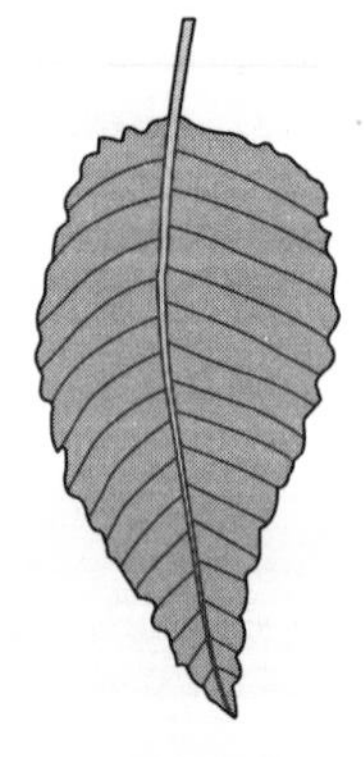

Shallow lobes

Teeth on lobes

A Key to Identifying Oak Trees			
Couplet	Description		Common Name
1a 1b	Short acorns Long acorns	Go to 2 Go to 5	
2a 2b	Leaves stay green all year Leaves fall off in autumn	Go to 3 Go to 4	
3a 3b	Tall tree Short shrub	*Quercus virginiana* *Quercus minim*	Southern live oak Dwarf live oak
4a 4b	Leaves have 5–9 deep lobes Leaves have 21–27 shallow lobes	*Quercus alba* *Quercus prinus*	White oak Swamp chestnut oak
5a 5b	Leaves have no teeth or lobes Leaves have teeth or lobes	Go to 6 Go to 7	
6a 6b	Narrow leaf Wide leaf	*Quercus phellos* *Quercus imbricaria*	Willow oak Shingle oak
7a 7b	3 lobes 7–9 lobes	*Quercus marilandica* *Quercus rubra*	Blackjack oak Northern red oak

6. All the organisms in the table
   (1) belong to different kingdoms
   (2) belong to the same species
   (3) belong to the same phylum
   (4) belong to different classes
7. What is the scientific name of a swamp chestnut oak?
8. What type of acorns would you find on a northern red oak?
9. You come across a tall oak tree with short acorns in the winter and notice that the leaves are still green. Based on this dichotomous key, what is the scientific name of this tree?
10. The scientific name of all of these oak trees starts with *Quercus*. What is the smallest classification group that all these trees have in common?
11. Based on this dichotomous key, identify two characteristics that willow oaks and shingle oaks have in common.

Reproduction

Cell Division

All cells come from other cells through the process of ***cell division***. In this process, one "parent" cell divides into two new "daughter" cells. This process is called mitosis. Each parent cell passes along to its daughter cells the set of "operating instructions" necessary for the cells to function properly. These instructions, or genetic information, are contained in threadlike structures called chromosomes. Chromosomes are found in the nucleus of the cell. (See Figure 6-3.)

The genetic information in the chromosomes also gives the cell, or the organism it belongs to, its characteristics, or traits, such as size and shape. All members of one species have the same number of chromosomes in each body cell. Chromosomes and their genetic information are passed on to the next generation during reproduction.

One-celled organisms reproduce through mitosis. When a one-celled organism undergoes mitosis, each new cell is a complete new organism.

Figure 6-3. *Cell division: Mitosis produces two new cells with the same number of chromosomes as the original cell.*

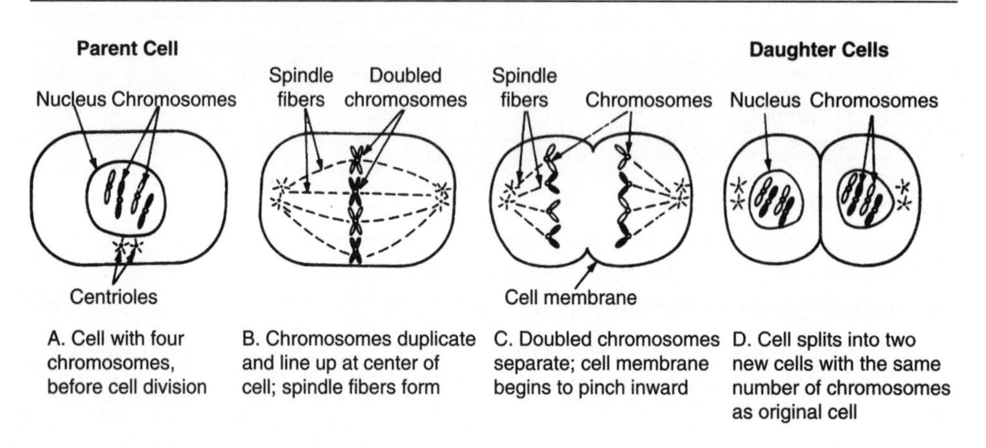

In an organism made up of many cells, cells must reproduce themselves to build new tissue for growth and to repair damaged tissue. They do this through mitosis.

Asexual Reproduction

Some organisms also produce offspring by mitotic cell division. In fact, any organism that reproduces asexually (with just one parent) does so through mitosis. The offspring that result are genetically identical to the parent.

Sexual Reproduction

Some organisms reproduce sexually, with two parents. Sexual reproduction involves the joining of two special reproductive cells, one from each parent. These *sex cells,* or *gametes* (sperm cells and egg cells), have only half the number of chromosomes found in body cells. Sex cells are formed by a type of cell division called *meiosis.*

During sexual reproduction, a sperm cell from the male parent joins with an egg cell from the female parent. This is called *fertilization*. Since each sex cell contains half of a normal set of chromosomes, when they join, they form one cell with a complete set of chromosomes (Figure 6-4).

Figure 6-4. *Fertilization occurs when an egg and a sperm unite.*

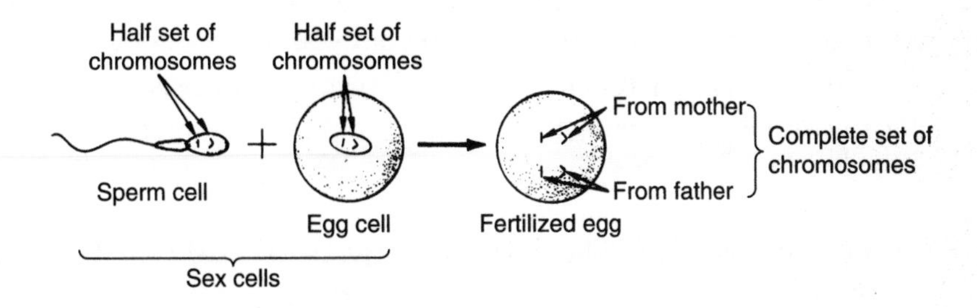

Process Skill 1

Interpreting Information in a Table

Many animals reproduce by sexual reproduction. During sexual reproduction, sperm from the male joins with an egg from the female. This process, called fertilization, may take place either within the female's body (*internal fertilization*) or outside of the female's body (*external fertilization*). After the egg is fertilized, it develops into an embryo. This process may also take place either inside the female (*internal development*) or outside the female (*external development*).

For example, a chicken lays an egg, which is already fertilized. The chicken then sits on the egg to keep it warm as the embryo in the egg develops. Chickens have internal fertilization and external development. The table below lists some different animals, their class and habitat, and the type of fertilization and development they undergo.

Study the table and answer questions 1 through 5 on page 182.

Animal	*Class*	*Habitat*	*Type of Fertilization*	*Type of Development*
Goldfish	Bony fish	Water	External	External
Robin	Bird	Land	Internal	External
Bee	Insect	Land	Internal	External
Cat	Mammal	Land	Internal	Internal
Frog	Amphibian	Water and land	External	External
Snake	Reptile	Land	Internal	External
Dolphin	Mammal	Water	Internal	Internal

Questions

1. Based on the table, what is required for external fertilization?
(1) internal development (3) water habitat
(2) land habitat (4) only fish can have external fertilization

2. Which of the following is required for internal development?
(1) land habitat (3) external fertilization
(2) water habitat (4) internal fertilization

3. A salamander is an amphibian. What would you predict about salamanders?
(1) They live both on land and in water.
(2) They have internal development.
(3) They live in water only.
(4) They have internal fertilization.

4. Alligators have internal fertilization and external development. Based on the table, which class do they most likely belong to?
(1) amphibian (2) reptile (3) fish (4) mammal

5. Based on this table, what two generalizations can be made about fertilization and development in mammals?

Development of the Fertilized Egg

The fertilized egg, called the *zygote*, forms new, identical cells through mitosis. You know, however, that multicellular organisms are made up of many different types of cells. (See Figure 7-1 on page 197.) At a certain stage in the development of the organism, the cells form different layers. Within these layers, the cells begin to change into the tissues and organs of the body. This process, called *differentiation*, is illustrated in Figure 6-5, which shows the development from the one-celled zygote to the multicellular embryo.

Figure 6-5. *Cleavage from zygote to embryo.*

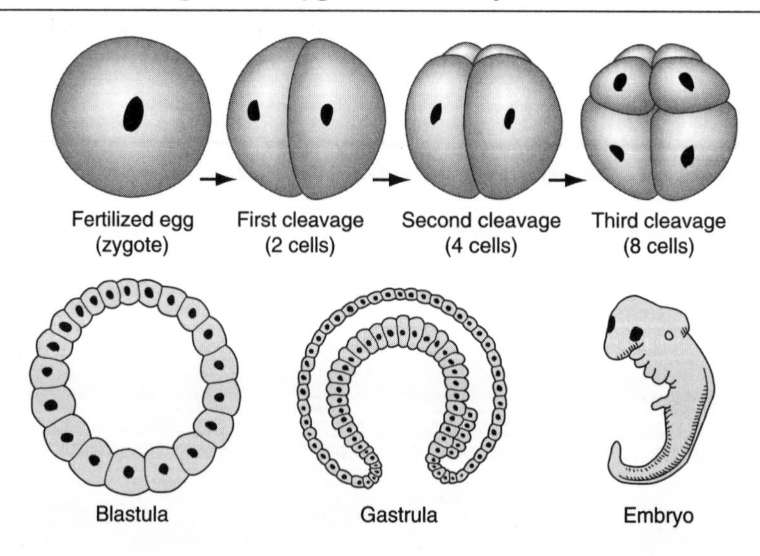

Eventually, an organism develops that resembles its parents. However, the new organism is not identical to either parent but has traits from both. In this way, sexual reproduction leads to variation in the next generation. Figure 6-6 shows a possible result of sexual reproduction in chickens.

Figure 6-6. *Sexual reproduction leads to variation in offspring.*

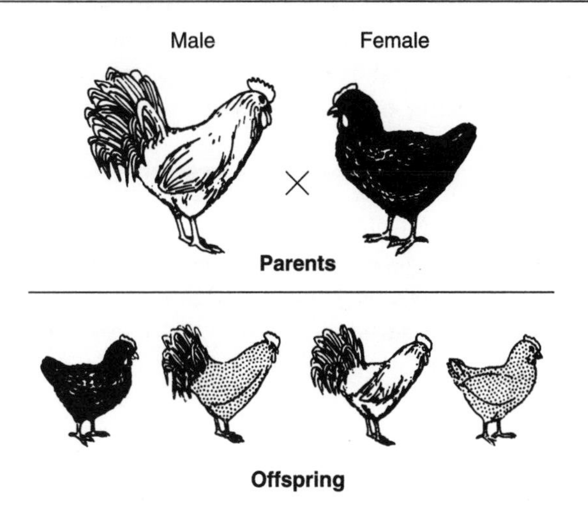

Sexual Reproduction in Plants

Plants, like animals, may reproduce sexually. Plants are rooted in the soil and cannot move from place to place; they have developed different structures and techniques to accomplish reproduction.

A flower is the reproductive organ of a plant. Pollen produced in the *anther* (see Figure 6-7) contains the male sperm cells. The female egg cells are located in the ovary at the base of the *pistil*. In plants, as in animals, the male and female reproductive cells must join in order for fertilization to take place. A plant depends on wind, rain, insects, birds, or other small animals to transfer the pollen from the anther to the pistil. This process is called pollination.

Figure 6-7. *Reproductive parts of a flower.*

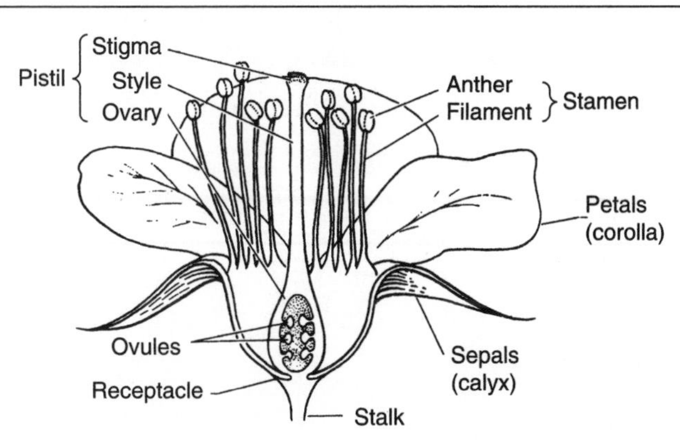

The fertilized egg develops into an embryo. The seed contains the embryo and provides food for its early development. The ovary of the flower develops into a fruit. Animals that eat the fruit help distribute the seed to new locations in their droppings. This is called *seed dispersal.* Wind is another agent of seed dispersal. Seeds with hooks stick to the fur of animals and are carried to new locations. Different types of seeds have developed to take advantage of various methods of dispersal, as illustrated in Figure 6-8.

Figure 6-8. *Methods of seed dispersal for different types of seeds.*

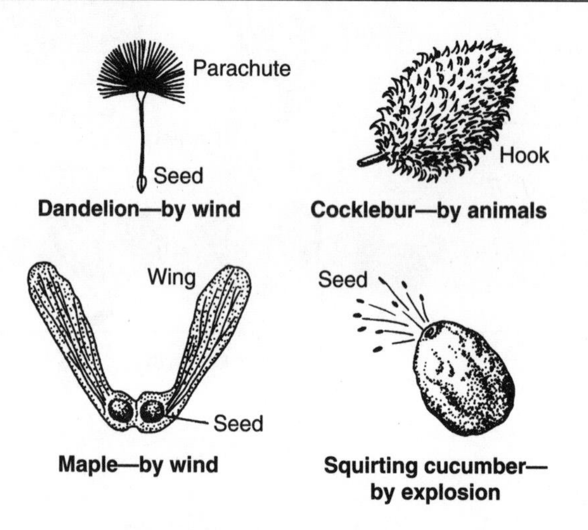

When conditions are favorable, the embryo in the seed begins to sprout into a young plant. The most important conditions for growth are temperature and moisture. The sprouting process is called ***germination***. During germination, the plant's roots, stems, and leaves begin to develop.

Part I

12. Asexual reproduction differs from sexual reproduction in that in asexual reproduction the offspring are genetically
    (1) different from either of their parents
    (2) identical to both parents
    (3) different from their single parent
    (4) identical to their single parent
13. The female sex cell is called
    (1) a sperm cell
    (2) an egg cell

(3) an embryo

(4) a zygote

14. Sexual reproduction is different from asexual reproduction because only sexual reproduction involves

(1) mitosis

(2) cell division

(3) duplication of chromosomes

(4) fertilization

15. A paramecium reproduces by a process called binary fission. During this process, the nucleus makes a copy of the genetic material and then the cell splits into two identical cells. This type of reproduction can best be described as

(1) fertilization

(2) germination

(3) differentiation

(4) mitosis

16. A male fruit fly has 4 chromosomes in each of its sperm cells. How many chromosomes are in each of its regular body cells?

(1) 8 (2) 2 (3) 16 (4) 4

17. Which process is illustrated by the following diagram?

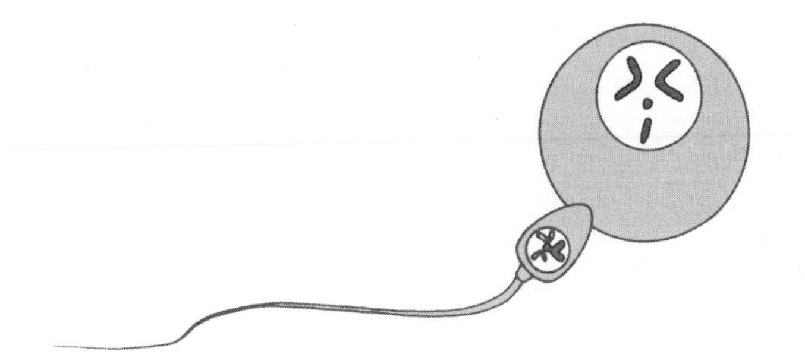

(1) meiosis

(2) mitosis

(3) fertilization

(4) cleavage

18. The reproductive organ of a plant is called the

(1) flower

(2) root

(3) stem

(4) leaf

Part II

Base your answers to questions 19 and 20 on the following diagram that illustrates a type of reproduction.

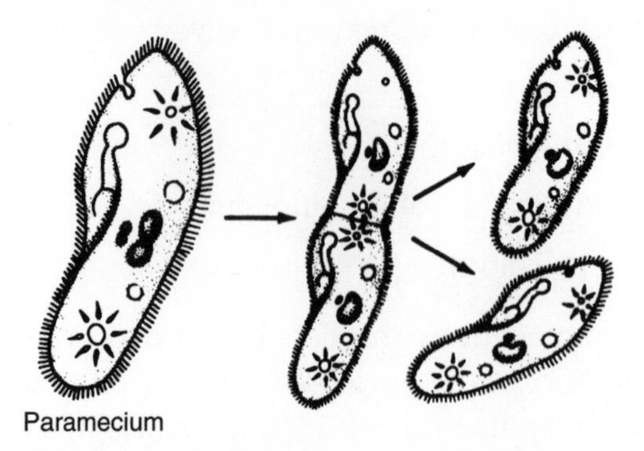

19. What type of reproduction is illustrated in the diagram?
20. Compare the genetic makeup of the two offspring to that of the parent.

---

21. Complete the following analogy: anther is to sperm as ovary is to ____.
22. Copy the table below into your notebook and indicate how many chromosomes are in each of the following types of cells found in elephants.

Type of Cell	*Number of Chromosomes*
Body	56
Sperm	
Egg	
Zygote	

Heredity

Inheritance of Traits

What type of thumb do you have? Did you know that there are two types of thumbs, "hitchhiker's" and "regular"? (See Figure 6-9.) The type of thumbs you have is determined by the genetic information you received from each of your parents. A piece of genetic information that influences a trait is called a ***gene.*** You received one gene containing information for the type of thumbs you have from each of your parents. These two genes can be the same or different.

Figure 6-9. *The two types of thumbs that can be inherited: hitchhiker's and regular.*

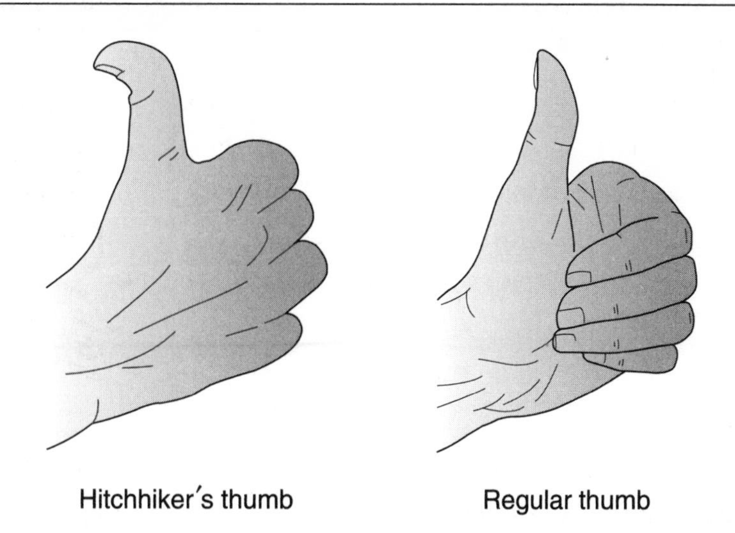

If both genes for a trait match, that trait will appear and you are "pure" for that trait. However, if you have genes for a trait that differ, one hitchhiker's and another regular, you are "hybrid" for that trait.

The type of thumb is just one of thousands of traits you inherit from your parents. Each trait is controlled by one or more pairs of genes. Therefore, each of your body cells contains thousands of different genes.

Mendelian Genetics

An Austrian monk named Gregor Mendel (1822–1884) performed experiments with pea plants to investigate how traits are inherited. He determined that traits are controlled by two genes (he called them "factors"). When these genes produce different versions of the same trait, one gene is dominant over the other. An individual with both genes (a hybrid) will exhibit only the version of the dominant gene. The gene for the trait that is not exhibited is called a recessive gene.

Mendel crossed pure tall pea plants with pure short pea plants. He observed that all the offspring were tall. He concluded that tall is the dominant trait. He also concluded that all of the offspring of this generation must be hybrids, containing the gene for tallness and the gene for shortness. When the hybrid tall pea plants were crossed with one another, most of the offspring were tall, but some were short. This proved that the hybrid tall pea plants still contained a gene (factor) for shortness. In the hybrid tall pea plants, the gene for shortness was hidden. In future generations, if the offspring received two genes for shortness, the offspring would develop into short pea plants. Only when the offspring received two short genes did they appear as short plants. Table 6-2 on page 188 indicates some common dominant and recessive traits studied in genetics.

Table 6-2. *Common Traits Studied in Genetics*

Organism	Trait	Dominant	Recessive
Human	Thumb shape	Regular	Hitchhiker's
Human	Earlobe	Free	Attached
Human	Blood type	A or B	O
Fruit fly	Wing	Normal	Vestigial
Fruit fly	Eye color	Red	White
Pea plant	Height	Tall	Short
Pea plant	Pea color	Yellow	Green
Pea plant	Seed shape	Round	Wrinkled

Using a Punnett Square

Fruit flies are often used to study genetics because they mature and reproduce quickly. Several generations can be observed in a short time.

One fruit fly characteristic studied is eye color. Fruit flies can have red eyes or white eyes. Red is dominant while white is recessive. To represent the gene, we use a capital letter for the dominant trait (*R* for red) and a lowercase of the same letter for the recessive trait (*r* for white). The capital letter is always written first. The possible gene combinations are outlined in Table 6-3.

Table 6-3. *Possible Gene Combinations*

Type of Gene	Representation	Appearance
Pure red	*RR*	Red
Hybrid red	*Rr*	Red
Pure white	*rr*	White

We can predict the possibility of an offspring having red eyes or white eyes if we know the types of genes of the parents. A diagram called a ***Punnett square*** can be used to predict the probability of an organism inheriting a given trait. The genes for one parent are placed at the top of the square and the genes for the other parent are placed at the side. Below is a Punnett square for two hybrid parents.

	R	*r*
R	*RR*	*Rr*
r	*Rr*	*rr*

Key:
R = gene for red eyes
r = gene for white eyes

Each box is filled in with the letter appearing above it and to its left. These boxes represent the possible combinations of genes. We see that of the four boxes, one will be pure dominant (*RR*), one will be pure recessive (*rr*), and two will be hybrid (*Rr*). This can be summarized as shown in Table 6-4.

Table 6-4. *Probabilities of Offspring of Hybrid Red-Eyed Fruit Flies*

Type of Genes	***Appearance***	***Probability***
RR = pure dominant	Red eyes	1/4 (25%)
Rr = hybrid dominant	Red eyes	2/4 (50%)
rr = pure recessive	White eyes	1/4 (25%)

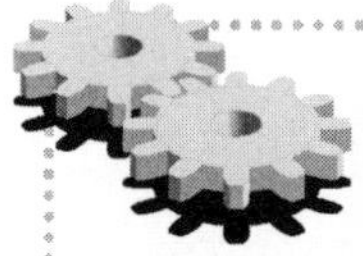

Process Skill 2

Interpreting the Results of an Experiment

Indira studied the traits of an unusual insect. She noticed that some of the insects had long antennae and some had short antennae. She separated the insects into two groups based on the length of their antennae and allowed them to mate only within each group. After the eggs were laid, the parents were removed. The eggs were allowed to develop. In one group, all the offspring had short antennae. Offspring in the other group had both types of antennae.

Questions

1. Indira can tell from this information that
(1) short is dominant (3) short is recessive
(2) long is recessive (4) there were two different species of insects

2. In describing the parents, it would be safe to infer that
(1) all those with long antennae were pure
(2) all those with long antennae were hybrid
(3) some of those with long antennae were hybrid
(4) some of those with short antennae were hybrid

3. Which procedure would make the experiment *less* reliable?
(1) Separate the insects to be mated before their antennae reach full size.
(2) Feed the long- and short-antennae groups the same diet.
(3) Maintain a constant environment for both groups.
(4) Repeat the experiment several times.

4. Predict the results of a cross between a pure long-antennae insect and a pure short-antennae insect.
(1) Some of the offspring will have short antennae.
(2) All of the offspring will have short antennae.
(3) Some of the offspring will have long antennae.
(4) All of the offspring will have long antennae.

Review Questions

Part I

23. Pea plants pure for purple flowers when crossed with pea plants pure for white flowers produce offspring that all have purple flowers. Which is the best conclusion to make about the inheritance of flower color in pea plants?

    (1) Color cannot be inherited. (3) Purple is recessive.

    (2) Purple is dominant. (4) White is dominant.

Use the following information to answer questions 24 and 25.

In humans, having dimples is a dominant trait.

24. What is the probability of two parents that are hybrid for having dimples producing a child with no dimples?

    (1) 0% (2) 25% (3) 75% (4) 100%

25. A parent without dimples and a parent who is hybrid for dimples have five children without dimples. What is the likelihood that their next child will have dimples?

    (1) 1/4 (2) 1/2 (3) 3/4 (4) 1/1

Part II

Use the information below to answer questions 26 through 29.

The Punnett square below represents a cross between a yellow and a green pea plant.

	y	y
Y	Yy	
Y		

Key:
Y = gene for yellow
y = gene for green

26. Draw a Punnett square and complete the results of the cross between these two pea plants.
27. What percentage of the offspring would be green?
28. Classify the genetic makeup of the parent plants.
29. Define the term "hybrid."

Use the information below to answer questions 30 through 33.

Fruit flies can have red or white eyes. When studying the inheritance of this trait, biologists use the letter *R* to represent the gene for red eyes and *r* to represent the gene for white eyes.

30. Which eye color is the dominant trait in fruit flies? Explain your answer.

31. What color eyes would a fly with the gene combination *Rr* have? Explain your answer.

32. Draw a Punnett square for a cross between two *Rr* fruit flies.

33. Based on your Punnett square, what percentage of the fruit flies would have red eyes?

Process Skill 3

Interpreting a Diagram

A diagram that shows how a trait is passed from generation to generation is called a pedigree. For questions 1 through 5, use the pedigree below to determine the genetic makeup of the individuals represented.

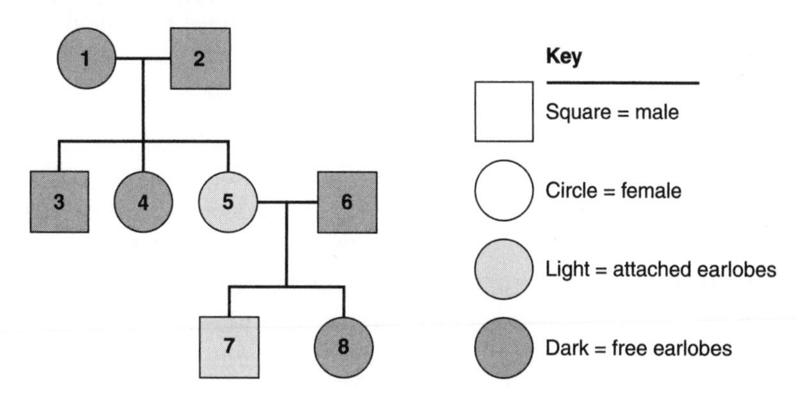

Questions

1. Individual 4 can best be described as
(1) a male with attached earlobes (3) a male with free earlobes
(2) a female with attached earlobes (4) a female with free earlobes

2. Based on this pedigree, it would be safe to assume that
(1) attached earlobes are dominant (3) attached earlobes are recessive
(2) free earlobes are recessive (4) earlobe type is not inherited

3. The genetic makeup of individual 2 can best be described as
(1) *FF* (2) *Ff* (3) *ff*

4. How do we know that individual 6 must be hybrid for the type of earlobe?

5. Use a Punnett square to predict the probability of the next child of individuals 5 and 6 having attached earlobes.

Mutations

Evolution accounts for the great diversity among living things. The process of evolution depends on genetic information being passed from one generation to the next through chromosomes. This occurs during reproduction.

If parents and offspring are so similar, why are there so many adaptations and such diversity among living things? Changes in genes can happen. Sometimes genetic material does not reproduce as expected. This may be caused by a natural process or by something in the environment. Such a change in the genetic material is a ***mutation***. The new genetic information will cause a change in the offspring. If this change is harmful to the organism, it will be less likely to survive and reproduce. If the change is helpful to the organism, it will be better able to survive and reproduce. The new genetic information will then be passed on to each new generation. If these changes increase the likelihood that the organism will reproduce, the changes will become more common within the population.

Scientists have discovered that mutations can be caused by environmental factors. Forms of radiation such as gamma rays, x-rays, and even ultraviolet light can cause mutations. In addition, certain chemicals have been shown to cause changes in genetic material.

Natural selection states that those organisms that are best adapted to their environment will survive and reproduce. After many generations, and many mutations, an organism may look and behave so differently from its ancestors that it has become a new species. This process, called ***evolution***, is often very slow and may take millions of years. However, bacteria and insects with very short life cycles may evolve relatively quickly. For example, penicillin is no longer effective against certain bacteria. Differences in genetic makeup may allow an individual resistant bacterium to survive treatment with penicillin. It will then reproduce asexually, forming additional resistant bacteria. These penicillin-resistant bacteria may soon become the main variety causing the infection. The infection can no longer be treated with penicillin. Scientists must develop new antibiotics to kill the new resistant varieties. Similarly, some insect populations have become resistant to certain insecticides.

Today, through advances in genetic engineering, scientists can cause a species to change to improve a particular characteristic. Plants and livestock with desirable traits such as resistance to disease, better taste, and higher yields are chosen in a process called ***selective breeding***. Selective breeding is a process in which individuals with the most desirable traits are crossed, or allowed to mate, with the hopes that their offspring will show the desired traits. For example, turkeys have been selectively bred to produce more white meat, racehorses are selectively bred for speed, and wheat is selectively bred to produce larger heads of grain. Through selective breeding, humans have replaced nature in determining which organisms survive and reproduce.

Genetic Diseases

Many mutations are not helpful. Diseases such as sickle-cell anemia and hemophilia, and conditions such as albinism (a lack of pigment) are caused by genetic changes that are passed down from generation to generation. Scientists have developed ways of checking genetic material for the presence of some of these harmful genes. Since most of these diseases are caused by recessive genes, both parents must carry the gene for the child to be at risk. A new health care field called *genetic counseling* helps parents understand and evaluate the risk that their children may inherit one of these diseases.

DNA

Chromosomes contain complex molecules called DNA, which is made of smaller molecules that are arranged in a particular sequence. This sequence of molecules is the code that determines the genetic information.

In a coordinated international effort, called the *Human Genome Project,* scientists from 18 countries decoded the messages contained in our chromosomes. The project began in 1990 and ended in 2003. The result of their work was the sequencing of 20,000 genes in the human genome. They hope to use this information to predict and even cure genetic diseases.

Genetic Engineering

Scientists have learned how to deliberately change the genetic material of an organism to benefit people. They have developed new strains of plants and animals that are resistant to disease and yield more food. This is done by actually changing the genetic material in these organisms in a process called ***genetic engineering***. Once these changes have been made, they are passed on to all future generations. For example, sheep have been genetically engineered to produce human insulin, a hormone that is used to treat diabetes. Some people feel that genetic engineering should not be attempted because the resulting organisms might cause more harm than good. If scientists can change plants and animals, might they change humans? There is disagreement over the whether it is right to genetically change humans.

Review Questions

Part I

34. To produce fast raceshorses, the owners of female racehorses sometimes pay thousands of dollars for the right to mate their females with the fastest male racehorses. This procedure is best described as

    (1) selective breeding (3) genetic engineering
    (2) mutation (4) adaptation

35. Sometimes, by exposing fruit flies to x-rays, scientists are able to produce fruit flies with completely new traits. These new traits can be passed on to future generations. These new traits are caused by

(1) mutations
(2) adaptations
(3) asexual reproduction
(4) selective breeding

36. Some diseases are caused by mutations. These diseases differ from infectious diseases because mutations

(1) can be cured by antibiotics
(2) can be passed on to the next generation
(3) are always fatal
(4) cannot be inherited

Base your answers to questions 37 and 38 on the diagram below.

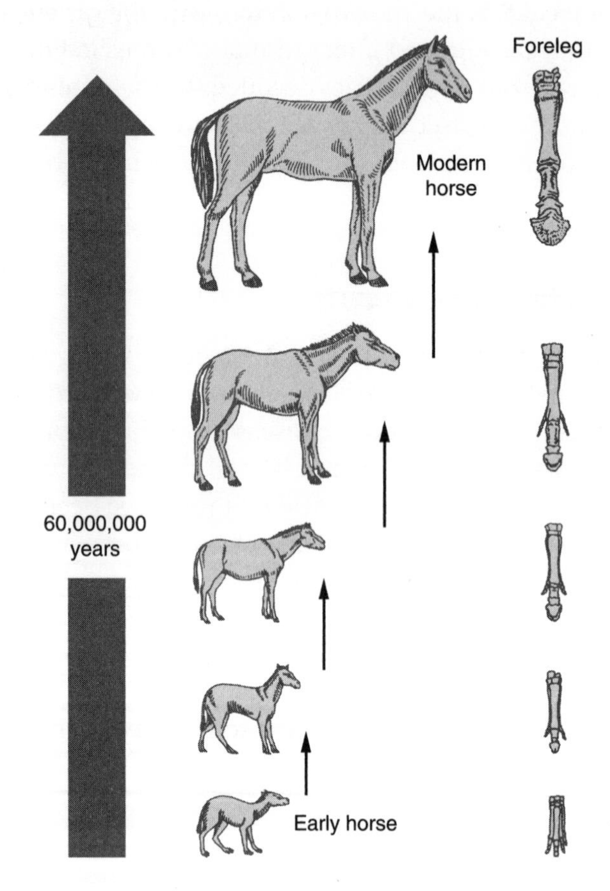

37. This diagram illustrates a process called

(1) selective breeding
(2) asexual reproduction
(3) genetic engineering
(4) evolution

38. The fossils of horses show that the size of the horse has changed gradually. These changes most likely occurred because
   (1) larger horses were more likely to survive and reproduce
   (2) smaller horses were more likely to survive and reproduce
   (3) genetic engineering has increased the size of horses
   (4) selective breeding has increased the size of horses

Part II

Base your answers to questions 39 through 41 on the paragraph below and your knowledge of science.

Antibiotics are medicines designed to kill harmful bacteria. If a doctor prescribes an antibiotic for you, it is important that you take the antibiotic for the entire time that the doctor tells you, even if you feel better after a few days. If you do not use enough medicine to kill *all* the bacteria, you may soon become sick again. Natural selection favors the strongest bacteria, those that can best resist the antibiotic. If any survive, they will pass this resistance on to the next generation. Since bacteria reproduce very quickly, you may soon become sick again.

39. Why might an antibiotic be less effective when the illness returns?

40. Why does evolution occur much more rapidly in bacteria than in most other organisms?

41. What process might have made some bacteria more resistant than others to the antibiotics?

Chapter 7 Human Systems

Vocabulary at a Glance

artery
blood
blood vessel
bone
brain
bronchi
capillary
cartilage
cellular respiration
gland
heart
hormone
involuntary muscle
joint
kidney
liver
locomotion
lung
lymph
lymph vessel
mammary gland
metabolism
muscle
nerve
neuron
organ
organ system
ovary
oviduct
respiration
respiratory system
sense organ
skin
sperm duct
spinal cord
testes
tissue
trachea
uterus
vagina
vein
voluntary muscle

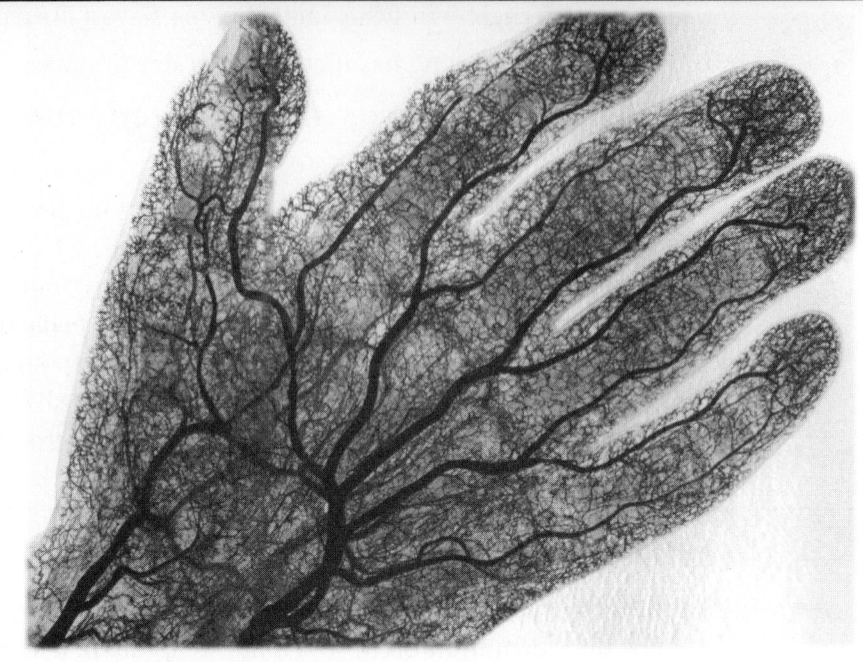

The circulatory system carries blood to every cell in the body.

Major Concepts

- Complex organisms, such as humans, show several levels of body organization.
- Cells are the basic units of life. Tissues are groups of similar cells that work together to carry out a life process. Organs are groups of tissues that work together to carry out a life process
- Organ systems are groups of organs that work together to carry out a life process. Different organ systems work together to carry out life processes.
- The human body includes the following types of tissues: blood, bone, muscle, nerve, and skin.
- The human body includes the following systems: skeletal, muscular, nervous, endocrine, digestive, circulatory, respiratory, excretory, and reproductive.
- All body systems work together; for example, the nervous system and endocrine system work together to regulate and control body activities.
- The circulatory system and respiratory system work together to bring oxygen in the blood to all body cells.

Organization, Support, and Movement of the Body

A human being is a complex organism, made up of a number of different body systems. Each system carries out a specific life process and thereby contributes to the operation of the whole body.

Human Body Systems Are Interdependent

All body systems depend on one another and work with one another to keep a person alive. For example, the respiratory system brings oxygen into the body. The circulatory system carries the oxygen throughout the body.

All the systems of the body work together to maintain a balanced internal state (homeostasis). Diseases as well as personal behaviors, such as poor diet and the use of toxic substances (alcohol, tobacco, and drugs), may interfere with maintaining this balance. Some effects on the body appear immediately, while others may not appear for years. During pregnancy, diseases contracted by the mother may also affect the development of the unborn child. Smoking and drinking alcoholic beverages during pregnancy may decrease the birth weight of the baby and cause other harmful effects.

Levels of Organization in the Human Body

1. *Cells.* Living things are made up of basic units called *cells*. The human body contains many types of cells, each designed to perform a different function. Figure 7-1 shows several kinds of cells.

Figure 7-1. *Different types of cells.*

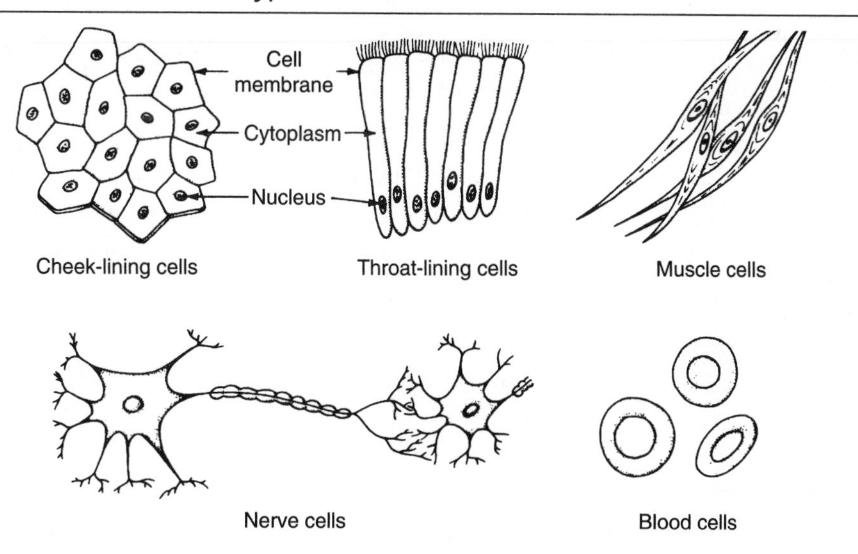

2. *Tissues.* A group of similar cells acting together forms a ***tissue***. Skin tissue covers the body. Muscle tissue produces body movements. Table 7-1 on page 198 lists some types of human tissues.

Table 7-1. *Types of Human Tissue and Their Functions*

Tissue	Function
Blood	Transports materials throughout the body
Bone	Supports and protects body and organs
Muscle	Helps body to move; aids in circulation, digestion, and respiration
Nerve	Carries messages
Skin	Covers and protects body; excretes wastes

3. *Organs.* A group of tissues working together forms an ***organ***. The heart is an organ that pumps blood throughout the body. It is composed mainly of muscle tissue, but also contains blood tissue and nerve tissue. Table 7-2 lists some important organs.

Table 7-2. *Important Organs and Their Functions*

Organ	Function
Heart	Pumps blood
Kidney	Removes wastes from blood
Lung	Exchanges gases with the environment
Stomach	Breaks down food by physical and chemical means
Brain	Controls thinking and voluntary actions

4. *Organ Systems.* A group of organs working together to carry out a specific life process makes up an ***organ system***. The circulatory system carries out the process of transport, moving materials throughout the body. Table 7-3 lists the human organ systems.

Table 7-3. *Human Organ Systems*

System	Function	Examples of Organs or Parts
Skeletal	Supports body, protects internal organs	Skull, ribs
Muscular	Moves skeleton, organs, and body parts	Arm and leg muscles
Nervous	Controls body activities; carries and interprets messages	Brain, spinal cord
Endocrine	Regulates body activities with hormones	Adrenal, pituitary glands
Digestive	Breaks down food into usable forms	Stomach, intestines
Circulatory	Carries needed materials to body cells and waste materials away from cells	Heart, arteries, veins
Respiratory	Exchanges gases with the environment	Lungs, bronchi
Excretory	Removes wastes from the body	Kidneys, skin
Reproductive	Produces hormones and offspring	Ovaries, testes

The Skeletal System

The human *skeletal system,* shown in Figure 7-2, supports and protects the body and its organs. The skeletal system includes the *skull, spinal column, breastbone, ribs, bones of the limbs* (arms and legs), and *cartilage.*

Figure 7-2. *The human skeleton.*

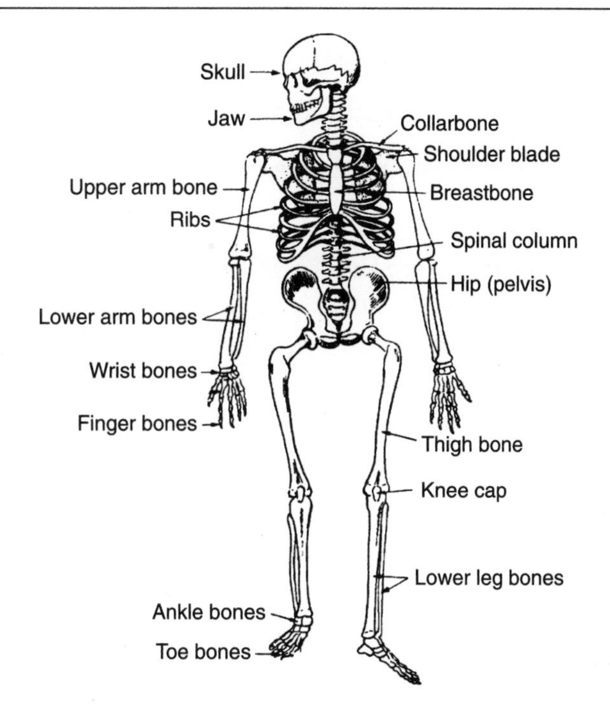

1. *Bones and Cartilage.* **Bones** are made of hard, strong material. ***Cartilage*** is a softer, more flexible tissue. Cartilage acts as a cushion between bones and provides flexibility at the ends of bones. Disks of cartilage separate the bones of the spinal column, cushioning them from one another.
2. *Joints.* Where one bone connects to another bone, a ***joint*** is formed. Most joints, such as the knee and elbow, allow the bones to move. However, some joints, like those in the skull, do not allow movement. Figure 7-3 shows three types of joints.

Figure 7-3. *Three types of joints.*

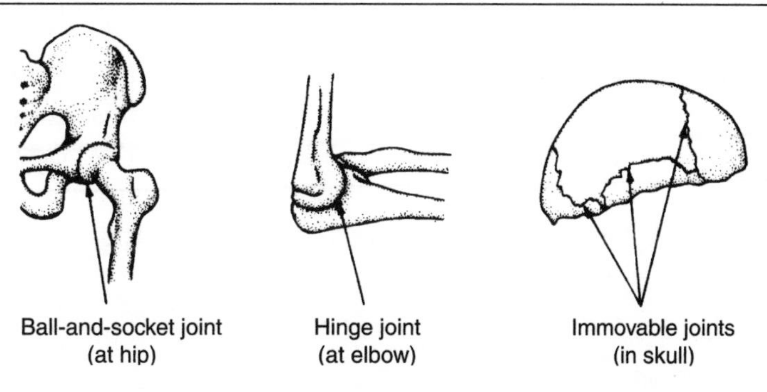

3. *Ligaments and Tendons.* At movable joints, the bones are held together by strips of tissue called *ligaments.* Bones are moved by muscles, which are attached to bones by *tendons,* cordlike pieces of tissue. A torn Achilles tendon is a common sports injury. This tendon joins the calf muscles of the leg to the heel bone.

The Muscular System

Muscles are masses of tissue that contract to move bones or organs. The *muscular system* contains two main kinds of muscles: *voluntary* and *involuntary.*

1. *Voluntary Muscles.* The *skeletal muscles,* which move bones, are examples of ***voluntary muscles***—muscles that you control. Skeletal muscles work with the skeleton to move body parts (see Figure 7-4), producing locomotion. ***Locomotion*** is the movement of the body from place to place. The muscles in the face and around the eyes are also voluntary muscles.

Figure 7-4. *Muscles, tendons, and bones of the arm allow movement.*

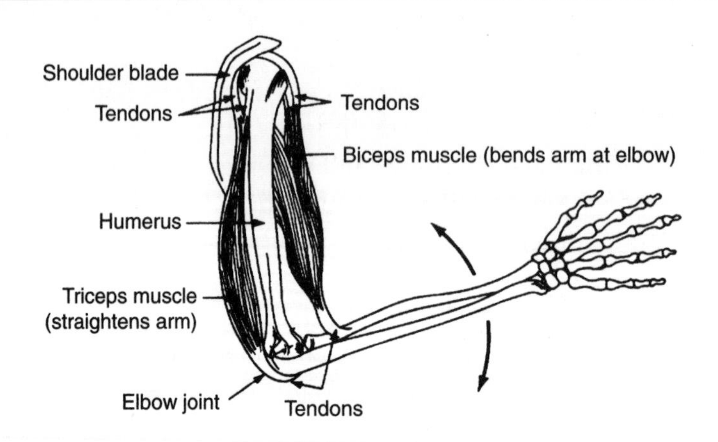

2. *Involuntary Muscles.* ***Involuntary muscles*** are not under our conscious control. There are two types of involuntary muscles: cardiac and smooth. *Cardiac* muscle, present only in the heart, pumps blood throughout the body. *Smooth* muscle, found in the respiratory, circulatory, and digestive systems, aids in breathing, controlling blood flow, and movement of food.

Review Questions

Part I

1. Which choice lists the levels of organization in the correct order, starting with the simplest?

   (1) cell, tissue, organ, organ system

   (2) cell, organ, organ system, tissue

   (3) tissue, cell, organ system, organ

   (4) organ, organ system, cell tissue

2. The cell illustrated below transmits messages. The tissue that contains this type of cell is called

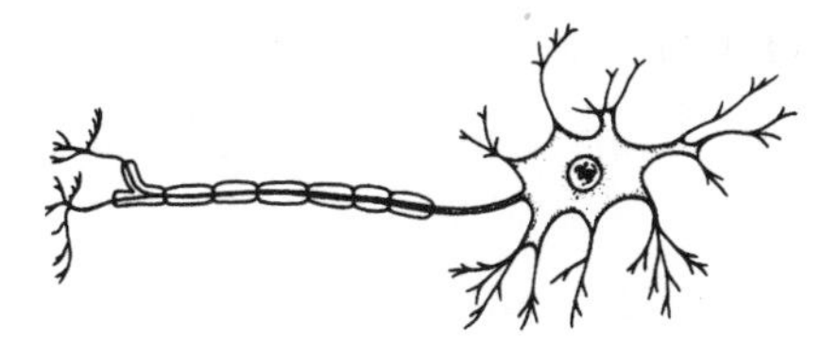

   (1) blood tissue | (3) muscle tissue

   (2) nerve tissue | (4) skin tissue

3. Tissue is composed of a group of

   (1) different organ systems working together

   (2) different organs working together

   (3) similar cells working together

   (4) similar organs working together

4. Complete the analogy: cells are to tissues, as tissues are to

   (1) organ systems | (3) organisms

   (2) organs | (4) bacteria

5. Which of these parts of the body contains the greatest number of cells?

   (1) blood tissue | (3) the arteries

   (2) the heart | (4) the circulatory system

6. Which activity is most closely associated with the respiratory system?

   (1) breathing | (3) eating

   (2) moving from place to place | (4) producing offspring

7. Which organ system is correctly matched with its role in maintaining homeostasis?

   (1) muscular—removes wastes from the body

   (2) respiratory—exchange of gases with the environment

(3) excretory—breaks down food into usable forms

(4) circulatory—moves organs and body parts

8. Which two systems of your body work together to move you from place to place?

(1) digestive and nervous
(2) reproductive and nervous
(3) muscular and skeletal
(4) excretory and respiratory

Part II

Use the following information to answer questions 9 and 10.

Humans are complex multicellular organisms. Life functions are carried out by different organ systems. These organ systems are made up of many different types of cells, tissues, and organs.

9. Describe how cells, tissues, organs, and organ systems are related to one another.

10. Identify which organ system is responsible for each of the following functions and name two organs that make up each of these body systems.

(a) carry materials to and from body cells

(b) support the body and protect internal organs

(c) move organs and body parts

(d) break down food into usable forms

(e) remove wastes from the body

Use the diagram below to answer questions 11 through 14. The diagram represents a human body system.

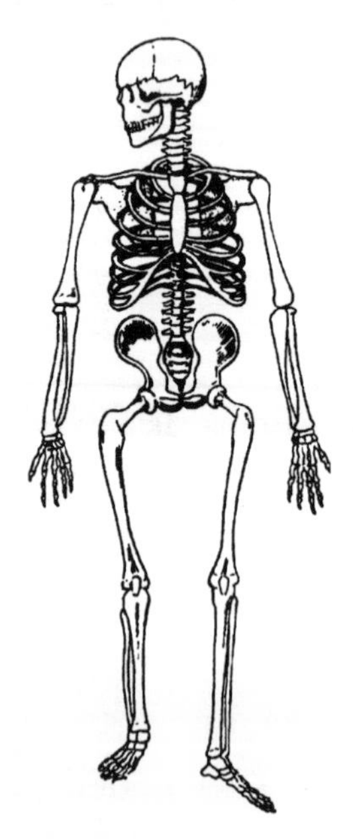

11. What is the name of this body system?

12. What are two functions performed by this body system?
13. Identify two types of tissue making up this body system.
14. Identify two organs making up this body system.

---

15. Complete the paragraph, using the terms "smooth," "cardiac," and "voluntary."

As the __(a)___ muscles of his stomach churned and digested food while he was returning home from his favorite restaurant, Paul noticed a pack of mean-looking dogs following him. The _(b)__ muscles in his heart pumped faster and faster. He realized that the _(c)__ muscles in the dogs' legs could move the dogs very quickly, so he knew not to run away. As the dogs moved closer, the _(d)__ muscle called the diaphragm caused Paul's breathing to speed up. Fortunately, the dogs passed by without even noticing him.

Regulation, Digestion, and Circulation

Regulation, digestion, and circulation are three of the processes carried out by systems in the human body. These processes keep us alive. Each process is carried out by different organ systems. The systems work together.

Regulation

The *nervous system* and the *endocrine system* work together to regulate body processes and actions. They provide us with a way of detecting and responding to *stimuli* (changes inside or outside the body). The nervous system (Figure 7-5) is made up of the *brain, spinal cord, nerves,* and parts of the sensory, or *sense organs*.

Figure 7-5. *The human nervous system.*

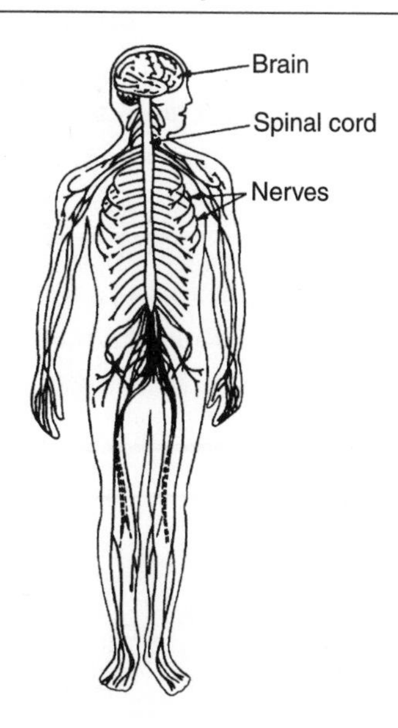

1. The ***brain*** receives and interprets *nerve impulses* ("messages") and controls thinking, voluntary action, and some involuntary actions, such as coordination, balance, breathing, and digestion.
2. The ***spinal cord*** channels nerve impulses to and from the brain and controls many automatic responses, or *reflexes*, such as pulling your hand away from a flame.
3. ***Nerves*** carry messages between the sense organs, the brain and spinal cord, and the muscles and glands.
4. The sensory, or ***sense organs***, which include the skin, eyes, ears, nose, and tongue, receive information from the environment.

Nerve cells, also called ***neurons***, receive and transmit nerve impulses (see Figure 7-6). Two types of neurons are the sensory and the motor neurons. *Sensory neurons* carry information from the sense organs to the brain or spinal cord. *Motor neurons* carry messages from the brain or spinal cord to muscles and glands, which respond to the messages.

Figure 7-6. *A typical neuron, or nerve cell.*

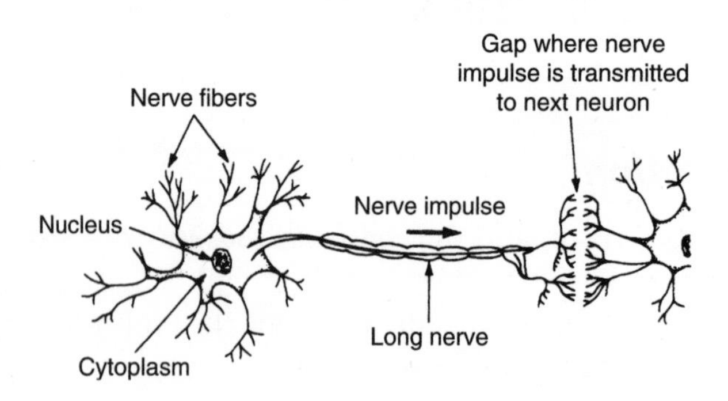

The endocrine system is made up of glands. A ***gland*** is an organ that makes and *secretes* (releases) chemicals. Endocrine glands secrete chemical messengers called ***hormones***. Figure 7-7 shows the major endocrine glands. When an endocrine gland secretes a hormone into the bloodstream, the blood carries the hormone to an organ, which responds in some way. For example, if you are suddenly faced with some danger, such as a snarling dog, the hormone *adrenaline* is released by your *adrenal gland*. The adrenaline makes your heart beat faster and makes your breathing more rapid. More sugar is released into your bloodstream to provide energy. These changes prepare your body to respond to the danger.

The Digestive System

Our cells need nutrients from food for energy, growth, and repair. The *digestive system* breaks down food into soluble nutrients that can then be absorbed into the bloodstream and carried to the cells.

Figure 7-7. *Major glands of the human endocrine system.*

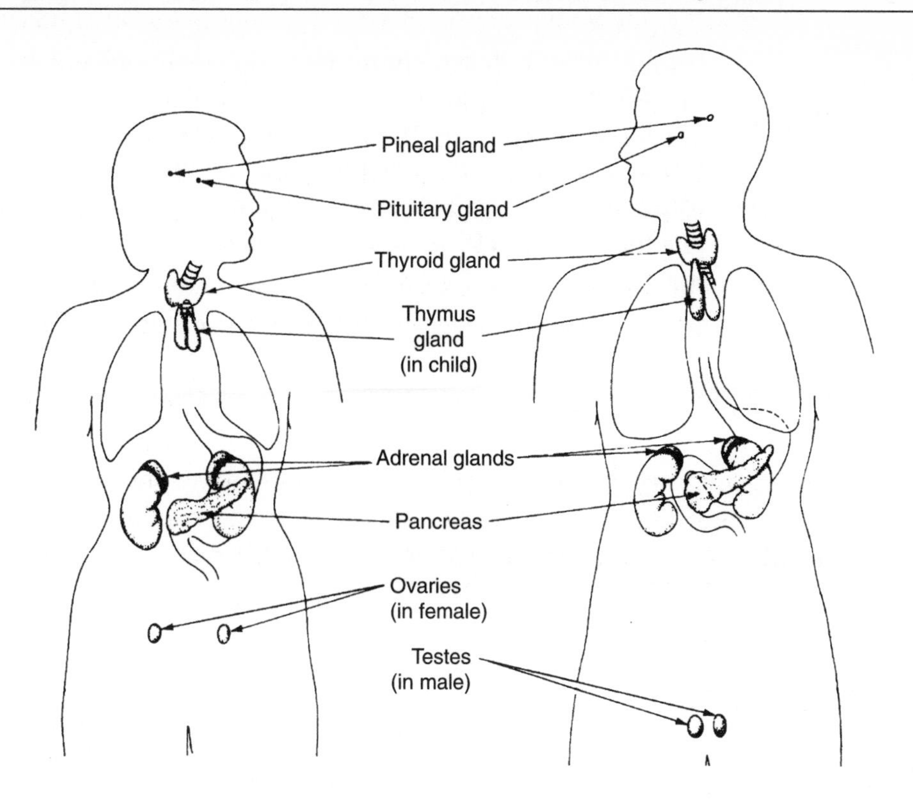

The digestive system, shown in Figure 7-8, consists of the digestive tract and the accessory organs. An accessory organ helps the functioning of another organ or system.

Figure 7-8. *The human digestive system.*

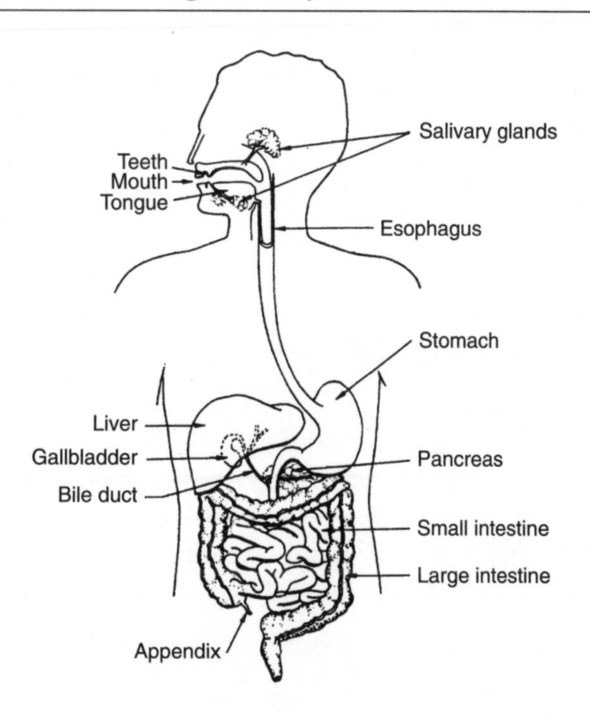

1. The *digestive tract* is a tube in which food is changed into a form that the body can use for growth, healing, and energy. The digestive tract begins at the mouth and continues through the *esophagus, stomach, small intestine,* and *large intestine.*
2. The *accessory organs* are the *pancreas, gallbladder,* and *liver.* They produce digestive juices that are released into the digestive tract. Bile and pancreatic juice are secreted by the liver and pancreas *into* the small intestine, where digestion occurs. The digestive juices help change food into forms the body can use. Table 7-4 lists the digestive juices, where they are produced, and what foods they digest.

Table 7-4. *Digestive Juices*

Organ	**Digestive Juice**	**Foods Acted On**
Mouth	Saliva	Starches
Stomach	Gastric juice	Proteins
Small Intestine	Intestinal juices	Sugars, proteins, fats
Pancreas	Pancreatic juice	Proteins, starches, fats
Liver	Bile	Fats

The digestive system breaks down food *physically* and *chemically.* (1) Food is physically broken down into small bits by chewing and by the action of muscles in the digestive tract. (2) The chemical breakdown of food releases nutrients that can be used by cells. Chemicals called *enzymes,* found in the digestive juices, accomplish this breakdown.

Digestion starts in the mouth and continues in the stomach and small intestine. When digestion has been completed, digested materials are absorbed into the bloodstream through the walls of the small intestine. Undigested materials, which make up the solid wastes called *feces,* pass on through the large intestine and are expelled from the body.

The Circulatory System

Nutrients absorbed into the blood must be carried to all body cells. This is the job of the *circulatory system*: to bring needed materials such as nutrients, water, and oxygen to the cells and to carry away wastes, like carbon dioxide, from the cells.

The parts of the circulatory system are the blood, the heart, the blood vessels (arteries, veins, and capillaries), lymph, and lymph vessels.

1. *Blood.* The ***blood*** is a liquid tissue containing plasma, red and white blood cells, and platelets. Plasma is the liquid part of the blood that carries dissolved nutrients, wastes, and hormones. The red blood cells, the white blood cells, and the platelets float in the plasma. Red blood cells contain the pigment *hemoglobin,* which is the chemical that carries oxygen to the cells. White blood cells fight infection. They surround the infecting organism and destroy it. Platelets cause clotting, which stops bleeding.
2. *Heart.* The ***heart*** (Figure 7-9) is a muscle that contracts regularly to pump blood throughout the body. The blood is pumped from the heart to the lungs, where it receives oxygen and gets rid of carbon dioxide. The blood then returns to the heart to be pumped to the rest of the body, as shown in Figure 7-10 on page 208.

Figure 7-9. *The human heart.*

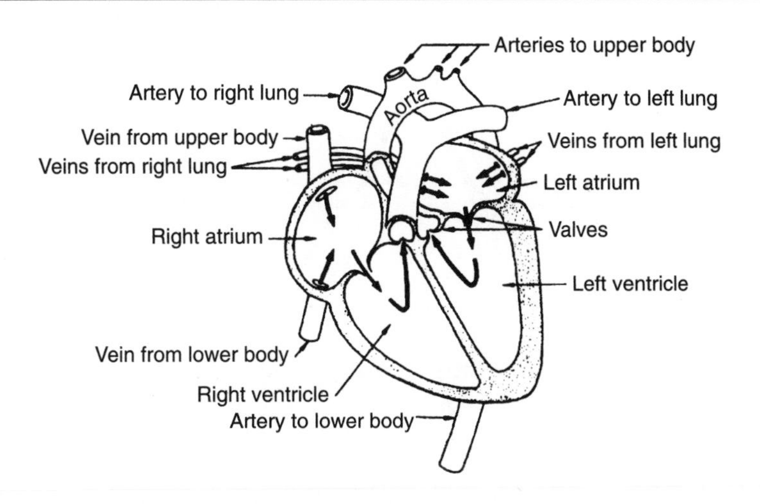

3. *Blood Vessels.* The blood flows through a network of tubes called ***blood vessels***. There are three types of blood vessels. ***Arteries*** carry blood away from the heart, while ***veins*** return blood to the heart. Connecting arteries to veins are ***capillaries***. (See inset in Figure 7-10.) Through the walls of the extremely small capillaries, materials are exchanged between the blood and the body's cells. Dissolved nutrients, water, and oxygen pass from the blood into the cells, and some wastes from the cells pass into the blood.
4. *Lymph.* Some of the watery part of the blood filters out through the walls of the capillaries into the surrounding tissue. This fluid, called ***lymph***, surrounds all the cells of the body. Lymph acts as a go-between in the exchange of materials between the blood and the cells. After receiving wastes from the cells, lymph is collected and returned to the bloodstream through ***lymph vessels***.

Figure 7-10. *Pathways of blood through the human circulatory system.*

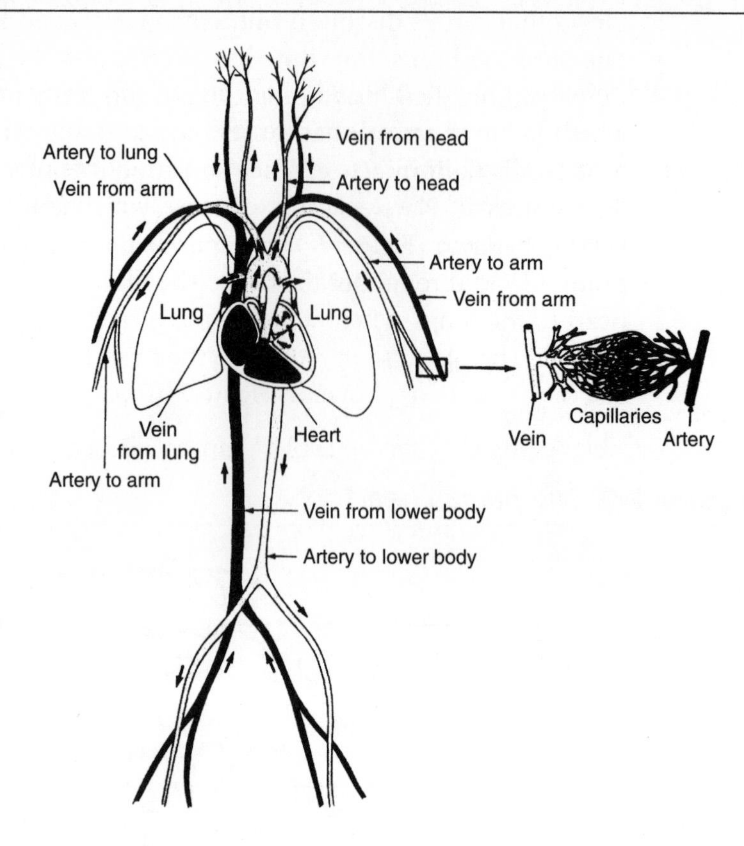

Take a virtual tour through the heart and circulatory system at *www.fi.edu/biosci/index.html* and click on the links in the paragraph that begins "Explore the Heart."

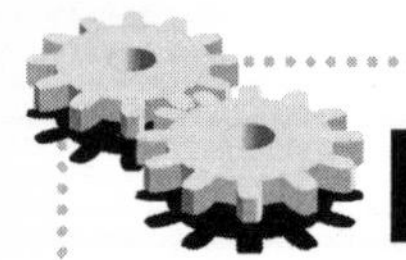

Process Skill

Interpreting a Diagram

The diagram below represents the circulatory system. It is not meant to show accurately the heart and lungs; instead, it shows the basic relationships among the main parts of the circulatory system and the sequence of events that take place within it.

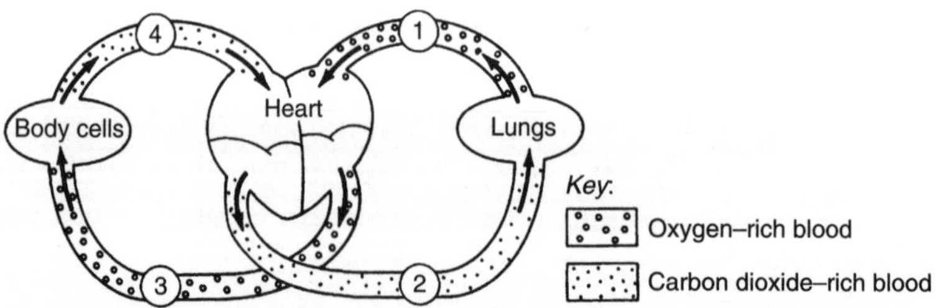

The circulation of blood is important to the process of respiration, since the blood carries fresh oxygen to the cells of the body and brings carbon dioxide to the lungs to be expelled as we exhale.

As you have learned, arteries are blood vessels that carry blood away from the heart. Which blood vessels in the diagram are arteries? The arrows indicate that blood vessels 2 and 3 carry blood away from the heart, so they are arteries. Blood vessels 1 and 4, which return blood to the heart, are veins. Study the diagram and then answer the following questions.

Questions

1. Blood rich in oxygen is found in blood vessels
(1) 1 and 2 (2) 2 and 3 (3) 1 and 3 (4) 2 and 4

2. Compared with blood vessel 1, the amount of carbon dioxide in blood vessel 2 is
(1) greater (2) less (3) the same

3. Which statement is true?
(1) All arteries carry oxygen-rich blood.
(2) All veins carry oxygen-rich blood.
(3) Arteries from the heart to the lungs carry oxygen-rich blood.
(4) Veins from the lungs to the heart carry oxygen-rich blood.

Review Questions

Part I

16. After it leaves the mouth, food next enters the
(1) stomach
(2) esophagus
(3) small intestine
(4) large intestine

17. The purpose of the digestive system is to
(1) break down food for absorption into the blood
(2) exchange oxygen for carbon dioxide in the lungs
(3) respond to stimuli
(4) carry nutrients to all parts of the body

18. The digestive system is made up of the digestive tract and accessory organs. Food travels through the digestive tract. Accessory organs produce digestive juices that are secreted into the digestive tract. Which organ is an accesory organ?
(1) mouth
(2) pancreas
(3) stomach
(4) small intestine

19. Which type of blood cell fights infection?

(1) red blood cell

(2) white blood cell

(3) blue blood cell

(4) platelet

20. Which life function is the direct responsibility of the circulatory system?

(1) excretion
(2) nutrition
(3) transport
(4) reproduction

21. The endocrine system produces chemical messengers that affect organs. These chemical messengers are called

(1) nutrients
(2) wastes
(3) impulses
(4) hormones

22. Which body system is represented in the diagram below?

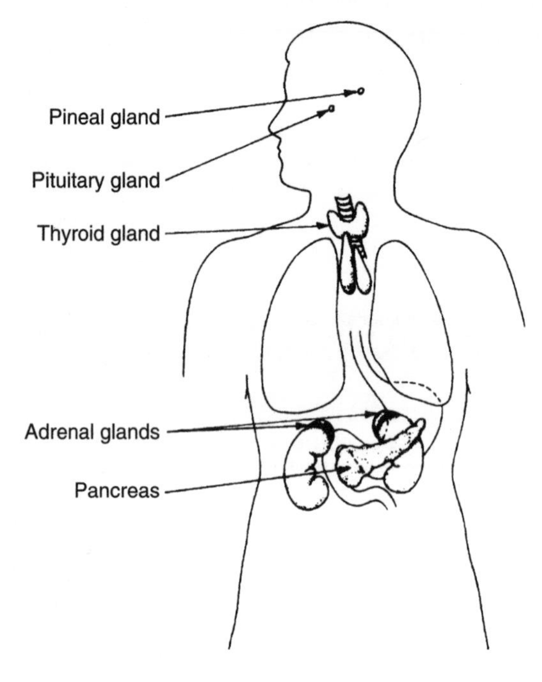

(1) circulatory

(2) endocrine

(3) skeletal

(4) nervous

23. Once you have taken a bite of an apple, which of the following represents the correct pathway of its nutrients?

(1) circulatory system → cell → digestive system

(2) cell → digestive system → circulatory system

(3) digestive system → circulatory system → cell

(4) circulatory system → digestive system → cell

Part II

Use the paragraph below to answer questions 24 through 26.

Regulation is the process of responding to changes. This is carried out by detecting changes and then sending messages to organs or muscles to respond to those changes.

24. Which two human systems are directly responsible for this life function?
25. Identify one organ in each of these systems.
26. Compare the way messages are carried in each of these systems.

Use the diagram below to answer questions 27 through 29. The diagram shows several organs in the human digestive system.

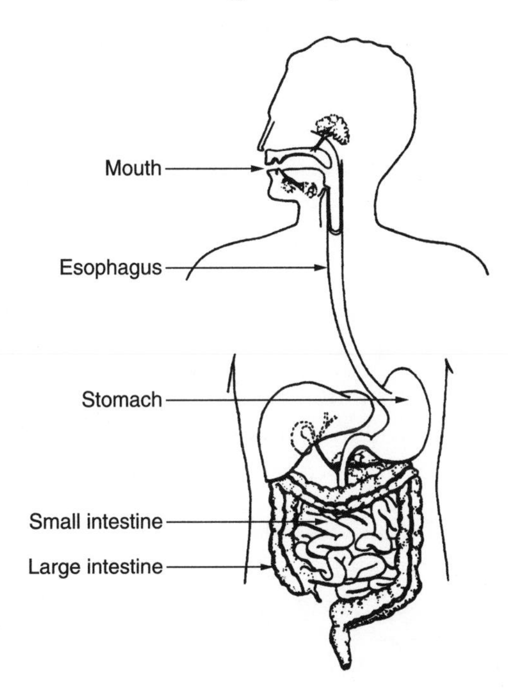

27. Identify two ways that food is changed as it passes through the digestive system.
28. Name two organs that participate in digestion but are not labeled in the diagram.
29. What is the main function of the digestive system?

Use the diagram below to answer questions 30 though 32. The diagram shows several organs of the human circulatory system.

Human Circulatory System

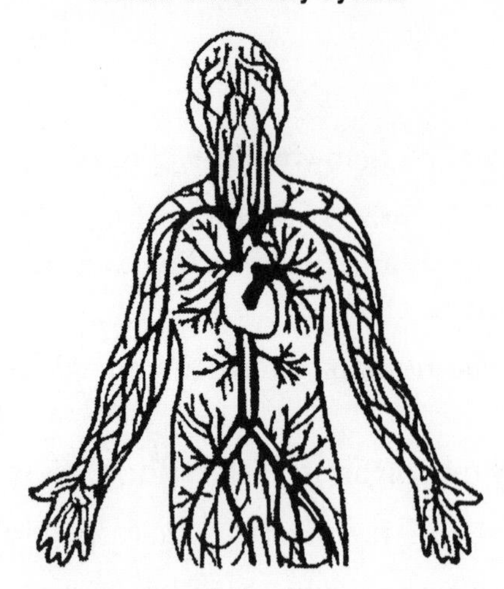

30. The heart is a very important part of the circulatory system. What does the heart do?

31. The diagram shows many blood vessels, leading to all parts of the body. Identify three different types of blood vessels and describe the function of each.

32. What is the function of hemoglobin and where is it found?

Respiration, Excretion, and Reproduction

The respiratory system provides oxygen to the cells. The excretory system removes wastes produced by the cells. The reproductive system ensures the continuation of the species.

The Respiratory System

The circulatory system gets the oxygen it brings to the cells from the ***respiratory system***. Cells use this oxygen in the process of ***cellular respiration,*** in which nutrients from digested food combine with oxygen to release energy and produce the waste materials carbon dioxide and water. This chemical process takes place in all body cells.

The *respiratory system,* illustrated in Figure 7-11, brings oxygen from the air to the blood, and brings carbon dioxide from the blood to the air. This process is called ***respiration***.

Figure 7-11. *The human respiratory system.*

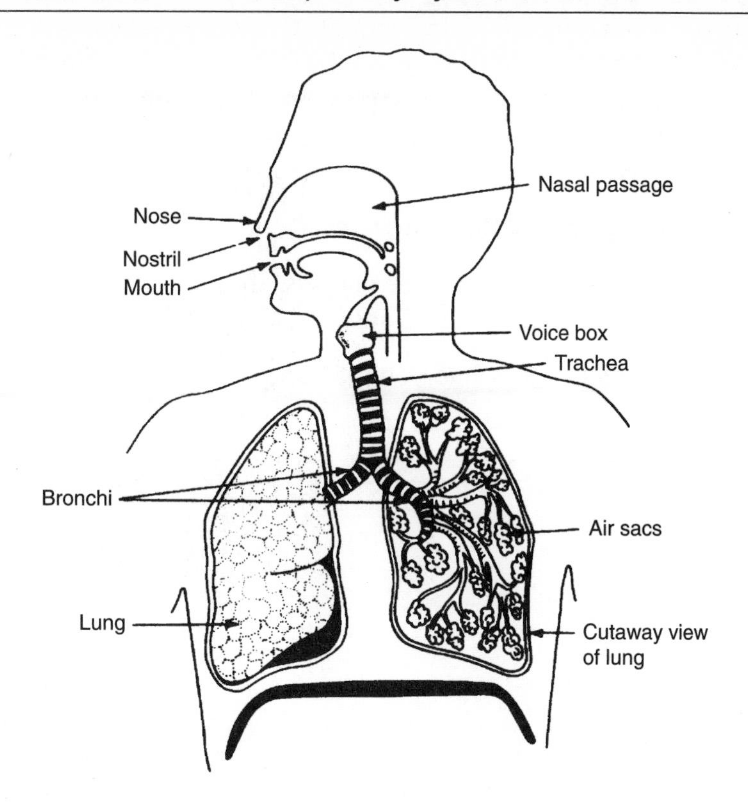

When you breathe in (*inhale*), your diaphragm contracts and air enters the nose or mouth and passes through the ***trachea***, or windpipe. The trachea branches off to each lung through two tubes called ***bronchi*** (singular, *bronchus*). The ***lungs*** contain millions of tiny *air sacs,* surrounded by capillaries. Here, respiratory gases are exchanged—oxygen enters the blood, while carbon dioxide leaves the blood and is breathed out (*exhaled*).

The oxygen that enters the blood in the lungs is carried to the cells of the body, where an exchange of gases again takes place. This time, oxygen leaves the blood and enters the cells, while carbon dioxide leaves the cells and goes into the blood. The carbon dioxide is carried to the lungs to be exhaled. This process is repeated continuously.

Laboratory Skill

Doing an Experiment

Why do you sometimes get out of breath? When your body needs to get rid of more carbon dioxide from your blood, your respiration rate changes. What might cause an increase in carbon dioxide in the blood? Exercise can change the amount of carbon dioxide in your blood. Jerry and Paul did an experiment to determine how exercise can affect the rate of respiration.

Part I—Resting breathing rate

Before taking any measurements, Jerry and Paul relaxed in a sitting position for two minutes. Using a stopwatch, they both determined how many breaths they took in 15 seconds. (They multiplied the number of breaths by 4 to determine the number of breaths per minute.)

Part II—Breathing rate after exercise

Paul and Jerry jogged in place for three minutes. They sat down and immediately counted their number of breaths for 15 seconds. The results are recorded below.

Results: Copy the table below into your notebook for your results:

	Number of Breaths in 15 Seconds	
	Paul	*Jerry*
Before exercise	5	6
After exercise	7	8

Questions

1. Using Paul's resting breathing rate, calculate his number of breaths per minute.
2. What effect does exercising have on the breathing rate?
3. Explain why exercising might change the breathing rate.
4. State a hypothesis that might be tested by this experiment.

The Excretory System

The activities of the body's cells produce waste materials that must be removed. These wastes are removed from the blood, and eventually from the body, by the *excretory system.* The excretory system consists of the *lungs, skin, kidneys,* and *liver.* Notice that lungs are part of two systems: respiratory and excretory.

1. The *lungs* rid the body of the waste products carbon dioxide and water vapor each time you exhale.
2. The ***skin*** gets rid of wastes when you perspire. Microscopic sweat glands deep in the skin excrete (release) *perspiration,* a liquid waste consisting mostly of water and salts. Perspiration leaves the body through the *pores,* which are tiny openings in the surface of the skin (Figure 7-12).

Figure 7-12. *Sweat glands in the skin expel wastes from the body through pores.*

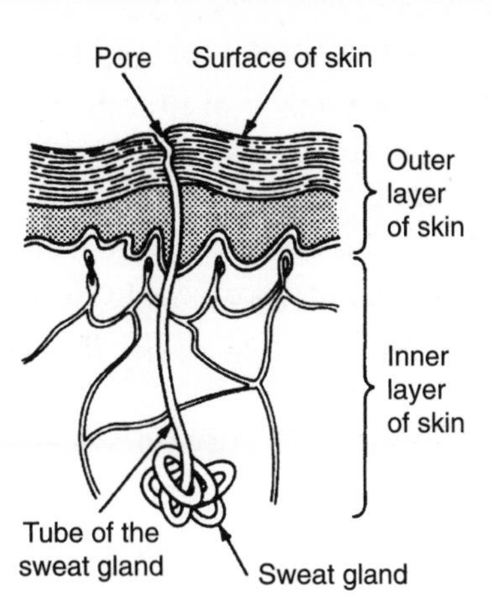

3. The two ***kidneys*** (Figure 7-13) help maintain the proper balance of water and minerals in the body. As blood flows through the kidneys, excess water, salts, urea, and other wastes are removed from the blood. These substances make up a fluid called *urine.* Urine is sent through a tube from each kidney to the *bladder,* where it is stored until excreted from the body.

Figure 7-13. *The human urinary system, part of the excretory system.*

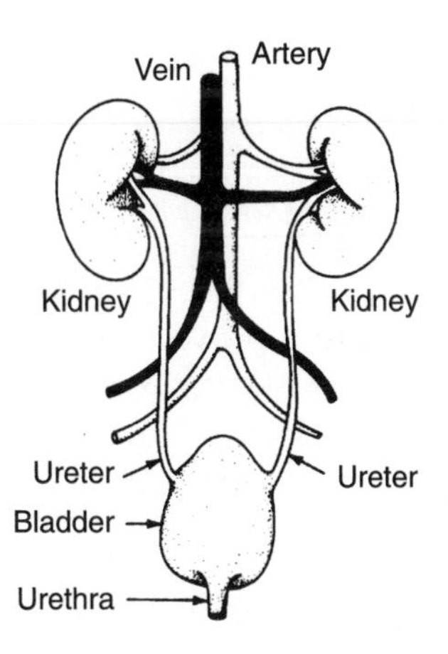

4. The ***liver*** produces *urea,* a waste resulting from the breakdown of proteins. Urea is taken by the blood to the kidneys, where it is filtered out of the blood and then expelled from the body in urine. The liver also removes harmful substances from the blood.

Metabolism

Your body carries out an amazing number of chemical reactions that keep you alive. These reactions break things apart and put things together. For example, you take in food, break it down, and use the energy stored in it; you build and repair tissues; you store fat. ***Metabolism*** is the total of all the chemical reactions that take place in the body.

Metabolism can be influenced by hormones, exercise, diet, and aging. Carbohydrate metabolism involves the hormone insulin. People who have diabetes either do not produce enough insulin or their bodies do not respond to it; therefore, they do not metabolize carbohydrates properly. Exercise increases metabolism because cells need more oxygen. When you eat too much, your metabolism stores the excess food as fat. As you reach old age, your metabolism slows.

The Reproductive System

The job of the reproductive system is the production of offspring. It is also responsible for hormone production. There are two human reproductive systems, male and female, as shown in Figure 7-14.

Figure 7-14. *The human reproductive systems: male (left) and female (right).*

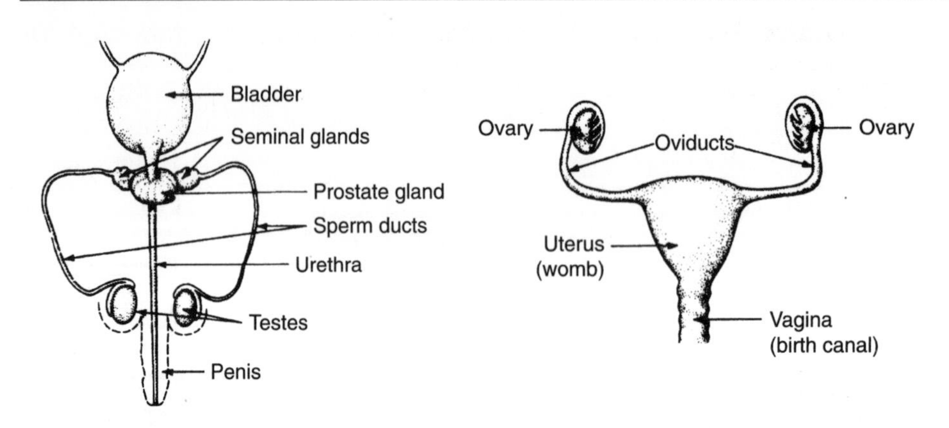

1. *Male.* The male reproductive system is made up of the *testes, penis,* and *sperm ducts*. The two ***testes*** (singular, *testis*) produce *sperm cells,* the male sex cells. These cells pass through tubes called ***sperm ducts,*** where they mix with a fluid to form *semen.* The semen is delivered through the *penis*.
2. *Female.* The female reproductive system is made up of the *ovaries, oviducts, uterus, vagina,* and *mammary glands.* The ***ovaries*** produce *egg cells,* the female reproductive cells. About once a month, beginning in adolescence, an egg cell leaves an ovary and travels

through one of the ***oviducts*** to the ***uterus***, or womb. If sperm cells are present in the oviduct, fertilization may take place.

During reproduction, sperm is released by the male into the female's reproductive system. When fertilization occurs, the fertilized egg attaches itself to the inner wall of the *uterus*, or womb. There it develops into a new offspring over a period of about nine months. At the end of this time birth takes place, and the offspring emerges through the ***vagina***, or birth canal. The newborn baby may be fed milk produced by the mother's ***mammary glands***. These glands are located in the female's breasts.

Review Questions

Part I

33. One purpose of respiration is to supply the cells with
    (1) carbon dioxide
    (2) nutrients
    (3) oxygen
    (4) water

34. The function of the circulatory system is to
    (1) carry materials to and from the cells
    (2) break down food into a usable form
    (3) regulate body activities
    (4) respond to stimuli

35. As oxygen from air sacs in the lungs moves into the blood, carbon dioxide from the blood moves into the air sacs. What two systems are involved in this process?
    (1) digestive system and circulatory system
    (2) respiratory system and nervous system
    (3) skeletal system and muscular system
    (4) respiratory system and circulatory system

36. What is the function of the reproductive system?
    (1) excretion
    (2) transport
    (3) nutrition
    (4) offspring

37. The kidneys, which remove dissolved wastes from the blood, are organs of the
    (1) endocrine system
    (2) excretory system
    (3) skeletal system
    (4) nervous system

38. Which of the following organs can be found in the male reproductive system?
    (1) ovary
    (2) oviduct
    (3) lungs
    (4) testes

39. Egg cells are produced in the

(1) sperm duct
(2) oviduct
(3) mammary glands
(4) ovaries

Part II

40. The figures below represent four human body systems. Identify the name and main function of each system.

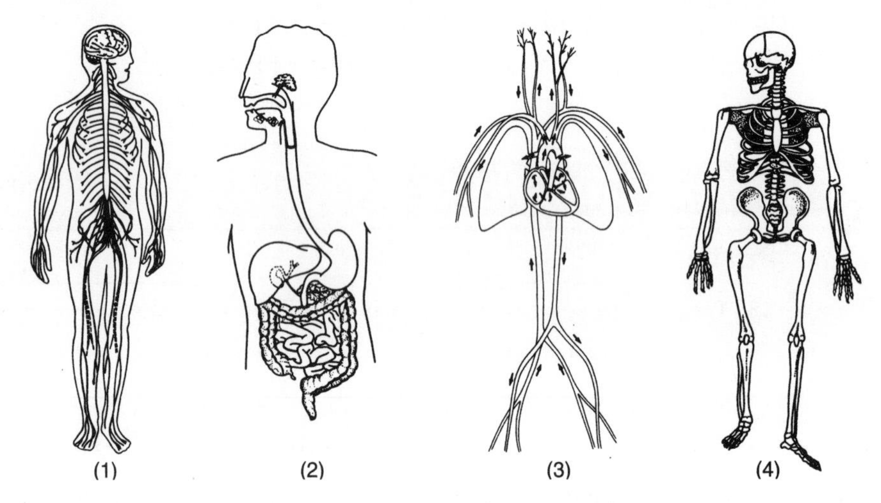

Use the diagram below to answer questions 41 through 43.

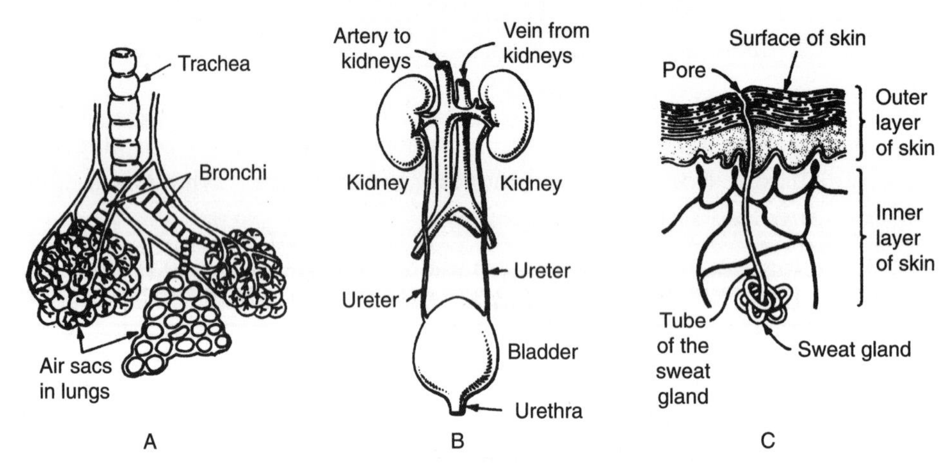

41. To which body system do all three structures belong?

42. Which organ filters the blood and removes urea?

43. What waste product is removed by all of these structures?

---

44. Carbon dioxide, water, and urea are wastes produced by the body. Describe the path of each of these wastes from the moment it leaves the cell. The path of water has already been done for you as a model to follow.

Water: cell → blood → skin

(a) Carbon dioxide

(b) Urea

45. Complete the following analogies:

(a) Female is to ovaries as male is to ______

(b) Ovaries are to egg as testes are to ______

46. Complete the concept map below by naming the excretory organs represented by A, B, C, and D.

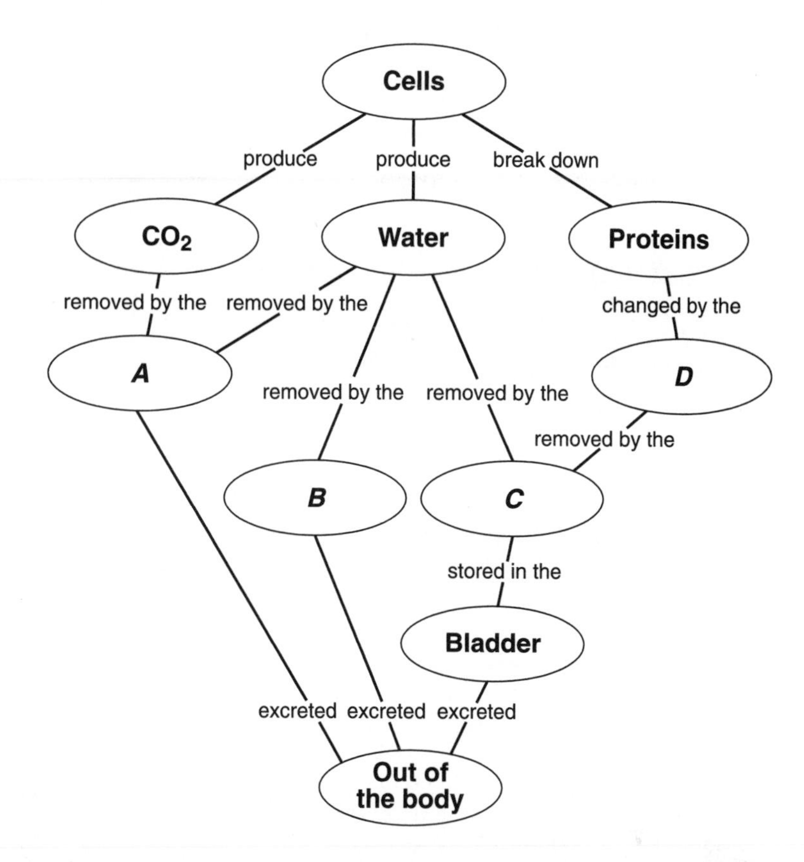

47. Use the Venn diagram below to organize the list of organs into the systems to which they belong. If an organ is part of two systems, place it in the area where the two systems overlap.

kidneys, urethra, stomach, esophagus, liver, lungs, skin, small intestine, gallbladder, nose, trachea

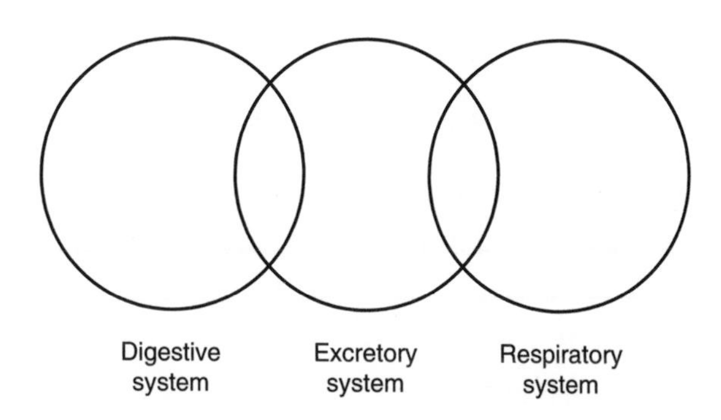

48. We breathe to provide our cells with oxygen. The oxygen is used to "burn" sugar to provide the cells with energy. The steps involved in this process are listed below, but their order has been jumbled. Place these steps in the correct order in your notebook.

(a) Sugar is burned to make carbon dioxide.

(b) Air enters the air sacs.

(c) Oxygen moves into the body cells.

(d) The diaphragm contracts.

(e) Carbon dioxide moves to the lungs.

(f) Oxygen enters the blood.

(g) Carbon dioxide enters the blood.

(h) Carbon dioxide leaves the blood.

49. The organs of a plant include the stems, leaves, roots, and flowers, as illustrated in the diagram below.

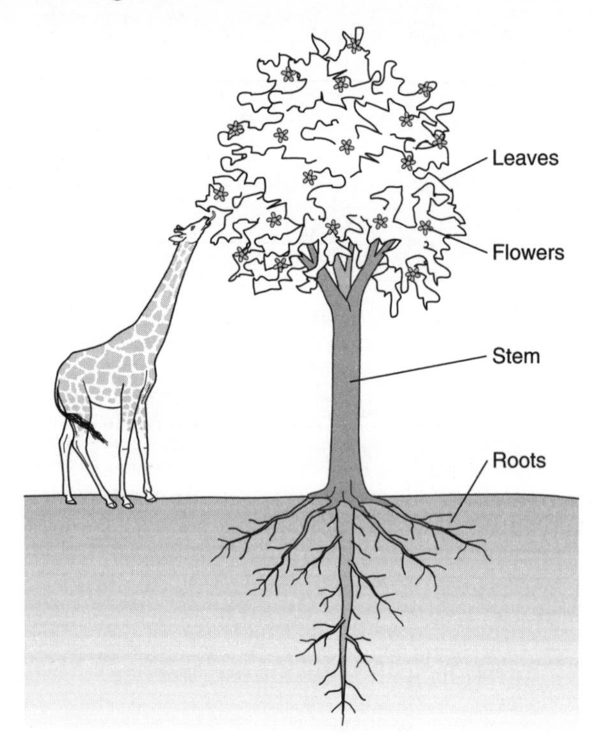

- The stem provides transport of materials between the leaves and the roots.
- The stem supports the leaves.
- The leaves provide nutrients in a usable form.
- The flowers produce offspring.
- The roots absorb water.

Identify which organ systems of a giraffe accomplish these same functions.

Chapter 8 Ecosystems

Meerkats are part of the ecosystem of the Kalahari Desert.

Vocabulary at a Glance

adaptation
behavioral adaptation
biodiversity
carnivore
commensalism
community
consumer
dormancy
ecological succession
ecosystem
endangered species
extinct
food chain
food web
fossil
habitat
herbivore
hibernation
migration
mutualism
omnivore
parasitism
physical adaptation
population
producer
response
stimulus
symbiosis

Major Concepts

- Living things and the nonliving things in their surroundings make up an ecosystem. Adaptations allow an organism to survive in a particular ecosystem.
- Green plants use sunlight to make their own food. They are called producers. Animals depend on other organisms for food. They are called consumers. All organisms get their energy directly or indirectly from the sun.
- Decomposers are organisms that break down the remains of dead plants and animals. Decomposers return nutrients to the environment.
- Producers, consumers, and decomposers may be linked in a food chain. Disturbing any part of a food chain affects other organisms in the food chain.
- There is a constant exchange of materials between an organism and its environment; these materials include food, water, oxygen, and wastes.
- Ecological succession is the natural process by which one community is replaced by another community in an orderly, predictable way.

Living things must react to changes in their environment. For example, when the air gets too hot, you perspire. When the light gets too bright, the pupils of your eyes get smaller. A change in the environment is a ***stimulus***. The way in which a living thing reacts to the change is a ***response***.

Sometimes an environment is harsh, which makes it difficult for an organism to survive. Living things have evolved in response to changes in their environments by developing ***adaptations***, characteristics that help them survive in their habitat.

Behavioral Adaptations

Some environments undergo very large changes in temperature, amount of sunlight, and water supply from season to season. Organisms that live in such environments have special behaviors called ***behavioral adaptations*** that help them adapt to these changes. These include migration, hibernation, and dormancy.

1. *Migration.* Have you ever heard that birds fly south for the winter? Many birds that live in places where the winters are cold fly to warmer regions as winter approaches. ***Migration*** involves moving from one environment to another.
2. *Hibernation.* Some animals survive the cold by finding a safe place (den or cave) and sleeping for most of the winter. Bears and bats use ***hibernation*** as a behavioral adaptation.
3. *Dormancy.* Other living things may adjust to extreme environmental changes by entering a state of ***dormancy***, becoming completely inactive. During the winter when a tree has lost its leaves, the tree is not dead—it is *dormant.* When spring comes, bringing warmer conditions, the tree grows new leaves. A seed may remain dormant for years, waiting for the proper conditions for growth.

Some organisms survive for only one season. Plants, called annual plants, go through their entire life cycle, including reproduction, during one season. Some insects such as the mosquito also live for only one season. Their eggs survive the winter and hatch the next summer.

Physical Adaptations

Living things have developed special characteristics called ***physical adaptations*** that allow them to survive under a given set of conditions. Organisms may be adapted for life in water, soil, or air. For example, a fish has gills so it can take in oxygen dissolved in the water. An earthworm's body shape helps it move through the soil. A bird has wings and light, hollow bones so it can fly. Many adaptations help an organism obtain food or escape predators in its environment. Figure 8-1 shows how certain birds are adapted for survival.

Figure 8-1. *Adaptations of birds to their environment.*

Earth's many environments include oceans, deserts, tropical rain forests, and the frozen Arctic tundra. Adaptations permit an organism to live in its own special environment, or ***habitat***. Organisms living in a dry, desert environment have adaptations that allow them to get and conserve water. For example, the cactus plant has specialized leaves, stems, and an extensive root system that helps it reach and store water.

Animals living in the icy Arctic have adaptations that help them to survive the region's very cold temperatures. For instance, polar bears have thick coats of fur; seals and whales have layers of protective fat called *blubber*. Both adaptations act as insulators to protect these animals from the extreme cold. Table 8-1 lists organisms from various habitats and their adaptations.

Table 8-1. *Organisms and Their Adaptations*

Organism	***Habitat***	***Adaptation***	***Function***
Giraffe	Grasslands	Long neck	Helps it reach leaves on trees, its main food
Arctic hare	Arctic	White fur in winter	Provides camouflage from predators
Monkey	Rain forest	Grasping tail	Acts as an extra hand, freeing hands and feet for other uses
Cactus	Desert	Waxy skin	Reduces water loss from evaporation

Extinction

An organism will survive only if it can adjust to changes in its environment. A species that cannot adapt to major changes in its environment will become ***extinct***. Extinction occurs when a species dies out. This usually happens when some essential part of the living or nonliving environment is removed. Common natural causes of extinction include natural disasters, climate changes, and habitat invasion by a predator. Humans also cause extinction by excessive hunting and pollution of land, air, and water.

Scientists know of thousands of species that once existed but are now extinct. Evidence of these extinct organisms is found in fossils. ***Fossils*** are the remains or traces of organisms that have lived in the past. Fossils include skeletons, shells, and impressions, such as footprints, preserved in rock. Some animals that have become extinct are dinosaurs, woolly mammoths, dodo birds, and passenger pigeons.

About 65 million years ago, roughly 70 percent of all animal species, including the dinosaurs, became extinct. Scientists are not sure what caused this mass extinction. Some have suggested that there was a sudden change in climate. Perhaps an asteroid or comet collided with Earth, raising huge clouds of dust that blocked the sun, cooling Earth. Only those species that were able to adapt to these drastic changes in the environment survived. The dinosaurs could not adapt successfully and died out. The elimination of the dinosaurs helped mammals to survive, evolve, and eventually become the dominant large animals that they are today.

Many species have become extinct due to the direct interference of humans. In 1598, Portuguese sailors discovered the dodo, a large, flightless bird on an island in the Pacific Ocean. (See Figure 8-2.) These birds had no natural enemies, so they did not fear the sailors and were easy to kill. Those that survived were killed by the dogs and pigs introduced to the island by the sailors. By 1681, the dodo was completely eliminated.

Figure 8-2. *The dodo (left) is extinct; the bald eagle was once threatened but was removed from the threatened species list in the summer of 2007. It is now a protected species.*

Learning from the mistakes of the past, we now have laws protecting species that are close to extinction. Many of these ***endangered species***, such as the California condor and Florida panther, have been rescued from extinction. The American bald eagle was also once thought to be in danger of extinction, but due to the efforts of conservationists, this eagle is no longer endangered. In fact, the bald eagle was removed from the threatened species list in the summer of 2007. This is a true success story. In May of 2008, the polar bear was declared a threatened species by the Interior Department, which said that the bears must be protected because of the decline in Arctic sea ice caused by global warming.

Process Skill 1

Reading for Understanding

Although many animal species in the United States became endangered during the previous century, only a few became extinct. The efforts of many people have been successful in increasing the population of several species, such as the bison and bald eagle. One species that may have disappeared, however, is the ivory-billed woodpecker. (See below.)

Unlike the passenger pigeon, which was hunted to extinction, the ivory-billed woodpecker seldom was hunted. Yet its numbers have decreased, perhaps to zero. Scientists blame the disappearance of this beautiful woodpecker on the loss of its habitat. The ivory-billed woodpecker eats beetles that are found in decaying hardwood trees. As human populations have increased in the southeastern United States, marshlands have been "cleared" and dead and decaying logs have been removed. Without its main food source, the ivory-billed woodpecker has disappeared from the United States. Every now and then, someone claims to have seen one, which creates great excitement among biologists. But, unfortunately, there is no positive evidence, so most biologists think this great bird is extinct.

Questions

1. Based on this passage, you could conclude that efforts to save endangered species are
(1) always successful (3) always unsuccessful
(2) usually successful (4) usually unsuccessful

2. Excessive hunting caused the extinction of the
(1) bald eagle
(2) bison
(3) passenger pigeon
(4) manatee

3. The major cause for the disappearance of the ivory-billed woodpecker seems to be
(1) water pollution
(2) destruction of habitat
(3) introduction of a new species
(4) introduction of chemicals into the environment

Communities and Ecosystems

A habitat usually contains many different types of organisms that interact and may depend on one another for survival. Within the habitat, all the members of a particular species make up a ***population***. All the different populations within a habitat make up a ***community***. When you set up an aquarium containing plants, catfish, and guppies, you create a small community.

To set up an aquarium, you must provide more than just the fish and the plants. You need water, a source of oxygen, and light. You must also maintain the proper temperature. These nonliving factors together with the living members of the community make up an ***ecosystem***.

The members of the community get the materials they need to survive from the ecosystem. In return, they give materials back, such as wastes and dead, decaying bodies. Materials are constantly being recycled within an ecosystem. Figure 8-3 shows how oxygen and carbon dioxide are recycled. Most of the oxygen in our environment is provided by green plants through the process of photosynthesis. Energy, however, is not recycled and must be provided by an outside source, such as the sun.

Figure 8-3. *Oxygen and carbon dioxide are recycled constantly in an ecosystem.*

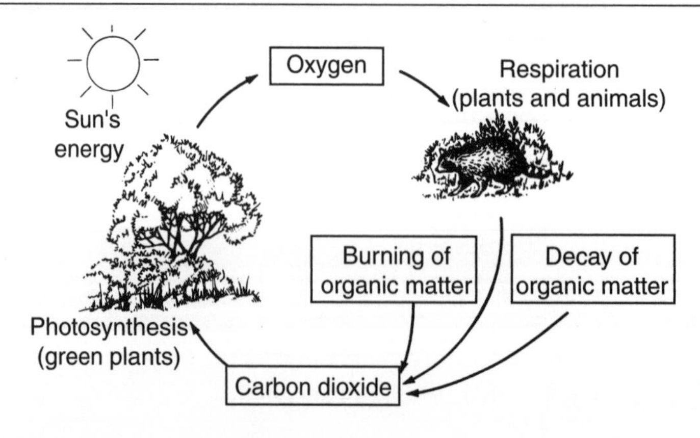

Review Questions

Part I

1. Organsims respond to changes in the environment. A change in the environment is a
   (1) stimulus
   (2) reaction
   (3) adaptation
   (4) migration
2. Which of the following is a physical adaptation?
   (1) A bird migrates in spring.
   (2) A bat hibernates in winter.
   (3) A cactus has waxy skin to hold in water.
   (4) A tree loses its leaves in the winter.
3. Many species of birds have webbed feet. Why might these different birds have the same type of feet?
   (1) They are adapted to live mainly on or near water.
   (2) They are adapted to live mainly on land and in trees.
   (3) They hibernate for the winter.
   (4) They are endangered.
4. Fur is a _______ adaptation to ______ a loss of heat.
   (1) behavioral prevent
   (2) behavioral, cause
   (3) physical, prevent
   (4) physical, cause
5. Some insect-eating birds fly to warmer areas in the winter due to a shortage of food. This behavior is called
   (1) a stimulus
   (2) hibernation
   (3) dormancy
   (4) migration
6. The complete and permanent loss of a species is called
   (1) adaptation
   (2) dormancy
   (3) extinction
   (4) migration

Use the information and the figure below to answer questions 7 and 8.

The figure below shows a saber-toothed cat and a woolly mammoth. Thick layers of fur made these animals well adapted to the cold environment of the ice age. However, both of these animals are now extinct.

7. Which of the following is most likely to cause the extinction of a species?
   (1) evolution
   (2) migration
   (3) gradual change in the environment
   (4) sudden change in the environment

8. Most of the evidence that extinct organisms such as the woolly mammoth and saber-toothed cat once existed is based on
   (1) photographs
   (2) eyewitness reports
   (3) fossils
   (4) pedigree charts

Part II

9. Explain the difference among a population, a community, and a habitat.

Base your answers to questions 10 through 16 on the reading below and your knowledge of science.

The bald eagle, penguin, and cormorant are birds that eat fish. However, they catch their prey in very different ways. The eagle flies overhead looking for fish and then dives to the surface where it reaches into the water with its sharp claws. A penguin is a very fast swimmer. It uses its wings to "fly" through the water. The penguin has heavier bones than most other birds, which help it to swim underwater. It also has a layer of fat that insulates it from the cold. Like the penguin, the cormorant swims and catches fish in its beak. However, underwater, unlike the penguin, the cormorant propels itself with its feet, not its wings. The cormorant uses its wings to fly through the air, not through the water.

The feathers of a penguin are waterproof, while the feathers of a cormorant are not. Because the feathers are not waterproof, the cormorant's wings will not trap air bubbles. This adaptation allows it to dive faster. After it hunts, the cormorant will spread its wings in the sun until they are dry.

10. Identify one way in which the eagle is adapted to its feeding habits.
11. Identify two ways in which the penguin is adapted to its feeding habits.
12. Identify one behavioral adaptation found only in the cormorant.
13. Which bird would be at a disadvantage when much of the water is covered with ice? Explain.
14. Which bird's adaptation allows it to catch larger fish? Explain.
15. What adaptation allows many species of penguins to live in very cold climates?
16. A person who can see objects at a great distance is often called an "eagle eye." Explain why this makes sense.

The Balance of Nature

Living things, organisms, interact with their environment. Organisms get food, water, and oxygen from the environment. They release wastes back into the environment. Living things depend on the environment, and the environment depends on the living things within it.

Producers and Consumers

All organisms need energy to survive. They get this energy from nutrients in food. During photosynthesis, green plants produce sugars (carbohydrates), our main source of food energy. Therefore, green plants are called ***producers***. Plant-eating animals, such as the zebra, grasshopper, and rabbit, are called ***herbivores***. The herbivores obtain energy-rich sugars when they eat and digest the plants

Meat-eating animals, called ***carnivores***, also get energy from plants, but indirectly. For instance, when a lion eats a zebra, it obtains nutrients from the meat of the zebra. The lion gets its energy from these nutrients. The zebra had gotten these nutrients from the plants it ate. Since both the lion and the zebra depend on other organisms for their food, they are called ***consumers.*** ***Omnivores*** are consumers too. Omnivores are animals that eat plants and other animals. Humans are omnivores. Every animal depends directly or indirectly on green plants for food and oxygen.

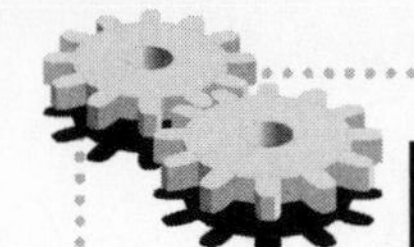

Process Skill 2

Interpreting the Results of an Experiment

Julia performed an experiment in which she planted 5 plants in sand and 5 plants in soil. All 10 plants were the same type. She gave them equal amounts of water and exposed them to equal amounts of sunlight. The experiment lasted for two weeks. The table below shows the growth of each of the plants.

Plants in Soil	*Increase in Height (centimeters)*	*Plants in Sand*	*Increase in Height (centimeters)*
1	2.0	6	0.5
2	1.9	7	0.6
3	2.2	8	0.4
4	2.1	9	0.7
5	1.9	10	0.6

Questions

1. What conclusion may be drawn from this experiment?
   (1) Plants grow just as well in soil as in sand.
   (2) Plants grow taller in sand than in soil.
   (3) Plants grow taller in soil than in sand.
2. What differences between the soil and the sand would explain the results of this experiment?
3. Which bar graph correctly represents the averaged results of this experiment?

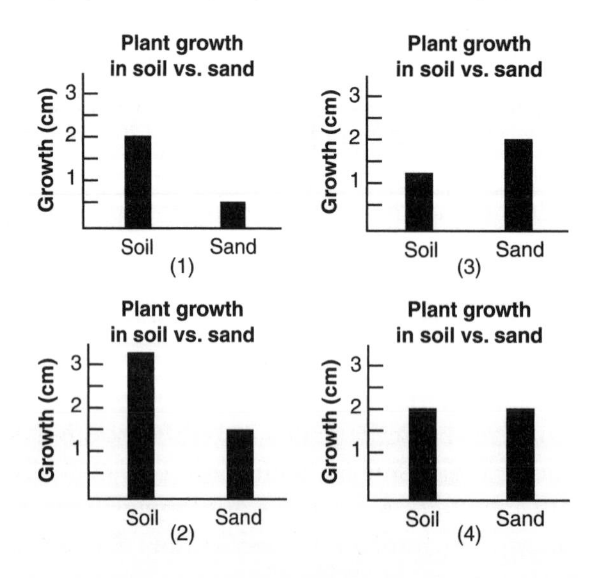

Food Chains

The nutrients in green plants get passed along from one organism to another in a sequence called a ***food chain***. Grass produces food during photosynthesis. A zebra eats the grass to get its nutrients. A lion, in turn, eats the zebra.

When the any organism dies, its body decays. Special organisms called *decomposers* break down the organism's remains and return nutrients to the soil. Plants, such as grass, can then use these nutrients.

Decomposers include fungi, such as mushrooms and molds, and some bacteria. Fungi and bacteria cannot make their own food and so depend on other living things for food. Decomposers are the last link in any food chain. Figure 8-4 shows an example of a food chain.

Figure 8-4. *A food chain.*

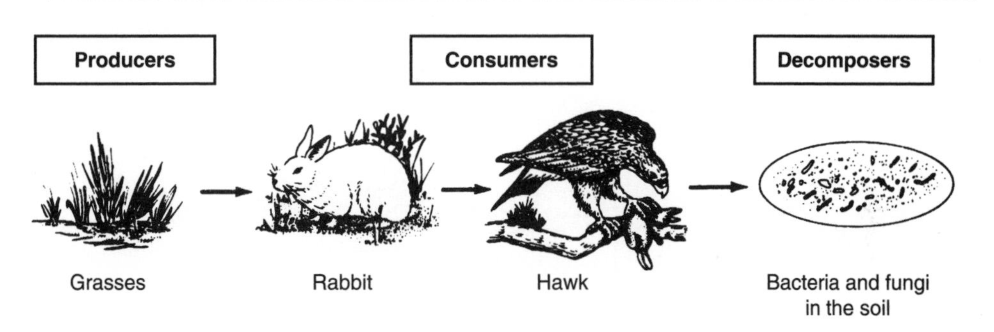

Suppose that, suddenly, there were no more zebras. How would the lion get its nutrients? A lion can also eat other animals. If something wiped out the zebra population, the lion would eat more of some other animals. Thus, the removal of one species, the zebra, would affect many other species.

Food Webs

Most ecosystems contain a number of food chains that are connected to form a ***food web***, as shown in Figure 8-5 on page 232. There is a delicate balance in an ecosystem among its producers, consumers, decomposers, and their environment. If this balance is disturbed, it could change the whole ecosystem. Visit the Web site *arcytech.org/java/population/facts_food chain.html* to learn more about food chains and food webs.

Food Pyramid

What happens to you when you exercise? Exercising causes you to "burn" a lot of food for energy. As you do this, your body temperature increases. You generate heat, which is passed to the environment. Every organism uses some of the energy it consumes and stores the rest. The energy that is used is passed to the environment in the form of heat. Only the stored energy is

Figure 8-5. *A food web consists of several interconnected food chains.*

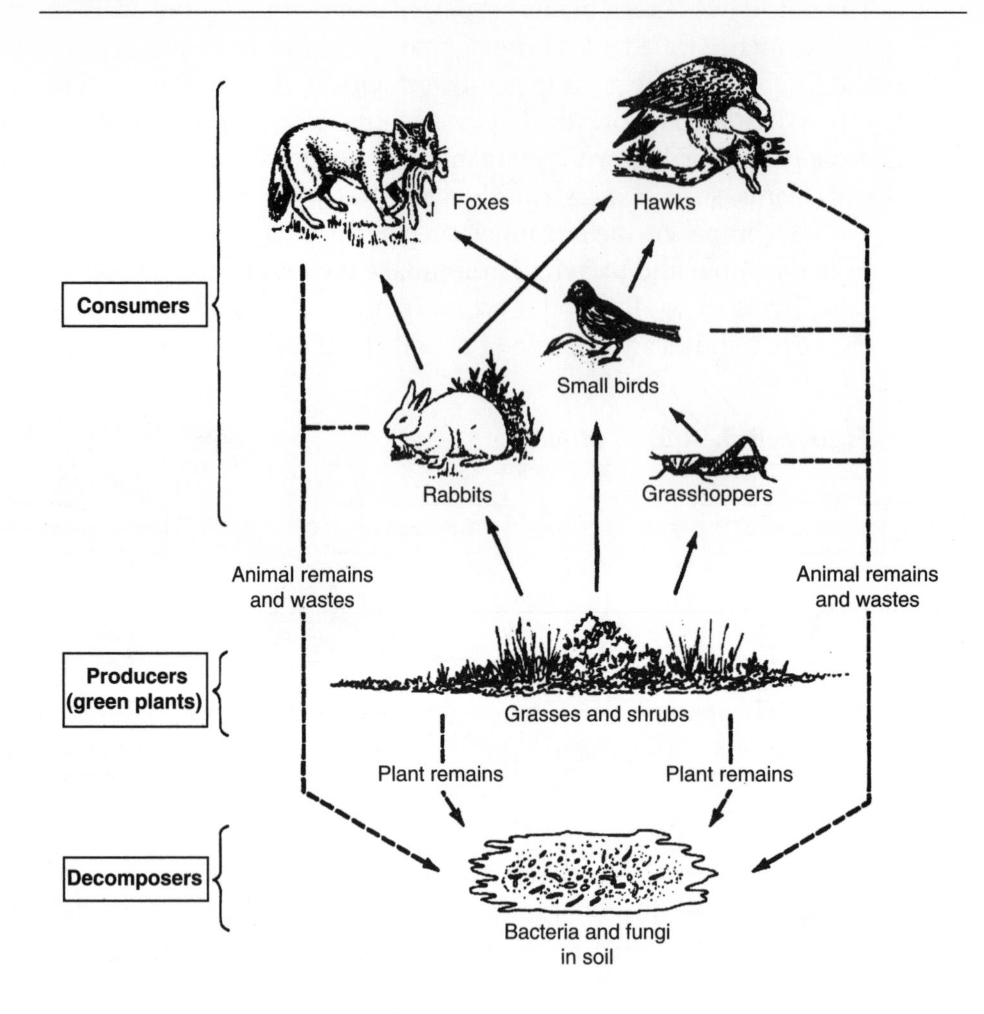

available to the next consumer in the food chain. This results in a decreasing amount of energy available at each step of the food chain. This can be represented by a pyramid, which gets smaller and smaller toward the top. Figure 8-6 illustrates an energy, or food, pyramid. A food pyramid always has green plants (producers) at its base. Green plants get their energy from the sun.

Symbiosis

When different organisms live together, they may interact in several ways. At least one organism always benefits from the relationship. The other organism, however, may or may not benefit. ***Symbiosis***, or a *symbiotic relationship*, occurs when one organism lives on or inside another one. For example, a flea living on the skin of a dog has a symbiotic relationship with the dog; this type of symbiosis is called ***parasitism*** because the flea benefits while the dog is harmed. On the other hand, bacteria living inside termites help the ter-

Figure 8-6. *A food pyramid shows the relationship of producers to consumers; the producers (plants) are always at its base.*

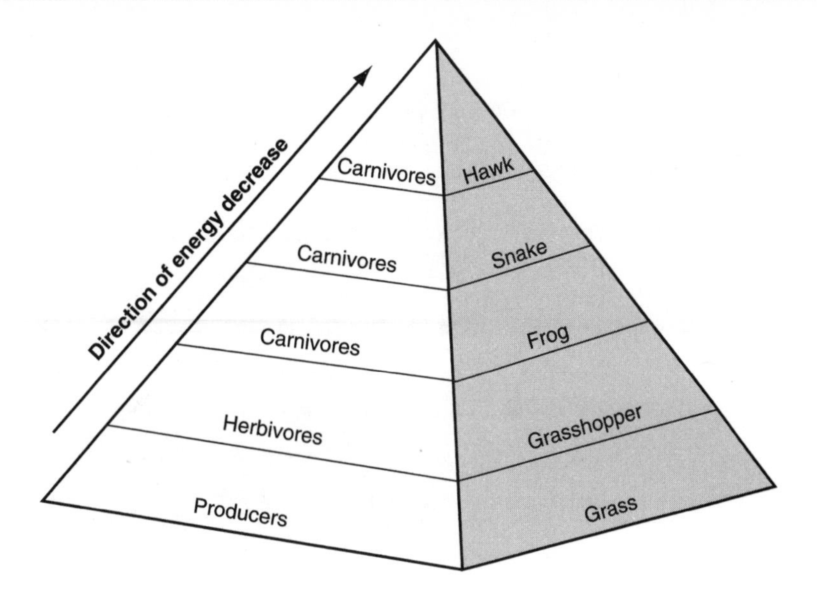

mites to digest wood. The bacteria as well as the termites benefit from this arrangement. This relationship is called ***mutualism***. In ***commensalism***, another symbiotic relationship, one organism benefits while the other is not affected. Orchids and the trees on which they grow have a commensal relationship. The tree only supplies support. The roots of the orchids absorb what they need from the air. Table 8-2 describes the three types of symbiotic relationships that may exist between organisms in a community.

Visit the Web site *www.factmonster.com/ipka/A0776202.html* to learn more about the way animals depend on one another.

Table 8-2. *Types of Symbiotic Relationships Between Organisms*

Relationship	***Description***
Mutualism	Both organisms benefit from the relationship.
Commensalism	One organism benefits, while the other is not affected.
Parasitism	One organism benefits, while the other organism is harmed.

Competition

Competition for food and space is an important part of the relationships among living things in an ecosystem. Both the moose and the snowshoe hare live in the same habitat and compete for food from the birch tree. The moose, by far the larger of the two animals, has an advantage over the hare in obtaining food. The moose can reach food high in the tree. In winter, when food is scarce, the hare is more likely to die of starvation because of competition for the limited food available. Plants also compete for space and sunlight. (See Figure 8-7.)

Figure 8-7. *Plants compete for resources, such as growing space and sunlight.*

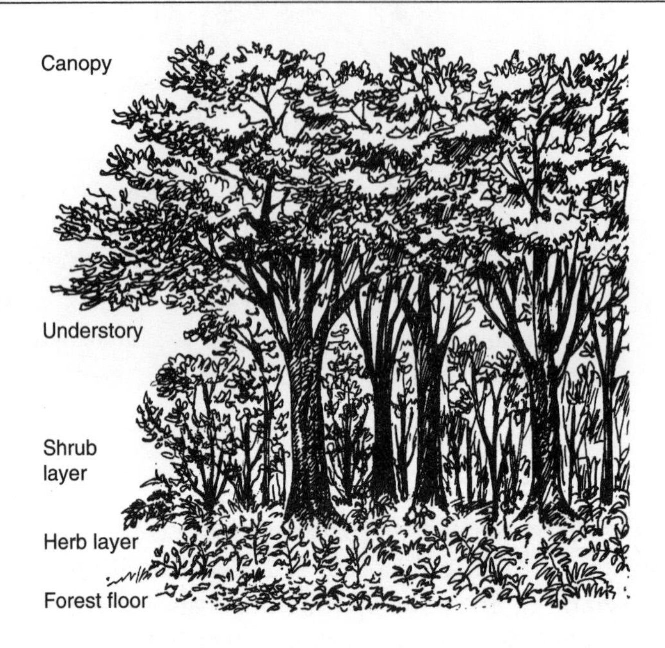

Upsetting the Ecosystem

Biodiversity describes the variety of life-forms that exist. It is important to all species that there are a large number of species living on Earth. Humans should preserve biodiversity because we depend on plants and animals for such things as food, medicine, and much of our industry.

Humans sometimes interfere with the balance of nature. For example, the early settlers in the northeastern United States killed off all the wolves in the region because the wolves killed farm animals. However, the wolves were the only natural enemies of the deer living in the area. Without the wolves to hold down their numbers, the deer population increased to the point where many deer starved to death in winter.

The actions of people are not the only things that can disturb the balance of nature. (See Table 8-3.) Sometimes, the delicate balance may be upset suddenly by natural events such as floods and forest fires. In 1980, for instance, a volcano called Mount St. Helens, in the state of Washington, erupted violently. The explosion destroyed almost 100,000 acres of forest. With time, the forest is returning to Mount St. Helens through a series of natural changes in the ecosystem. In the summer of 2000, many areas in the Northwest and Texas were badly damaged by forest fires. In time, these areas also will regrow.

Table 8-3. *Causes of Change in Ecosystems*

Gradual Changes	Sudden Change
Succession	Volcanic eruptions
Climate change	Forest fires
Human population growth	Human actions
Continental drift	Floods, meteorite impacts

Ecological Succession

After a forest fire or volcanic eruption has destroyed an ecosystem, the soil becomes enriched with minerals from the decaying remains of the plants and animals that had lived there. Soon, small new plants sprout. These become homes and food for insects and small animals. Eventually, these plants die and are replaced by other, larger plants. Each new community changes the environment, making it more suitable for the next community. Finally, a community emerges that is not replaced.

Before the volcano erupted on Mount St. Helens, there was a forest of spruce and fir trees and the animals of that community along its slopes. This community was destroyed by the eruption. Eventually, those plants and animals will return as each community prepares the environment for the next stage. The natural process by which one community is replaced by another in an orderly, predictable sequence is called ***ecological succession***. Figure 8-8 illustrates the ecological succession of a barren area into a forest.

Figure 8-8. *Ecological succession on land.*

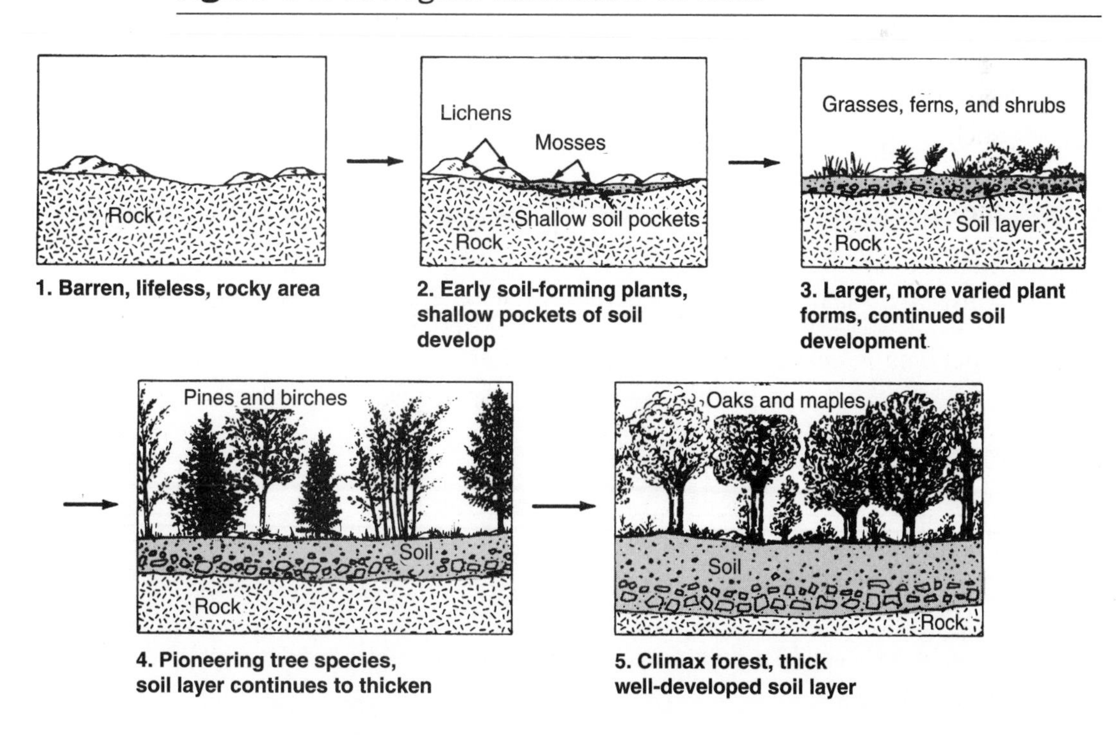

Conserving Natural Resources

A forest is an important *natural resource.* It supplies wood and oxygen, conserves soil and water, and provides a habitat for wildlife and recreation for people.

Forests that are destroyed can be replaced, although replacement takes a long time. This means the forest is a *renewable resource,* a resource that can be replaced. When plants and animals die and decay, their nutrients are returned to the soil. Therefore, soil is a renewable resource. Water, too, is a renewable resource, since it is constantly recycled through the environment (Figure 8-9). Other examples of renewable resources are wind and sunlight.

Figure 8-9. *Water is recycled constantly through the environment.*

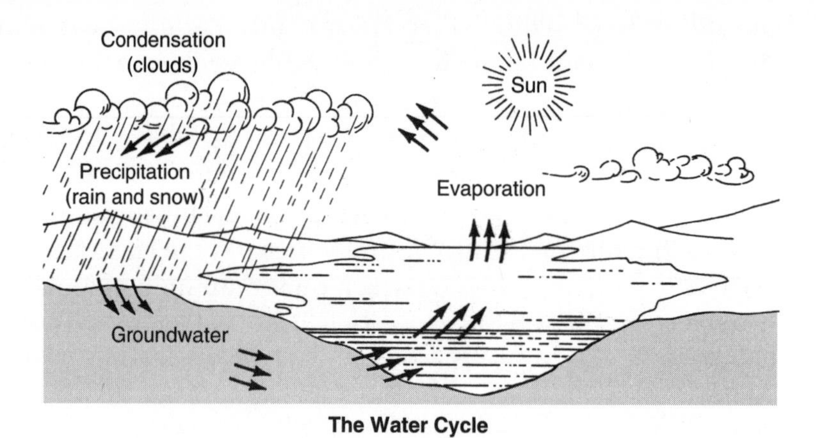

The Water Cycle

Aluminum, like other minerals, is not replenished by nature. Minerals and other materials that are not naturally replaced are *nonrenewable resources.* To guarantee enough of these valuable materials in the future, we must conserve and recycle them today.

Although nature does recycle water, soil, and forests, humans often use them up faster than nature can replace them. It is important, therefore, to conserve these resources as well, or we may have shortages of them someday.

Review Questions

Part I

17. What do all organisms need to survive?
    (1) soil
    (2) blood
    (3) carbon dioxide
    (4) energy

18. Green plants are called producers because they
    (1) produce oxygen gas
    (2) obtain energy from animals
    (3) manufacture their own food
    (4) can be decomposed by bacteria

19. Herbivores obtain energy directly from
    (1) the sun
    (2) plants
    (3) animals
    (4) decaying organisms and wastes

20. Through photosynthesis, plants change energy from the sun into sugars. These sugars are transferred to animals when they eat the plants. Animals that eat only plants are called
    (1) herbivores
    (2) carnivores
    (3) omnivores
    (4) producers

Use the diagram below to answer questions 21 through 23. The diagram represents a food chain.

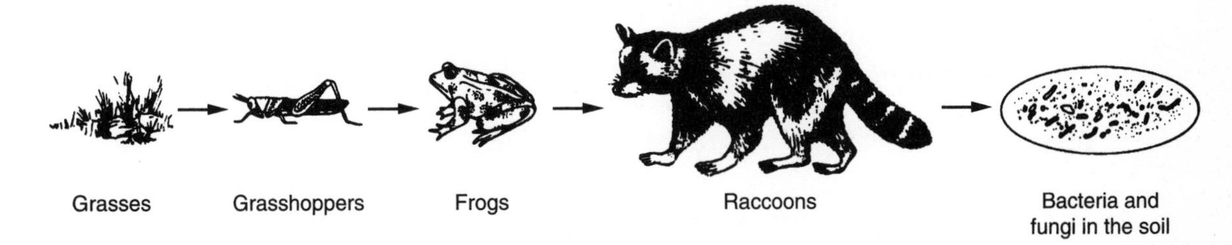

21. Which organisms are the producers in this food chain?
    (1) grasses
    (2) grasshoppers
    (3) raccoons
    (4) bacteria and fungi

22. Which organisms return nutrients to the soil where they can be used by other organisms?
    (1) grasshoppers
    (2) frogs
    (3) raccoons
    (4) bacteria and fungi

23. What is the role of the frog in this food chain?
    (1) It is a producer.
    (2) It is a decomposer.
    (3) It is a herbivore.
    (4) It is a carnivore.

Base your answers to questions 24 through 26 on the diagram below and your knowledge of science. The diagram shows a food web.

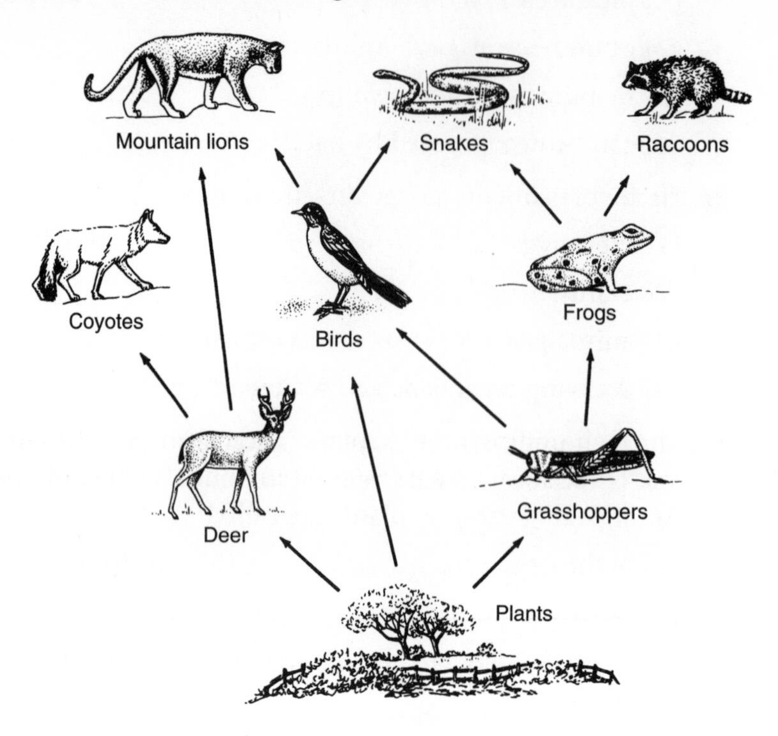

24. This food web illustrates how energy flows through an ecosystem. Which organisms do not provide energy for any other animals in this food web?

(1) plants
(2) snakes
(3) deer
(4) frogs

25. The organisms that return nutrients to the soil are not shown in this food web. These organisms are called

(1) producers
(2) carnivores
(3) decomposers
(4) omnivores

26. Which animal is an omnivore?

(1) deer
(2) mountain lion
(3) bird
(4) grasshopper

27. A natural disaster completely destroys a forest environment. The first new organisms to appear in the recovering environment would probably be the

(1) pine trees
(2) maple trees
(3) small shrubs
(4) mosses and grasses

28. Which term includes the other three?
    (1) mutualism
    (2) commensalism
    (3) symbiosis
    (4) parasitism

29. In parasitism,
    (1) both organisms are harmed
    (2) both organisms benefit
    (3) one organism benefits, and the other is not affected
    (4) one organism benefits, and the other is harmed

30. As the population of old shrubs decreases in a changing ecosystem, the population of new trees increases. The old community of shrubs
    (1) destroys the ecosystem
    (2) prepares the ecosystem for the new community
    (3) is the climax community
    (4) could not change the ecosystem

31. A natural resource that can be replaced by nature is called a renewable resource. Which of the following is not a renewable resource?
    (1) water
    (2) wood
    (3) soil
    (4) gold

Base your answer to question 32 on the paragraph and your knowledge of science.

Lice are organisms that live among human hairs. They are commonly found in children between the ages of 3 and 12. They feed on very small amounts of blood taken from the human scalp. Lice do not spread disease, but they are annoying. Their bite causes the scalp to itch. Often, scratching the bite can lead to irritation or infection. Doctors prescribe medicated shampoo to get rid of lice.

32. The relationship between lice and humans can best be described as
    (1) mutualism
    (2) parasitism
    (3) commensalism
    (4) metabolism

33. Using the diagrams below, the correct order of the stages of succession is

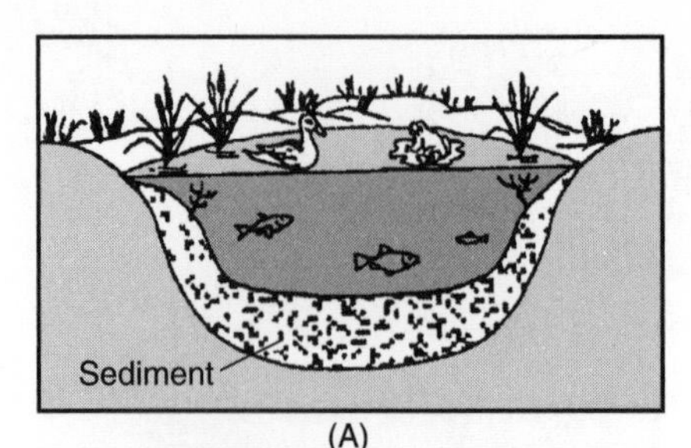

(A)

(B)

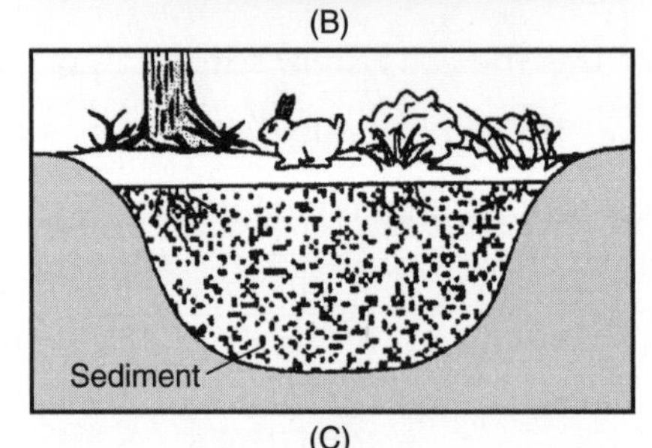

(C)

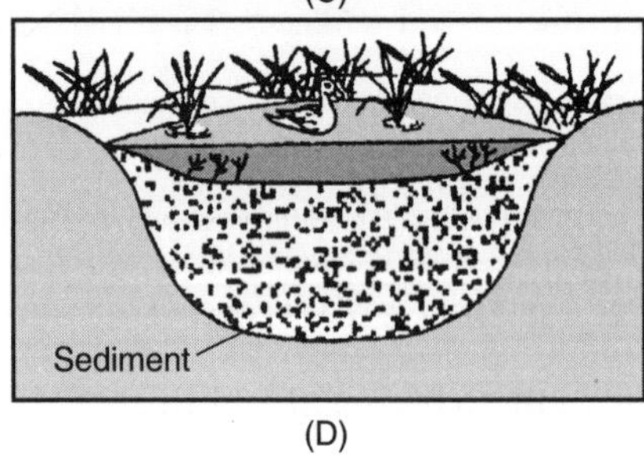

(D)

(1) B→A→D→C

(2) A→D→C→B

(3) C→B→A→D

(4) D→A→C→B

Part II

Base your answers to questions 34 through 36 on the following diagram, which shows a food web.

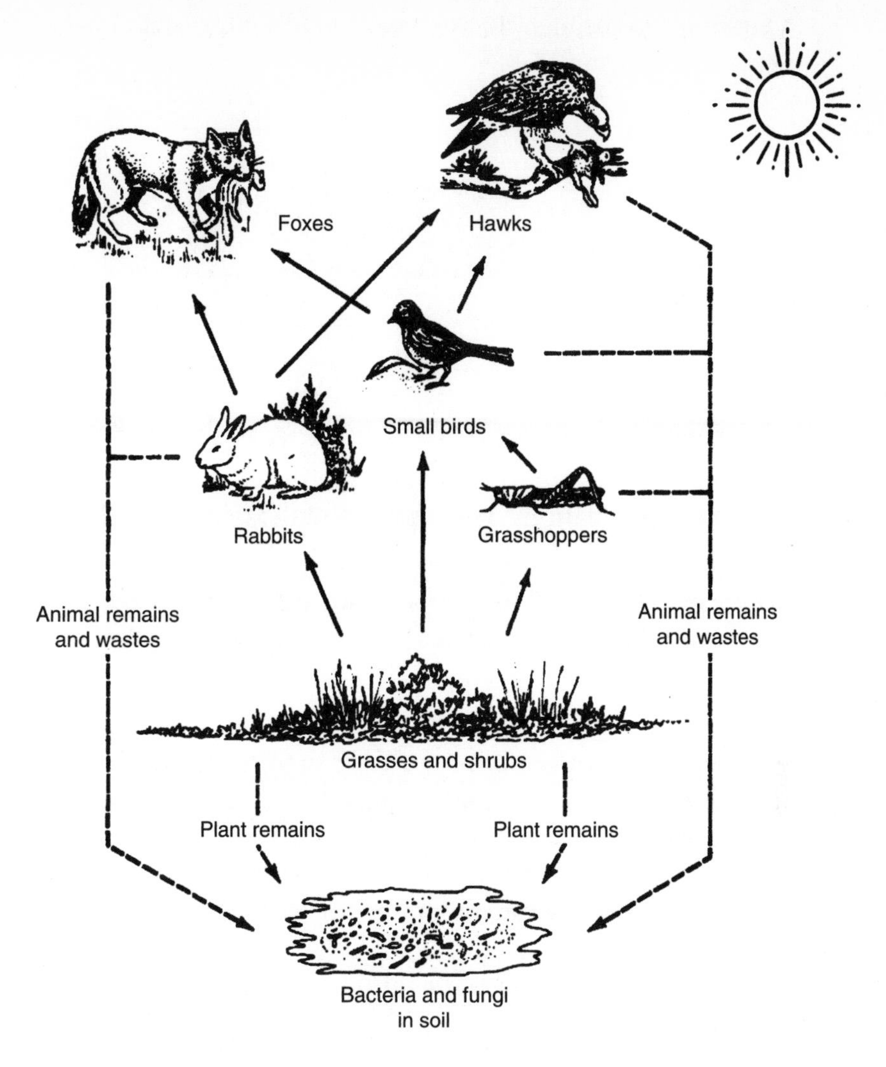

34. Which two carnivores compete for the same food sources?

35. Why are the grasses and shrubs considered "producers"?

36. If an insecticide were used to kill the grasshoppers, what would happen to the population of small birds? Explain your answer.

---

37. When asked to describe the three types of symbiosis, a student simply handed in the following three pairs of pictures. What type of symbiosis does each picture represent? Explain your answer.

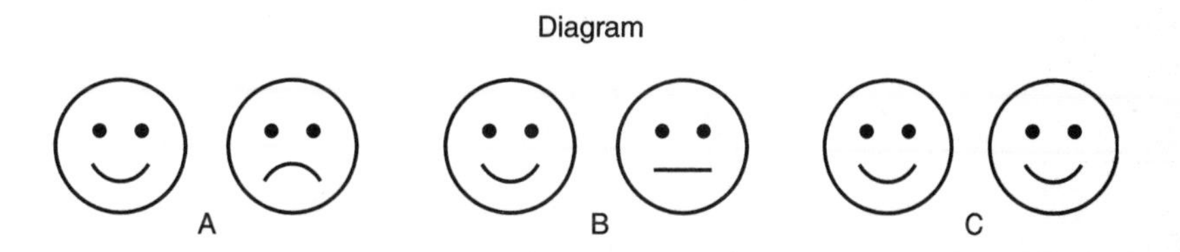

Questions 38 through 40 are based on the information below that represents a food pyramid.

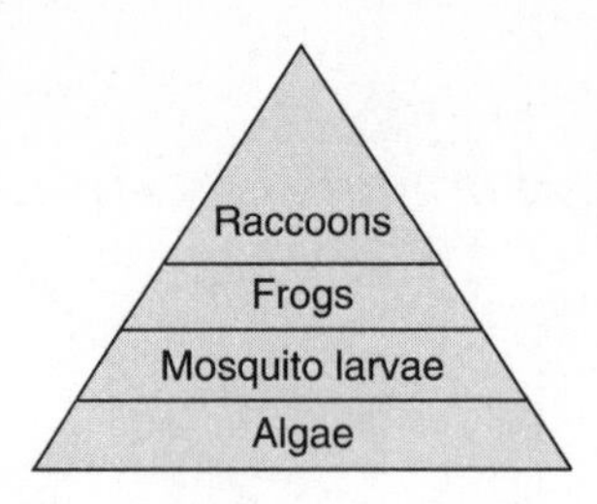

38. Which organism in the pyramid contains the greatest amount of energy?

39. What role do the mosquito larvae play in this food pyramid?

40. A disease kills most of the frogs in the ecosystem. What would happen to the populations of raccoons and mosquitoes?

Base your answers to questions 41 and 42 on the following reading passage and your knowledge of science. The reading passage describes an urban food web.

Prospect Park, in Brooklyn, New York, is home to a great many organisms that depend on one another for their survival. There are hundreds of trees and plants that provide the seeds eaten by the squirrels, mourning doves, and sparrows that live in the park. Rabbits, grasshoppers, and caterpillars eat the leaves of the plants. At least two different species of hawk are found in Prospect Park. The sharp-shinned hawk eats sparrows and grasshoppers, while the larger red-tailed hawk is able to kill and eat the larger mourning doves, as well as rabbits and squirrels.

41. Which two bird species compete for food?

42. Complete the diagram below by filling in the names of the eight animals mentioned in the reading passage in their appropriate place.

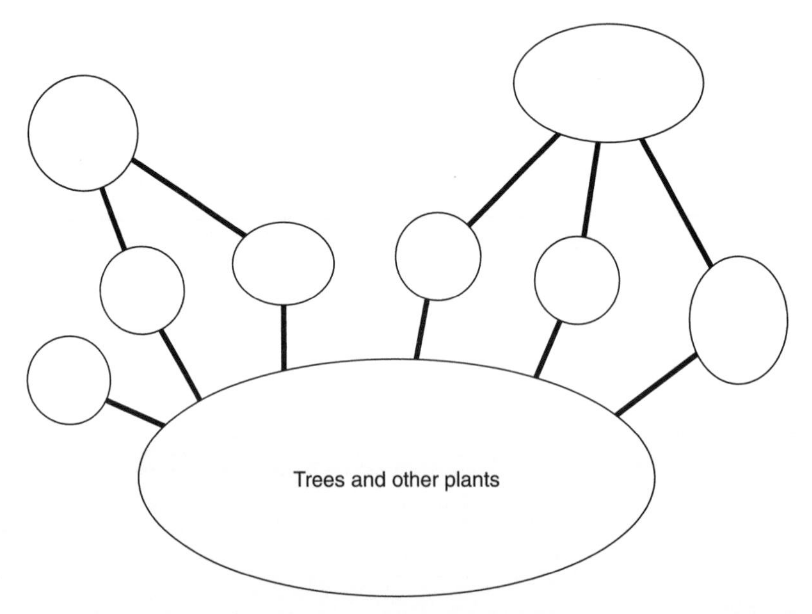

Chapter 9 Earth's Surface

Interesting rock formations are found on Earth's surface.

Vocabulary at a Glance

bedrock
chemical property
contour line
crust
igneous rock
latitude
longitude
map
metamorphic rock
mineral
physical property
profile
rock
sedimentary rock
soil
topographic map

Major Concepts

- Physical properties, such as hardness, streak color, and cleavage, and chemical properties, such as a reaction to acid, help identify minerals.
- Rocks contain one or more minerals. The three types of rocks—igneous, sedimentary, and metamorphic—are classified according to how they formed. The rock cycle shows how various processes can change rocks from one type to another.
- Fossils are the remains or traces of organisms that lived long ago. They tell us about ancient environments and climate. Fossils are almost always found in sedimentary rocks.
- Studying rock formations gives clues to the history of crustal activity in an area.
- Topographic maps are flat models that show the land surface. Contour lines show the shape and form of the land.
- We use a grid system (coordinate system) of west-east lines of latitude and north-south lines of longitude to determine positions on Earth. With a compass and a topographic map, we can accurately determine direction.

Bedrock and Soil

Earth's rocky outer layer is the ***crust***. The surface of the crust is made of bedrock, rock fragments, and soil, as shown in Figure 9-1.

Figure 9-1. *Typical section of Earth's surface showing the bedrock covered with rock fragments and soil.*

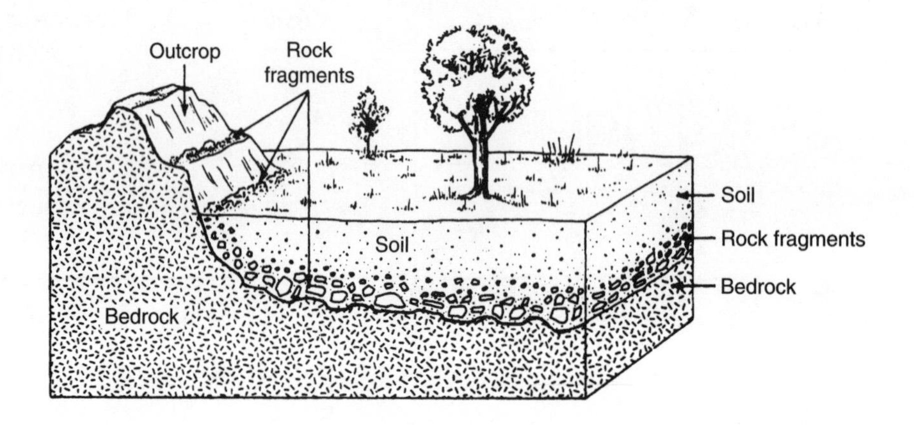

Bedrock is the solid rock part of the crust. Bedrock is visible at Earth's surface as an *outcrop*. Rock fragments (sediments) are pieces of broken-up bedrock. The pieces can range in size from giant boulders to tiny grains of sand.

Soil is a mixture of small rock fragments and *organic matter* (materials produced by living things, such as decaying leaves and animal wastes). Water and air are also important parts of soil. Soil and large rock fragments make up most of Earth's surface, with the bedrock hidden underneath.

Minerals

Rocks are made of minerals. ***Minerals*** are naturally occurring solid, inorganic (nonliving) substances. There are many minerals in Earth's crust, but only a few rock-forming minerals make up most of the crust. Feldspar is the most common mineral. Some other common minerals are quartz, mica, and calcite. Minerals have *physical* and *chemical properties* that are used to identify them.

1. ***Physical properties*** of minerals include streak color, hardness, luster, cleavage, and color.

   Streak color is the color of the powdered form of the mineral. You determine streak color by scratching the mineral on the unglazed part of a ceramic tile (streak plate). Some minerals leave no streak color.

Hardness indicates how a mineral resists being scratched. Minerals are assigned a number between 1 and 10 to indicate their hardness. A mineral with a hardness of 1 is the softest and 10 the hardest. Mohs' scale of hardness is shown in Table 9-1. It lists a representative mineral for each hardness 1–10. Hardness of some common objects is also listed. They are set apart by an asterisk (*). A mineral can be scratched only by another mineral or object with a higher number on the hardness scale.

Table 9-1. *Mohs' Scale of Hardness*

Mineral	Hardness	Mineral	Hardness
Talc	1 (softest)	Glass	5.5*
Gypsum	2	Feldspar	6
Fingernail	2.5*	Steel file	6.5*
Calcite	3	Quartz	6
Penny	3*	Topaz	7
Fluorite	4	Corundum	9
Apatite	5	Diamond	10 (hardest)
Iron nail	5*		

Luster refers to how a mineral looks when it reflects light. Some common mineral lusters are metallic, glassy, greasy, and earthy.

Cleavage is a mineral's tendency to break along smooth, flat surfaces. The number and direction of these flat surfaces are clues to a mineral's identity. Minerals that have cleavage break into characteristic shapes, as shown in Figure 9-2. Not all minerals have definite cleavage; some fracture unevenly when broken.

Figure 9-2. *Two minerals that have cleavage.*

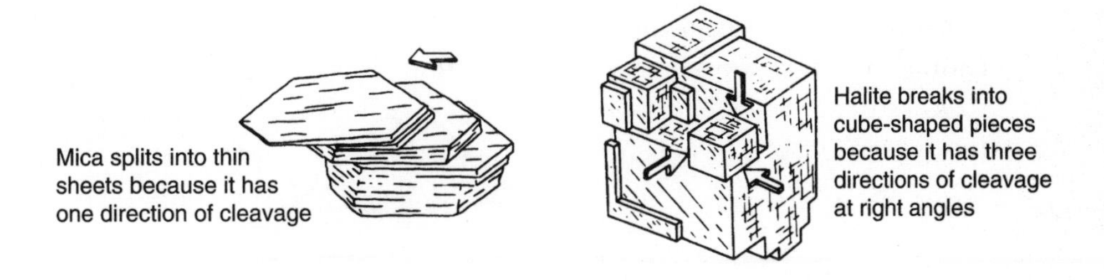

Color is not always a good way to identify a mineral. Some minerals, such as quartz, come in different colors. On the other hand, samples of different minerals may have the same color. Color is best used with other properties to identify a mineral.

2. Minerals also have ***chemical properties***, such as how they react with an acid. For example, calcite, the main mineral in limestone and marble, bubbles, or fizzes, when hydrochloric acid is placed on it. The fizzing is the result of a chemical reaction between the calcite and the acid, which produces bubbles of carbon dioxide gas. Most chemical tests are difficult to administer in the classroom.

Rocks

The ***rocks*** that form Earth's crust are natural materials composed of one or more minerals. Like minerals, rocks are identified by their physical and chemical properties. Rocks are classified into three types—igneous, sedimentary, and metamorphic—depending on how they formed.

1. ***Igneous rocks*** are formed by the cooling and hardening of hot, liquid rock, associated with volcanic activity. Melted rock material is *magma* when underground and *lava* when it pours onto Earth's surface. Extrusive igneous rocks form when lava cools quickly on Earth's surface. These volcanic igneous rocks that form from rapid cooling of lava contain tiny, sometimes invisible crystals. Basalt is a dark-colored volcanic rock made of crystals too small to be seen with the unaided eye. In fact, lava on Earth's surface may cool so rapidly that obsidian, an igneous rock with a glassy texture, forms.

   Intrusive features form when magma cools slowly underground. Igneous rocks that form underground develop large, coarse crystals. Granite is a light-colored igneous rock that contains large, easily visible mineral grains. Figure 9-3 shows processes that produce

Figure 9-3. *Igneous rocks form from hot molten rock—magma underground and lava on Earth's surface.*

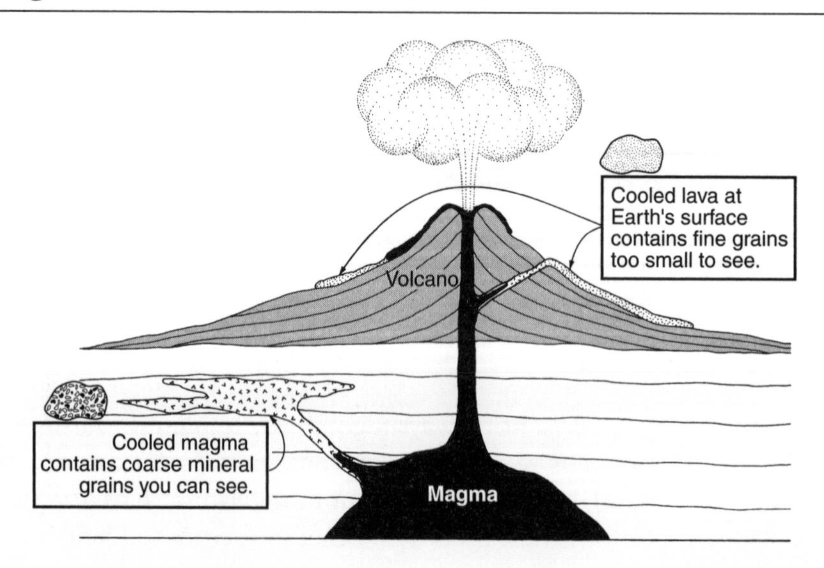

igneous rocks. Igneous rocks are identified by their color and by the size of the mineral grains (crystals) they contain.

Magma cools at different rates, depending on how far below Earth's surface it is. The closer the magma is to the surface, the smaller the grain (crystal) size of the minerals formed in the rock. This occurs because overlying rock is a good insulator and does not let heat escape. Table 9-2 lists the characteristics of some common igneous rocks.

Table 9-2. *Some Common Igneous Rocks*

Rock Name	Grain Size	Color	Where Formed
Obsidian	Glassy, no grains	Dark	Extrusive
Basalt	Less than 1 mm (fine)	Dark	Extrusive
Rhyolite	Less than 1 mm (fine)	Light	Extrusive
Granite	1–10 mm (coarse)	Light	Intrusive
Pegmatite	Greater than 10 mm (very coarse)	Light	Intrusive

2. ***Sedimentary rocks*** are made of particles called *sediments* that pile up in layers. Sediments are small rock or seashell fragments that are carried by wind or water and eventually deposited. Sedimentary rocks usually form near or under water. When entering an ocean or lake, a river or stream loses energy, slows, and drops the sediments it is carrying. It drops the largest sediments first and, as the water's flow continues to slow, it drops smaller and smaller particles. As sediments accumulate for millions of years, they become buried and harden into sedimentary rocks. (See Figure 9-4.) Table 9-3 lists some common sedimentary rocks.

Figure 9-4. *Most sedimentary rocks form from particles settling out of water.*

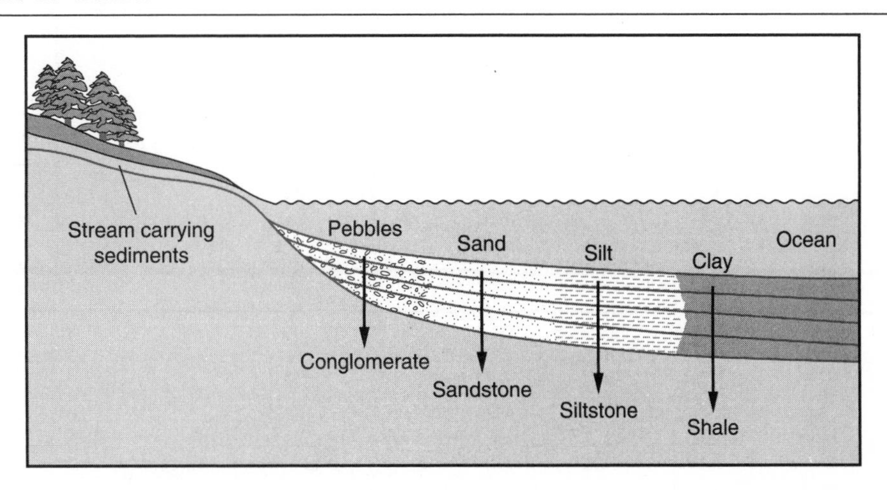

Table 9-3. *Some Common Sedimentary Rocks*

Rock Name	Description/Particles
Shale	Sheets of tightly packed clay particles
Siltstone	Powdery grains of cemented silt particles
Sandstone	Cemented grains of sand
Conglomerate	Visible cemented rounded pebbles
Limestone	Calcite from tiny seashell particles

3. ***Metamorphic rocks*** form when either igneous or sedimentary rocks are changed by heat, pressure, or both. This can happen when magma heats rocks it touches or when forces deep underground squeeze rocks for long periods of time. The high temperatures and pressures this creates change the appearance and mineral composition of the rocks, changing them into metamorphic rocks. (See Figure 9-5.)

Figure 9-5. *Metamorphic rocks form from heat, pressure, or both deep underground.*

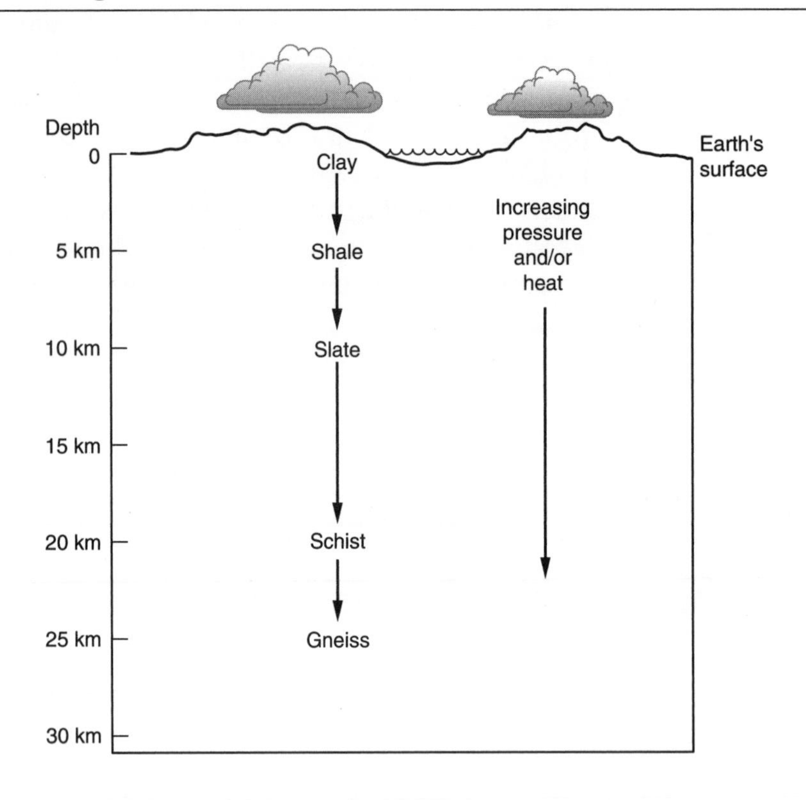

Marble and slate are metamorphic rocks formed from sedimentary rocks. Marble is formed from limestone. Slate is formed from shale. Gneiss (pronounced "nice") is a metamorphic rock that can form from granite, an igneous rock. Table 9.4 lists the characteristics of some common metamorphic rocks.

Table 9-4. *Some Common Metamorphic Rocks*

Original Rock	**Change in Rock**	**Metamorphic Rock**
Shale (sedimentary)	Compacted, does not easily split	Slate
Shale (sedimentary) or granite (igneous)	Flattened grains, light and dark bands	Gneiss
Sandstone (sedimentary)	Harder and denser	Quartzite
Limestone (sedimentary)	Harder and denser, crystals form	Marble

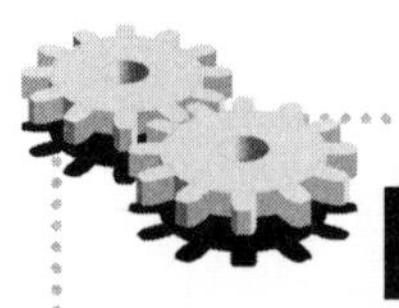

Laboratory Skill

Identifying Minerals Using a Flowchart

Each mineral has a set of physical and chemical properties that can be used to identify it. Properties are best determined from a mineral that has a fresh, clean surface. If a specimen's surface is dirty or oxidized, you may need to break the specimen to obtain a clean surface. Mineral identification tables organize these properties to make it easier to identify the mineral.

The following steps describe the process of identifying an unknown mineral.

1. Streak color: Scratch the mineral on the unglazed part of a ceramic tile. If you see a streak on the tile, record the color of the streak. If no streak appears on the tile, the specimen does not have a streak color.
2. Color: Determine the color of the specimen. Minerals that are black, dark green, brown, red, or blue are considered dark colored; otherwise they are light colored.
3. Hardness: Using simple tools such as a glass plate and a carpenter's nail (the hardness of each is between 5 and 6), determine if the mineral is hard (can it scratch the glass plate?) or soft (can it be scratched by the nail?).
4. Cleavage or fracture: This is best determined by breaking the specimen and observing how it breaks. Generally, if the specimen shows flat surfaces, corners, or edges, it has cleavage; otherwise, it has fracture.
5. After you determine the properties of the mineral, use the correct table, A or B, on pages 250–251 to identify your specimen. If the specimen has no streak, use Table A. If it has a streak, use Table B. To use Table A, determine whether the specimen is light or dark colored; hard or soft; has cleavage

or fracture; and where it fits into the colors and properties listed. To use Table B, first determine the streak color. Then decide whether it is hard or soft, has cleavage or fracture, and if it fits into the colors and properties listed.

6. Matching the specimen with known minerals, pictures of minerals, and more extensive descriptions can help assure final identification. Pictures of minerals can be found at *www.webmineral.com/specimens.shtml.*

Table A. *No Streak*

Mineral Color	Hard/ Soft	Cleavage/ Fracture	Common Colors/Properties	Mineral Name
Light colored	Hard	Cleavage	Gray, white, pinkish	Feldspar
		Fracture	White; looks waxy	Milky quartz
			Pink; looks glassy to waxy	Rose quartz
	Soft	Cleavage	Very soft, soapy feel	Talc
			White, soft, scratch with fingernail	Gypsum
			Clear, salty taste (do not taste)	Halite
			Colorless; thin sheets peel easily	Muscovite mica
			White to gray; bubbles in weak HCl acid	Calcite
		Fracture	Tan, earthy	Bauxite
Dark colored	Hard	Cleavage	Black, elongated grains; hardness 5–6	Hornblende
		Fracture	Gray to black; glassy to waxy	Smoky quartz
			Red, brown, yellow; dull or waxy	Jasper
			Black, gray; dull or waxy	Flint
			Red, brown, green; looks glassy	Garnet
	Soft	Cleavage	Black, brown; thin sheets peel easily	Biotite mica
		Fracture	Green to black; soapy feel	Serpentine

Table B. *Shows Streak*

Streak Color	Hard/Soft	Cleavage/Fracture	Common Colors/Properties	Mineral Name
Black/dark green/gray	Hard	Cleavage	No common minerals	
		Fracture	Black, magnetic (it sticks to magnet)	Magnetite
			Brassy, yellow, looks metallic	Pyrite
	Soft	Cleavage	Metallic silver; cubes	Galena
		Fracture	Black to gray; marks paper; feels greasy	Graphite
Red/brown	Hard	Cleavage	No common minerals	
		Fracture	Red to brown; hardness 5–6	Hematite
	Soft	Cleavage	Yellow, brown; looks glassy to waxy	Serpentine
		Fracture	Yellow-brown; earthy luster	Limonite
			Silver-gray; tiny flakes; looks metallic	Specularite
Green	Soft	Fracture	Bright green; looks earthy, with azurite	Malachite
Blue	Soft	Fracture	Bright blue; looks earthy, with malachite	Azurite

Questions

1. How can you distinguish between halite and calcite?
2. An unknown mineral sample scratches a glass plate but cannot be scratched by a carpenter's nail. For identification purposes, how is the mineral's hardness described?
3. Mary has two unknown minerals (A and B) that she wants to identify. She observes the following characteristics of the minerals:

Characteristics	*Mineral A*	*Mineral B*
Color	Dark brown	Black
Streak	Brown streak	No streak
Hardness	Scratches a glass plate	Scratches a glass plate
Description	No cleavage	Waxy luster

What are the names of minerals A and B?

Process Skill 1

How Do Rocks Change from One Type to Another Type?

Natural processes can change the three types of rock—igneous, sedimentary, and metamorphic—into a new type of rock.

- Igneous rocks form when heat deep within Earth's crust melts rock into a liquid that then cools and hardens.
- Sedimentary rocks form when weathering and erosion of rock produce sediments that become buried and cemented together.
- Metamorphic rocks form when heat, pressure, or both within Earth's crust affect rock and change its chemical composition and structure.
- There is no preferred path in the rock cycle. For example, igneous rock can change into sedimentary, metamorphic, or even a different igneous rock, depending on what happens to it.

Basically, over long periods of time, old rock material is recycled into new rock material. These changes and processes make up the rock cycle shown in the diagram below. Study the diagram and answer the questions that follow.

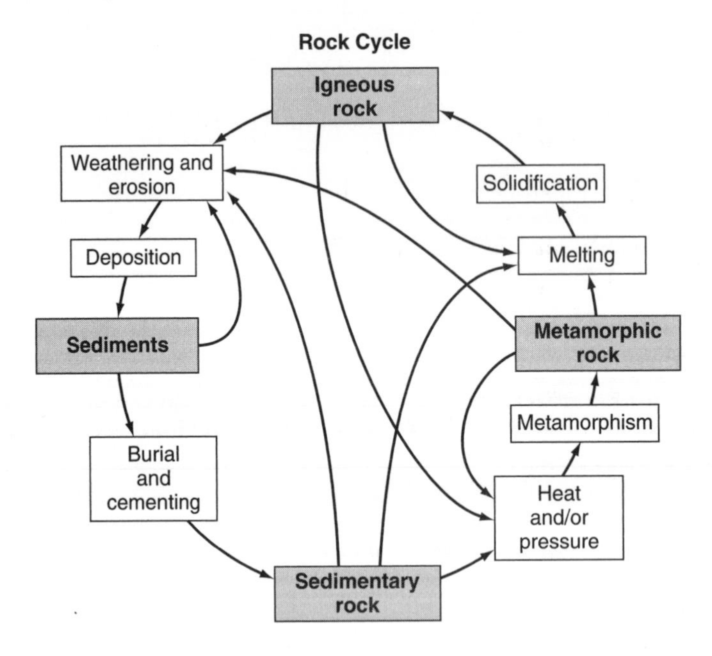

Questions

1. The processes necessary to change a metamorphic rock into a sedimentary rock are
(1) weathering and erosion, deposition, burial and cementation
(2) melting and solidification

(3) heat and pressure
(4) none of the above

2. Which of these sequences *cannot happen*?
(1) igneous rock → heat and/or pressure → metamorphic rock
(2) igneous rock → weathering and erosion → deposition → burial → sedimentary rock
(3) igneous rock → melting and solidification → igneous rock
(4) all three sequences are possible

3. Using the rock cycle diagram, how many years does it take an igneous rock to become a sedimentary rock?
(1) 100 years (3) 1,000,000 years
(2) 500 years (4) cannot be determined

4. What type of rock is formed by the solidification of a liquid rock mixture?

5. Give two possible series of processes that could change sedimentary rocks into igneous rocks.

Review Questions

Part I

1. When driving through the mountains, you may see exposed bedrock. These rock exposures are called
(1) mountain cliffs (3) outcrops
(2) giant boulders (4) rock fragments

2. The table indicates the hardness of six minerals. Which mineral is hard enough to scratch calcite but will not scratch quartz?

Mineral	*Hardness*
Quartz	7
Fluorite	4
Calcite	3
Corundum	9
Gypsum	2
Diamond	10

(1) gypsum (3) fluorite
(2) corundum (4) diamond

3. Which rocks form deep underground from heat, pressure, or both?
(1) extrusive igneous rocks
(2) volcanic rocks

(3) sedimentary rocks

(4) metamorphic rocks

4. Which process shows a chemical property that helps you identify a mineral?

   (1) reflecting light from the surface of a mineral

   (2) placing a drop of acid on the mineral causing it to fizz, or bubble

   (3) breaking a mineral

   (4) observing the color of a mineral

5. Sandstone is a sedimentary rock. This means it was formed by

   (1) cooling and hardening of magma

   (2) great heat or pressure, or both

   (3) particles settling in water

   (4) cemented clay particles

6. The rocks that formed this mountain were produced by

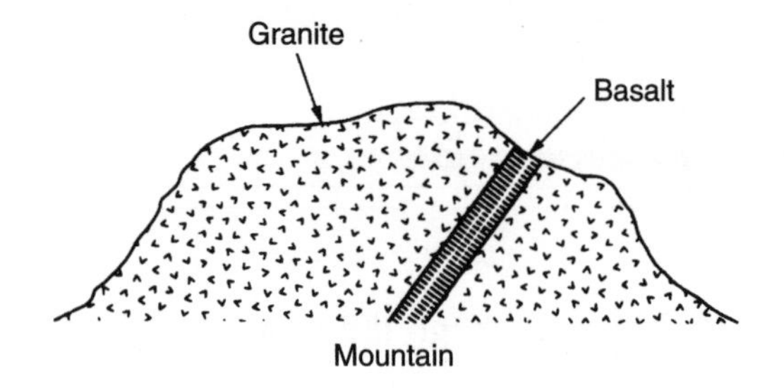

   (1) volcanic activity

   (2) particles settling in deep water

   (3) particles settling in shallow water

   (4) heat, pressure, or both underground

7. The diagrams show magnified crystals of four igneous rocks. Which igneous rock most likely formed from magma that cooled very quickly and closest to Earth's surface?

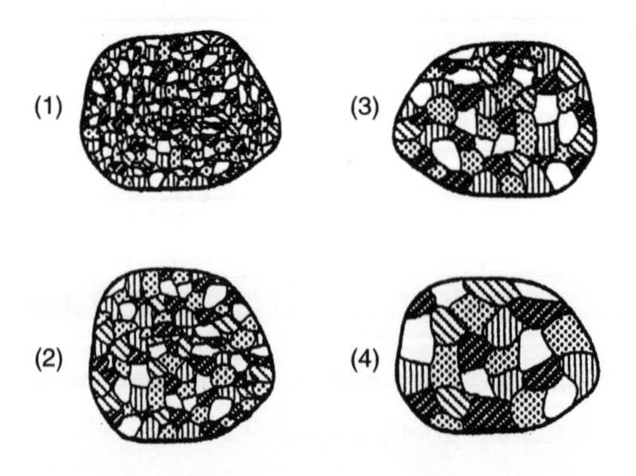

8. The diagram below is the rock cycle.

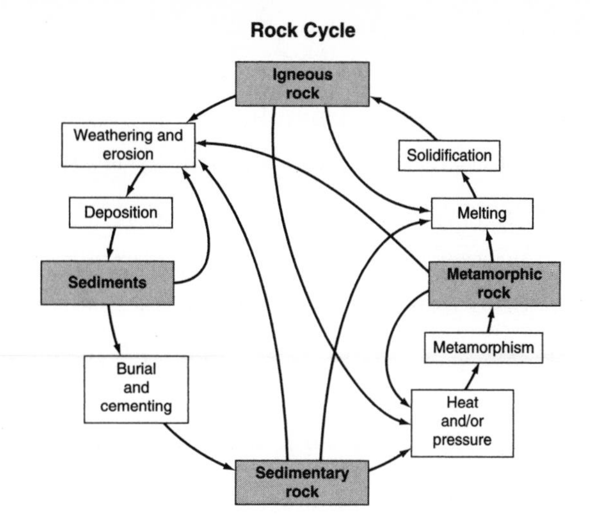

Which processes form igneous rocks?

(1) melting and solidification

(2) sedimentation and evaporation

(3) crystallization and cementation

(4) compression and precipitation

Part II

Base your answers to questions 9 through 12 on the descriptions of the four different rock specimens listed below. Use the mineral identification table to determine the name of each specimen.

Specimen 1: Does not react to acid, white color, and can easily be scratched by specimen 3.

Specimen 2: Reacts to acid, white color, and can easily be scratched by specimen 3.

Specimen 3: Does not react to acid, white color, none of the other specimens will scratch it.

Specimen 4: Does not react to acid, white color, can be scratched only by specimen 3.

Mineral Identification Table

Mineral	*Hardness*	*Acid Test*	*Common Color*
Quartz	7	No reaction	White
Calcite	3	Reaction	White
Gypsum	2	No reaction	White
Feldspar	6	No reaction	White

9. What is the name of specimen 1?

10. What is the name of specimen 2?

11. What is the name of specimen 3?

12. What is the name of specimen 4?

---

13. Describe how you determine a mineral's streak.

14. Describe how you determine a mineral's hardness.

15. Describe how you determine a mineral's cleavage.

Base your answers to questions 16 through 18 on the table below and your knowledge of science. The table shows the characteristics of the minerals quartz and calcite.

	Common Color	*Hardness*	*Cleavage/ Fracture*	*Other*
Quartz	White	7	Fracture	
Calcite	White	3	Cleavage	Bubbles with acid

16. Given only a sample of each mineral, how could you tell which one is quartz and which one is calcite?

17. Which mineral can scratch a piece of glass (hardness of glass is 5.5)?

18. Which mineral would most likely have flat surfaces and straight edges if you broke it?

Base your answers to questions 19 and 20 on the diagram below and your knowledge of science. The diagram shows a cross section of a volcanic mountain.

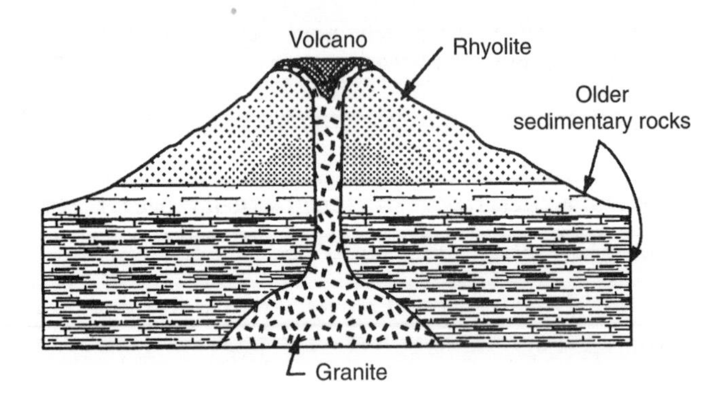

19. Rhyolite and granite have the same chemical composition and are formed from the same type of liquid rock. Why is the grain size of rhyolite less than 1 mm, while the grain size of granite is greater than 1 mm?

20. Describe what will happen to the volcano if it stops erupting.

Base your answers to questions 21 through 23 on the diagram on page 257 and your knowledge of science. The diagram shows some layers of rocks.

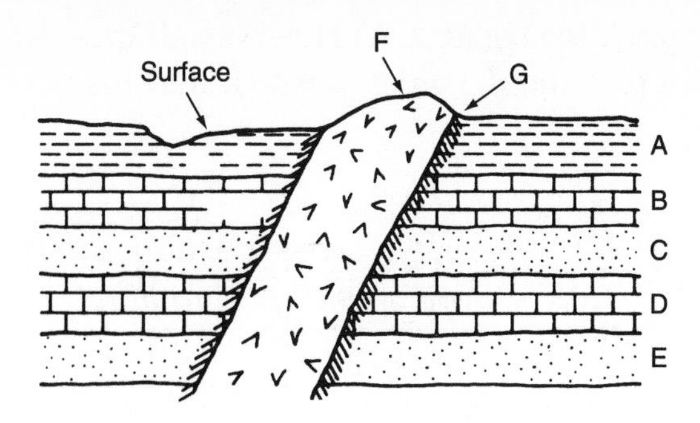

21. What type of rocks are layers A, B, C, D, and E?
22. Rock F is an igneous rock. How did rock layer F form?
23. What type of rock would you expect to find at G, the heated zone where the igneous rock touched the other rocks?

Earth History

Interpreting Rocks

Scientists have pieced together much of Earth's history by studying rocks all over the world. The rocks in an area contain information about that area's past. For example, sedimentary rocks indicate that an area was probably once covered by water. Fossils in sedimentary rocks tell of ancient living things and the environments in which they lived.

Scientists can interpret clues in rocks that tell the order in which they were formed. Horizontally layered sedimentary rocks are easiest to interpret. The bottom layers were laid down first and are therefore the oldest. The top layers were laid down last and are therefore the youngest. (See Figure 9-6.)

Figure 9-6. *In a stack of sedimentary rocks, the oldest layers are at the bottom and the youngest at the top.*

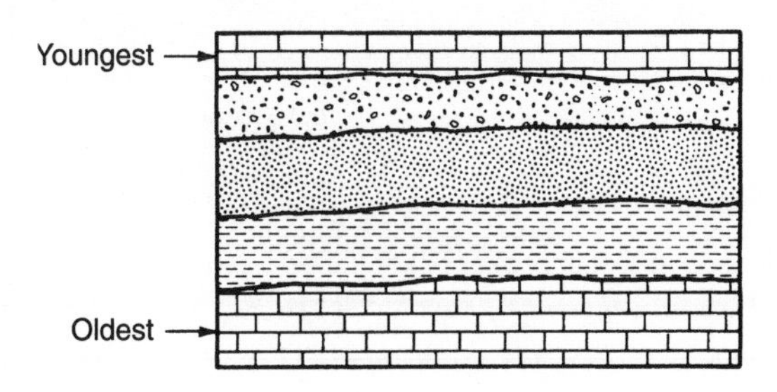

This simple situation is often changed by later events. Folding and faulting of rock layers sometimes cause these layers to be *overturned* (turned over). This makes it difficult to tell which rocks are the oldest and which are the youngest. (See Figure 9-7.) However, all these features are clues to events in Earth's past and the order in which they happened.

Figure 9-7. *A geologic cross section showing a complex history of folding, erosion, volcanism, and faulting.*

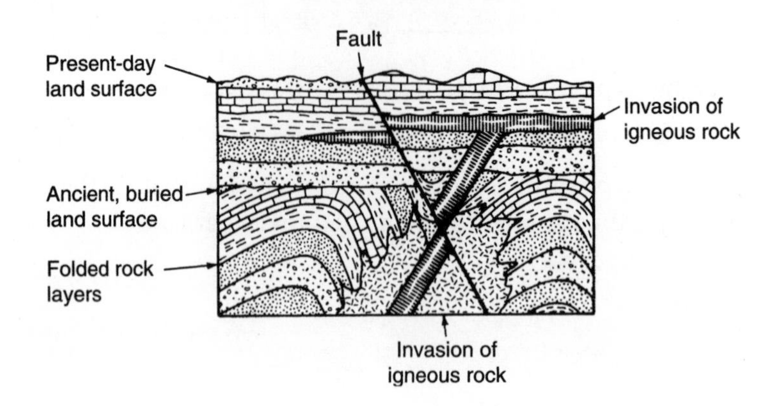

Fossils

Fossils are the remains or traces of organisms that lived long ago. Figure 9-8 shows several types of fossils. Fossils are formed when a dead plant or

Figure 9-8*. Five fossils found in New York State.*

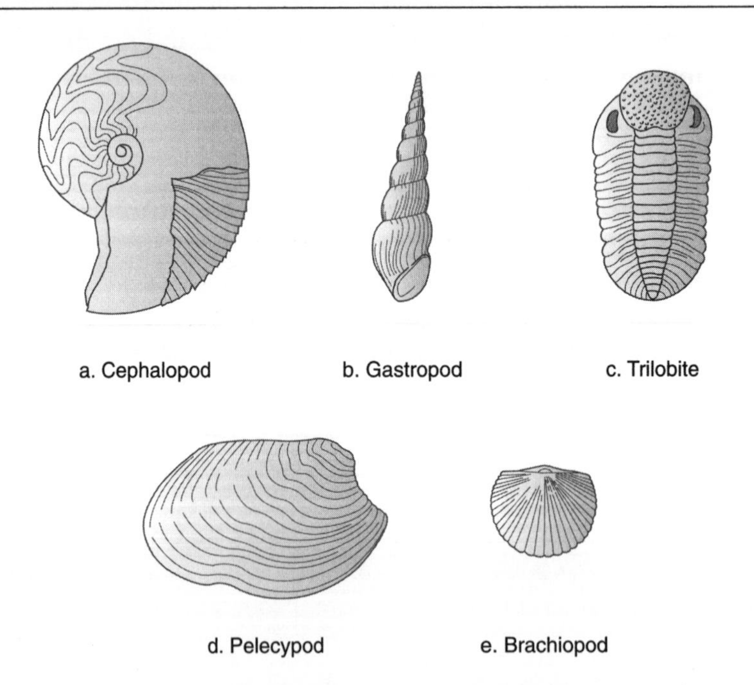

animal, or some trace, like a footprint in mud, is covered by sediment that later hardens into rock. Almost all fossils are found in sedimentary rock.

Scientists have learned much about Earth's past by studying fossils. Fossil evidence has helped scientists trace the evolution of life from simple ancient organisms to complex present-day life-forms. Fossils also provide clues to ancient environments. For example, corals live only in warm, shallow, sun-filled waters. Finding fossil corals in central New York State suggests that the area was once covered by a warm, shallow sea.

Fossils can sometimes be used to match up rock layers that are far apart. Finding the same group of fossils in rock layers at separate locations indicates that those layers formed at the same time. (See Figure 9-9.)

Figure 9-9. *Fossils can be used to match distant rock layers. They can be used to determine that rock layers at separate locations formed at the same time.*

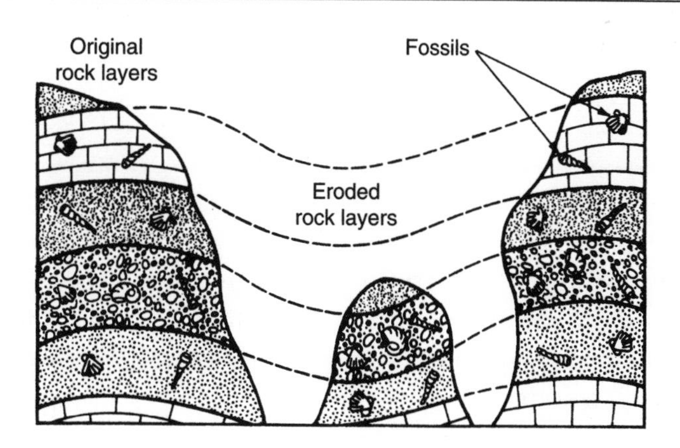

Dating Rocks

The relationships among rock layers, fossils, folds, faults, and intrusions of igneous rock can indicate the order of events in Earth's past. However, they cannot reveal the actual age of the rocks. To determine the age of rocks, scientists use a technique called *radioactive dating.*

Most rocks contain small amounts of radioactive substances that change (break down or decay) into nonradioactive substances at a known rate. For example, radioactive *uranium* changes into lead at a known rate. By measuring and comparing the amounts of uranium and lead in a rock, the age of the rock can be determined.

Using this technique, scientists have been able to assign dates to major events in Earth's history, such as periods of mountain building, the formation of oceans, and the appearance of various life-forms. Scientists estimate that Earth is about four and a half billion years old.

Process Skill 2

Determining the Sequence of Geologic Events

Many events from Earth's past are recorded in rocks. By examining rock features, such as bends (folds), large cracks (faults), igneous rock intrusions (magma injected into pre-existing rock), and eroded surfaces, the sequence of events that produced present-day rock structures can often be determined.

For example, observe the rock structure in Diagram 1 below. Notice that the igneous rock cuts across the sedimentary rock layers. For this to happen, the sedimentary layers must have already been there. Therefore, the sedimentary rocks were formed first.

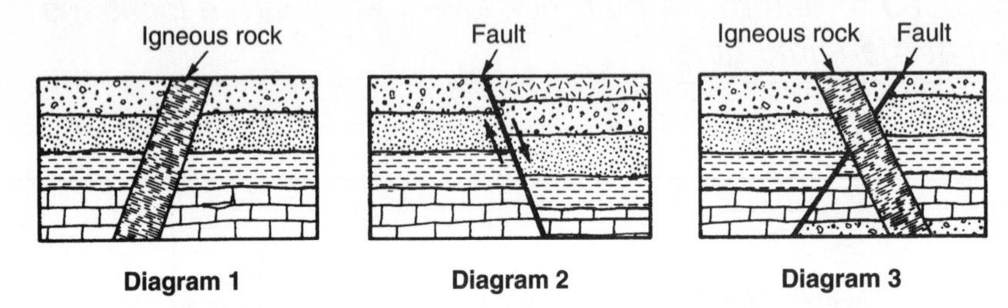

In Diagram 2, the fault has moved the sedimentary rock layers so that they do not match up across the fault. This means that the sedimentary layers were formed first. The direction in which matching layers have been shifted indicates that the rocks on the right side of the fault have moved downward in relation to the rocks on the left side, as shown by the arrows.

Diagram 3 is more complex. The fault has moved the sedimentary layers, so the sedimentary rocks must have formed before the fault. The igneous rock cuts across the sedimentary layers, so the sedimentary rocks must have formed before the igneous rock. But the fault has not moved the igneous rock, so the igneous rock must have formed after the fault. The order of formation here is: sedimentary rocks, fault, and igneous rock. Examine Diagram 4 and answer the questions below.

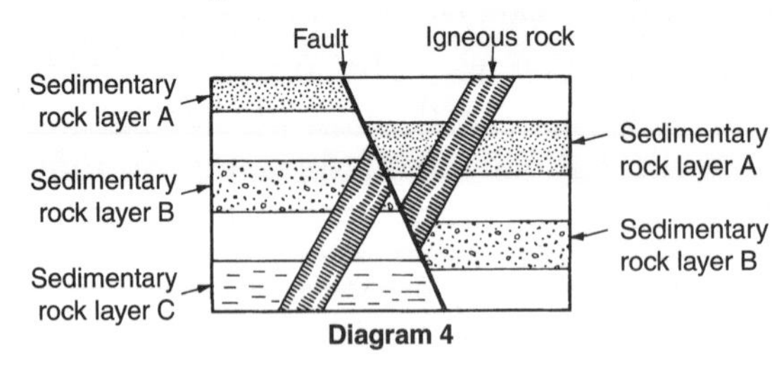

Questions

1. Which rock layer formed first? How do you know?

2. Which event occurred second: faulting or igneous rock intrusion? How do you know?

3. List in order the three geologic events indicated by this rock diagram.

Review Questions

Part I

24. Fossils in rock formations can help geologists determine

    (1) how the rocks formed

    (2) the absolute age of the rocks

    (3) the type of rocks

    (4) information about past environments

25. Becky identified a rock outcrop near her school as gneiss, a metamorphic rock. What does this suggest about the region's past?

    (1) Underground volcanic activity once took place.

    (2) Surface volcanic activity once took place.

    (3) The area was once underwater.

    (4) Great heat or pressure once affected the rocks.

26. Similar fossils are found in two locations about 50 km apart. What does this tell us about the two locations?

    (1) The rocks must be identical.

    (2) The rocks formed at the same time.

    (3) The rock locations were once closer together.

    (4) The rocks formed at the same depth.

27. The diagram shows layers of sediments deposited in a body of water. How do you know layer D formed first?

    (1) Layer D is shale.

    (2) Layer D is sedimentary rock.

    (3) Layer D is metamorphic rock.

    (4) Layer D is on the bottom of the layers.

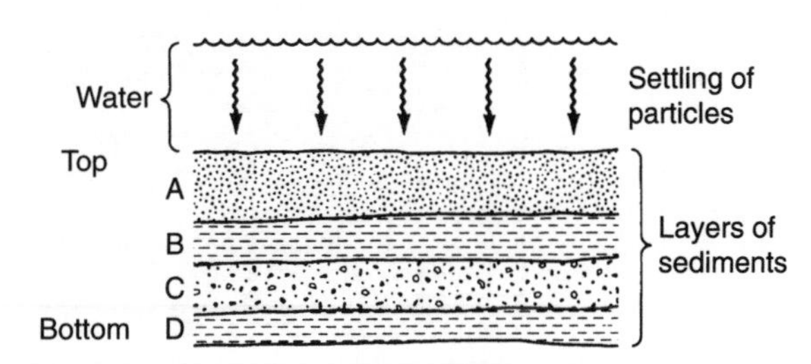

28. A rock that contains fossil seashells was most likely formed as a result of

    (1) volcanic activity

    (2) sedimentation

    (3) heat and pressure

    (4) magma cooling

29. Evan found a coral fossil in the bedrock near his home. Which statement about the land around his home is most likely correct?

(1) The land was once under deep, cold water.

(2) The land was once under deep, warm water.

(3) The land was once under shallow, cold water.

(4) The land was once under shallow, warm water.

30. To determine the age of a rock, it must contain

(1) radioactive material (3) fossils

(2) quartz (4) more than one mineral

Part II

Base your answers to questions 31 through 33 on the diagram below and your knowledge of science. A rock has fossils of three different ancient species: cephalopods, gastropods, and brachiopods. The table shows the time ranges during which these organisms lived.

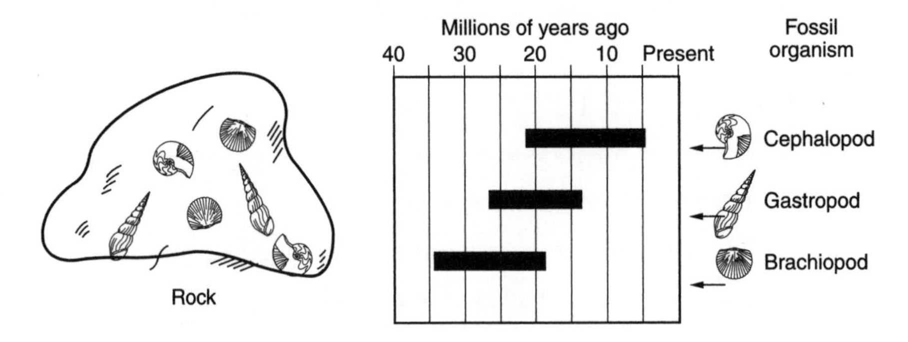

31. About how many years did the brachiopods live?

32. What happened to the brachiopods 18 million years ago?

33. When did the rock form?

Base your answers to questions 34 through 36 on the diagram below and your knowledge of science. The diagram shows a geologic rock cross section.

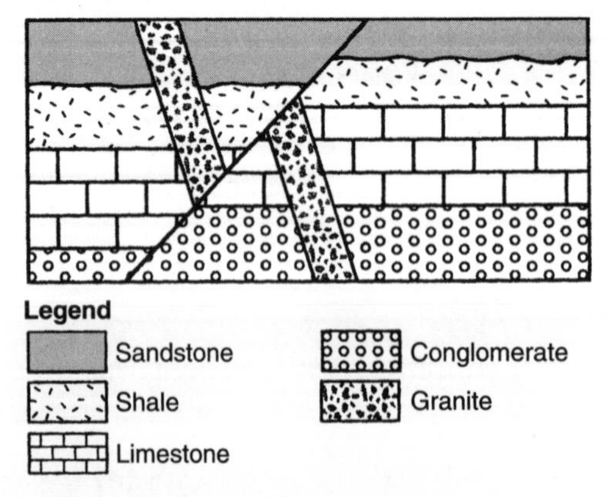

34. Which rock layer formed first? How do you know?

35. What evidence indicates that the granite formed after the sandstone?

36. Copy the words below into your notebook. Place the events in the order that produced this geologic cross section. Place 1 in front of the first event, 2 in front of the second event, and 3 in front of the third event.

_____ faulting

_____ deposition of sedimentary rocks

_____ igneous rock intrusion

Base your answers to questions 37 through 39 on the fossil table below and your knowledge of science. The fossil table shows when three fossils lived. Assume that if a fossil is not shown, it did not exist at the time.

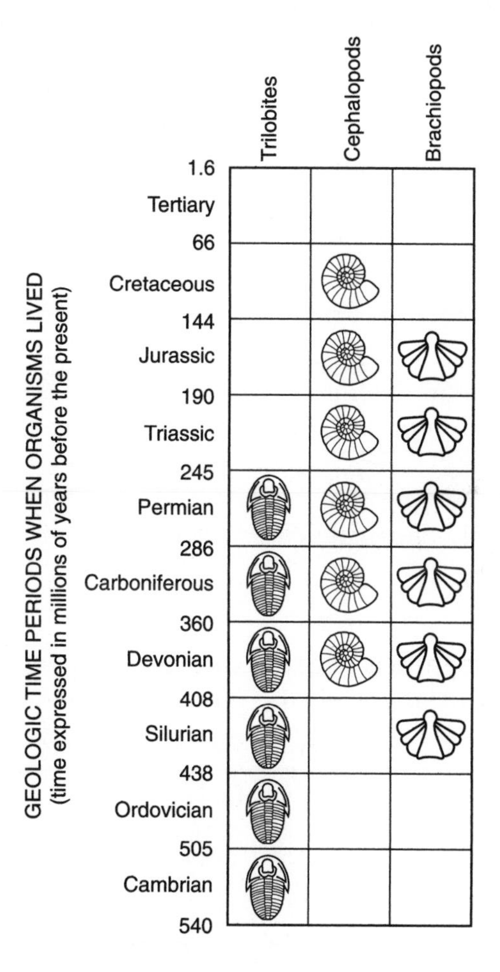

37. How many years did the trilobites exist?

38. What happened to the brachiopods 144 million years ago?

39. A rock was found that contained trilobites and brachiopods but not cephalopods. When did the rock form?

Visualizing Earth's Surface

A *map* is a flat model that shows a portion of Earth's surface. Different types of maps are used to show different types of information about Earth's surface. Weather maps show weather conditions and provide information to airplane pilots, boaters, and skiers. Geologic maps show rock formations and provide geologists with information about the location of oil and mineral deposits. Road maps show the location of roads, towns, or cities, and help us in everyday travel.

Latitude and Longitude

Earth is a sphere, and a sphere can be cut horizontally or vertically into circles. Each circle contains 360 degrees of arc. These circles form the coordinate system we know as latitude and longitude. The distance between the North Pole and the equator represents one-quarter of the distance around Earth, or 90°. Lines of ***latitude*** run west-east and are measured in degrees north and south of the equator. The equator is 0° latitude. As you travel north of the equator, the latitude increases until you reach the North Pole (90° north latitude). As you travel south of the equator, the lines of latitude increase until you reach the South Pole (90° south latitude). (See Figure 9-10.)

Figure 9-10. *Latitude lines are west-east lines that are measured from the equator (0° latitude) to the North Pole (90°N) and South Pole (90°S). Longitude lines are north-south lines that are measured from the prime meridian (0° longitude) to the 180° line halfway around Earth.*

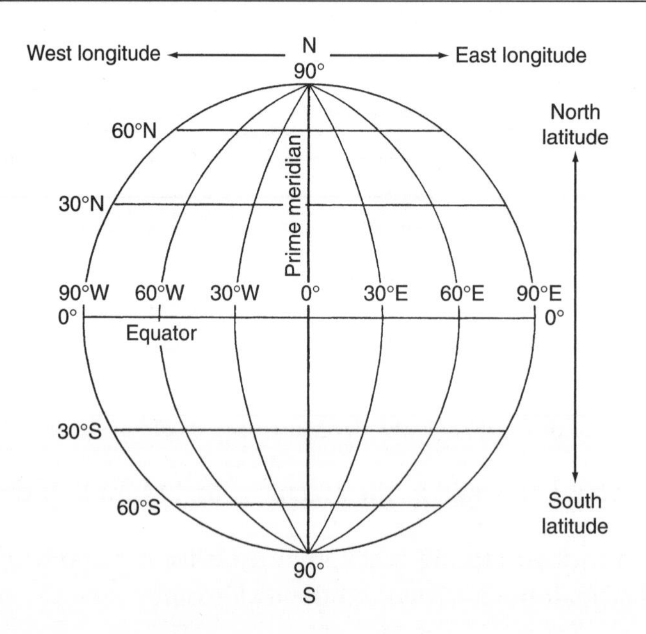

Lines of ***longitude*** run north-south and are measured in degrees west and east. The 0° longitude line, the prime meridian, is an imaginary line that connects the North and South Poles and passes through Greenwich, England. From the prime meridian, you can travel east one-half the distance around Earth in the east longitude hemisphere to the 180° longitude line, or you can travel west one-half the distance around Earth to the 180° longitude line. For maps that show large areas, such as continents or oceans, latitude and longitude measurements are given in degrees.

Topographic Maps

Topographic maps (see Figure 9-11) show the form and shape of the land's physical features. They are drawn from a bird's-eye view—in other words, they are drawn as if the mapmaker were above the landform looking down. The physical features include mountains, valleys, hills, plains, rivers, and lakes. Topographic maps are used by engineers when planning roads, by hikers planning a hike, and by anyone interested in knowing the shape of the land.

Figure 9-11. *A topographic map of an island. The contour interval is 10 meters. Based on the scale, the distance from point x to point y is 1.5 kilometers.*

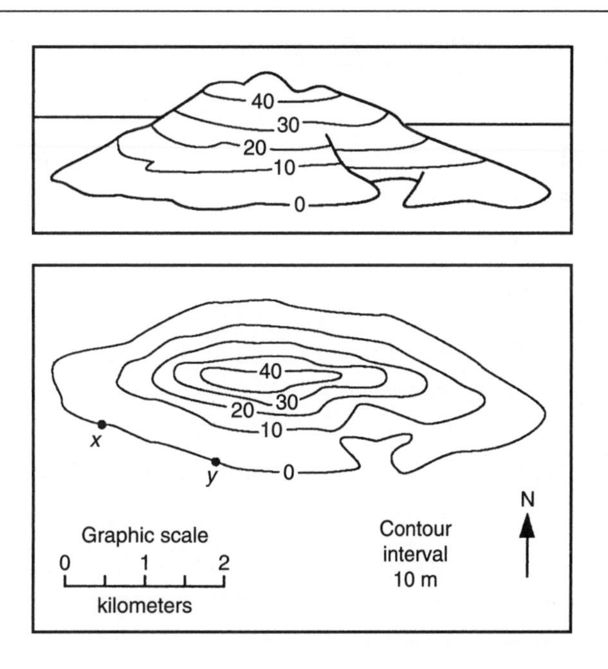

Topographic maps show the shape of the land with ***contour lines***, lines that connect points of equal elevation. Contour lines show land elevations above or below sea level. To make reading elevation easy, every fifth contour line is an index contour line. It is darker, and its elevation is labeled.

Certain rules are observed on contour maps. One rule is that contour lines of different elevations may never cross. Crossing different contour lines would mean that a certain place is at two different elevations, which is not possible.

Topographic maps use colors and symbols to show features. Blue is used for bodies of water such as lakes and rivers. Roads, buildings, railroad tracks, and cemeteries are shown in black. Green is used for areas of vegetation such as woodlands. Contour lines are brown. Different types of symbols are used on topographic maps to show many natural and constructed features. Figure 9-12 shows some of these symbols.

Figure 9-12. *Some symbols used on topographic maps.*

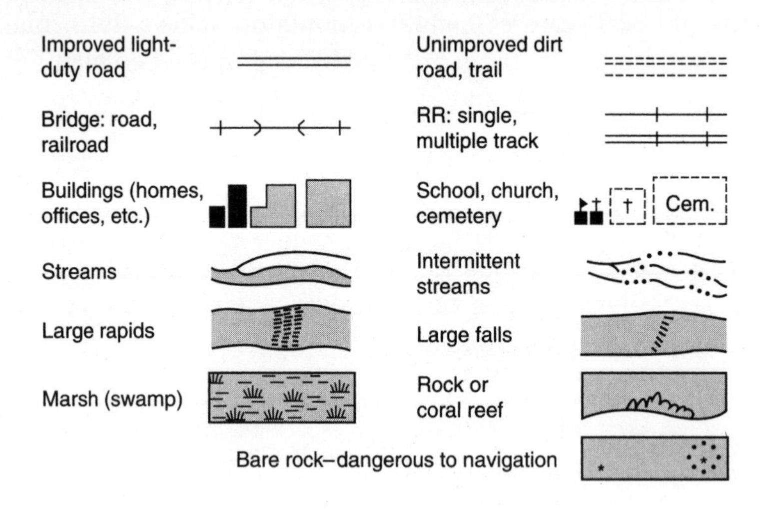

Landscapes

Landscapes are shown by the shape and spacing of contour lines. (See Figure 9-13.) For example:

- Closely spaced contour lines indicate a steep slope (a).
- Widely spaced contour lines indicate a gentle slope (b).
- Circular contour lines indicate a hill or mountain (c).
- Contour lines that cross a stream or river make a V shape that points upstream (d).
- U-shaped bends in contour lines indicate a wide, deeply eroded valley (e).
- Hachure marks (short inward-pointing lines) on a contour line indicate a depression (f).

Figure 9-13. *From the shape of a set of contour lines, you can recognize landscapes.*

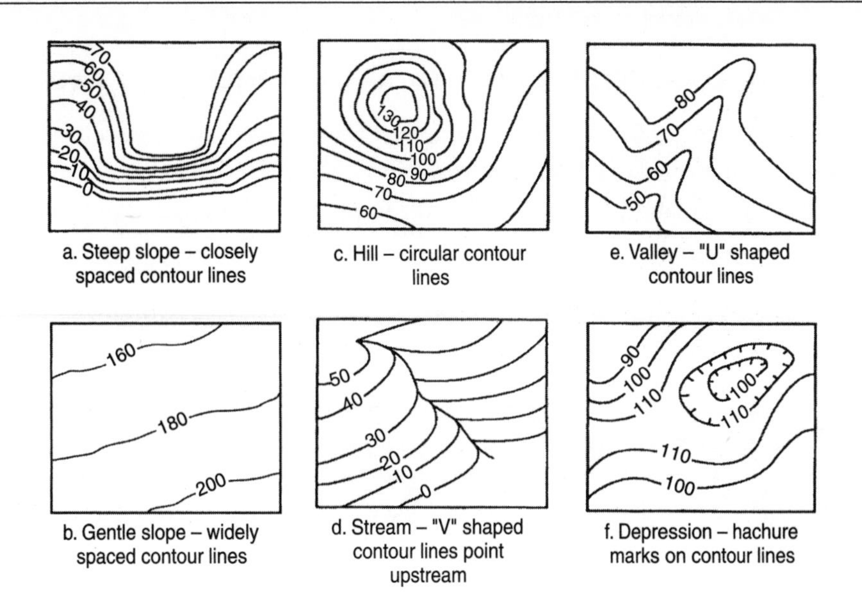

Profiles

A ***profile*** is a side view or cross section of an area on Earth's surface. A profile shows how the land goes up and down as if you were walking in a straight line on Earth's surface. (See Figure 9-14.) By drawing a profile between two points on a map (called a baseline), you get a true picture of the shape of the land you would cross if you traveled from one point to another point. For example, in Figure 9-14, if you were to walk from point M to point N in a straight line, you would have to climb over a hill.

Figure 9-14. *Making a profile from baseline MN on a topographic map. The profile shows the shape of the land.*

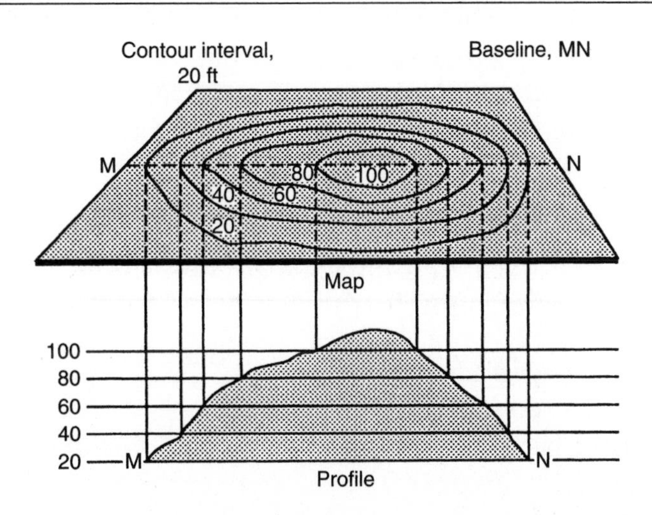

Process Skill 3

How Is a Compass Used to Determine Direction?

A compass is a magnetized metal needle that freely pivots above a circular dial that is labeled with the major geographic direction points. The most commonly labeled points are north, northeast, east, southeast, south, southwest, west, and northwest. (See Figure A below.)

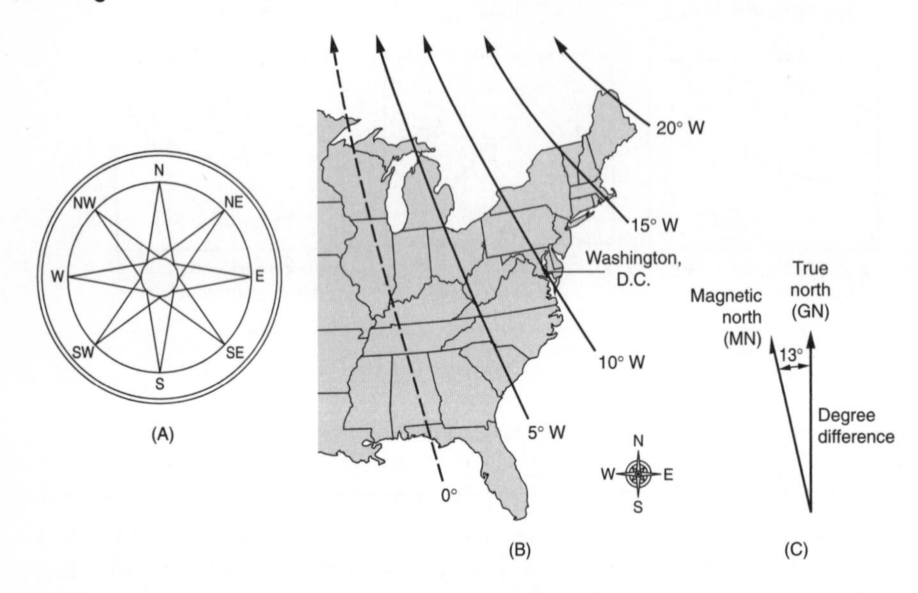

Earth's iron-rich core acts much like a giant bar magnet. At one end of the giant magnet is magnetic north. When allowed to swing freely and rest, the compass needle points to magnetic north (MN). True north (usually labeled GN for geographic north) is located at the North Pole. Magnetic north and true north are at different locations; therefore, they are in different directions. In New York, magnetic north is between 8 and 15 degrees *west* of true north. (See Figure B.) When accuracy of direction is required, this difference must be taken into account.

In New York, you can determine true north by using a compass and a topographic map. First, from the margin of a local topographic map, find the difference between true north and magnetic north. (See Figure C.) Now use your compass to locate magnetic north. Move eastward from magnetic north the number of degrees indicated in the margin on the topographic map. This is the direction of true north.

Questions

1. Explain where in the United States a compass needle points to true north.
2. What is the degree difference between magnetic north and true north in Washington, D.C.?
3. In New York, magnetic north is always
   (1) west of true north
   (2) east of true north
   (3) south of true north
   (4) in the same direction as true north

Review Questions

Part I

Questions 40 through 43 refer to the topographic map below.

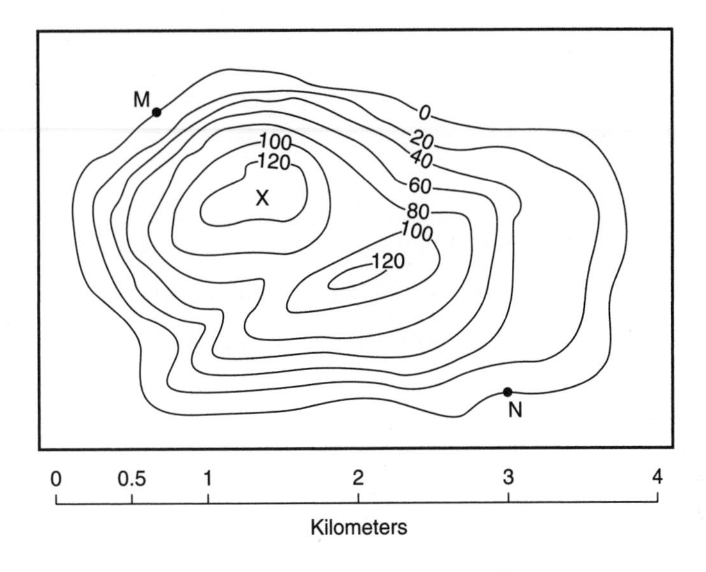

40. The contour interval of the map is

(1) 10 meters
(2) 20 meters
(3) 50 meters
(4) 100 meters

41. The best estimate of the elevation of point X is

(1) 100 meters
(2) 120 meters
(3) 140 meters
(4) 130 meters

42. The distance between points M and N is most nearly

(1) 1 kilometer
(2) 2 kilometers
(3) 3 kilometers
(4) 5 kilometers

43. If you walked in a straight line from point M to point N, you would walk

(1) on level ground
(2) uphill, downhill, uphill, and downhill
(3) uphill and downhill
(4) downhill, uphill, and downhill

44. Using the small topographic map below, which statement is correct?

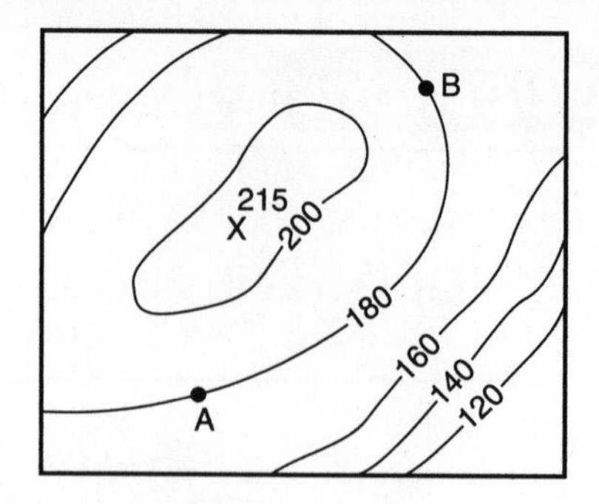

(1) Point A is higher than point B.

(2) Point B is higher than point A.

(3) Points A and B are at the same elevation.

(4) The relationship between points A and B cannot be determined.

45. What type of contour lines indicates a hill?

(1) circular (2) widely spaced (3) closely spaced (4) V-shaped

46. In New York State a compass needle points

(1) to true north

(2) west of true north

(3) east of true north

(4) to the North Pole

Part II

Base your answers to questions 47 through 49 on the diagram below and your knowledge of science. The diagram represents a portion of Earth's latitude and longitude grid.

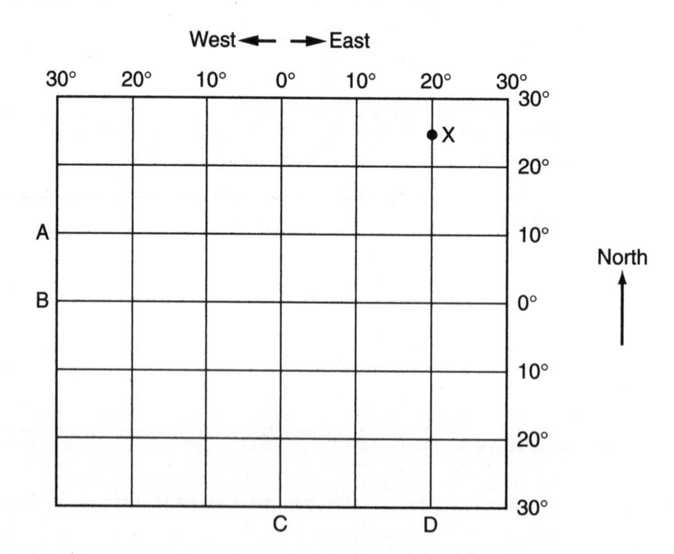

47. What type of line is line D?

48. What is the latitude and longitude of point X? (Your answer must include a value, unit, and direction of each.)

49. What line (letter) represents the equator?

---

50. Draw a topographic map of an island. Use a contour interval of 50 meters. Place a 225-meter hill on the island. Place an X at the top of the hill. Include a graphic scale that shows 2 cm equals 1 km.

Chapter 10 Processes That Change Earth's Surface

Vocabulary at a Glance

atmosphere
core
earthquake
faulting
folding
hydrosphere
landform
lithosphere
longitudinal wave
mantle
mountain
plain
plateau
plate tectonics
primary wave
seafloor spreading
secondary wave
volcano
weathering

Glaciers, like this one on Mount Kilimanjaro, change the shape of Earth's surface. Continental glaciers once covered New York State.

Major Concepts

- External processes, such as weathering and erosion, wear down Earth's surface. Internal processes, such as mountain building, faulting, and volcanism, build up Earth's surface.
- Weathering breaks down rocks. Physical weathering breaks rocks into smaller pieces and chemical weathering changes the chemical composition of the rocks.
- Erosion moves rock material from one place to another.
- Interactions of the lithosphere, hydrosphere, and atmosphere change Earth's surface.
- Plate tectonics explains that the crust is made up of plates that move and interact, causing earthquakes and volcanoes. Most earthquakes and volcanoes are located on the boundaries of tectonic plates.
- Major landforms, on the surface or under the ocean, are produced by plate tectonics.
- Continental drift was supported at first by the fit of the continents, matching fossils, and ages and types of rock formations at the boundaries of the plates. Today, plate tectonics and ocean-floor features strongly support that the continents were once together.

Interacting Earth Systems

Planet Earth is made up of three spheres: a rock sphere, or ***lithosphere***; a water sphere, or ***hydrosphere***; and a gaseous sphere, or ***atmosphere*** (Figure 10-1). On Earth's surface, these three spheres meet and affect one another by exchanging energy and matter.

Figure 10-1. *The atmosphere, hydrosphere, and lithosphere meet near Earth's surface.*

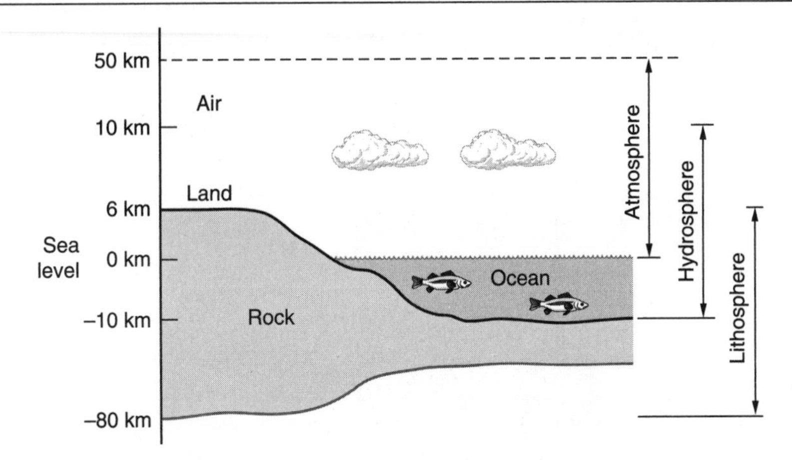

Energy is exchanged constantly among the lithosphere, atmosphere, and hydrosphere. Table 10-1 gives examples of energy exchange among Earth's spheres.

Table 10-1. *Examples of Interactions Among Earth's Spheres*

Atmosphere → Lithosphere
Atmospheric gases and moisture produce physical and chemical weathering, causing rocks to crumble.

Atmosphere → Hydrosphere
Wind blowing across water surfaces produces waves. The stronger the wind, the larger the waves.

Lithosphere → Hydrosphere
Volcanic and earthquake activity on the ocean floor produce tsunami waves. Strong vibrations of the seafloor are transferred to the water, causing large ocean waves to form.

Hydrosphere → Lithosphere
Ocean waves breaking along beaches carry sand particles, causing coastline erosion.

Hydrosphere → Atmosphere
Warm ocean currents travel north, warming air they meet; cold ocean currents travel south, cooling air they meet.

Lithosphere → Atmosphere
Volcanic activity sends ash particles high into the atmosphere. These particles block the sun's radiation, producing cooler temperatures.

External Processes

Internal and external processes are constantly at work shaping and changing Earth's surface. *Internal processes* cause the land to rise above sea level. *External processes* wear the land down to sea level. Figure 10-2 illustrates these processes and their effects on Earth's surface features.

Figure 10-2. *Earth's surface is shaped by the interaction of internal and external processes.*

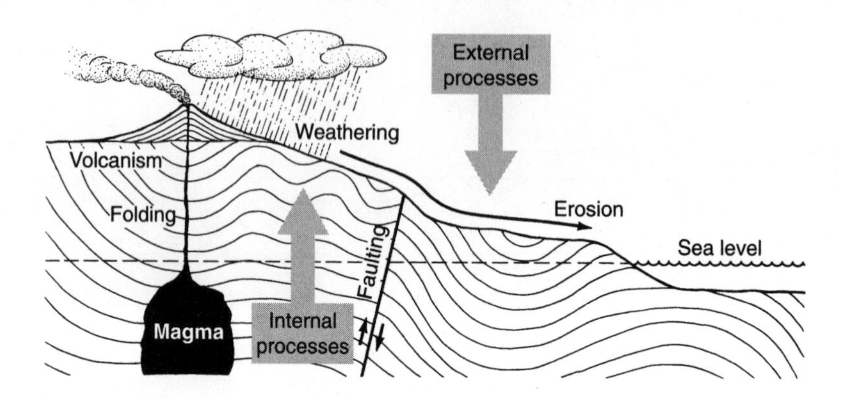

Weathering

External processes include weathering and erosion. Together, these processes wear down Earth's surface.

Weathering is the breaking down of rocks into smaller pieces. The most important agents of weathering are rain, ice, and atmospheric gases. Both physical and chemical agents can cause weathering. In *physical weathering*, rocks are broken into smaller pieces by physical agents. For example, water seeps into cracks in a rock and freezes. As the water freezes, it expands and breaks the rock into smaller pieces. (See Figure 10-3.) The expanding roots of plants growing in cracks can also split rocks. The smaller pieces have the same chemical composition as the original rock.

Chemical weathering is the breaking down of rocks through changes in their chemical makeup. These changes take place when rocks are exposed to air or water. For example, when rainwater combines with carbon dioxide in the air, a weak acid is formed. This acid dissolves calcite in rocks such as limestone and causes the rocks to crumble. Also, oxygen and water react chemically with iron-containing minerals in a rock. The iron is changed into rust, which crumbles away.

By breaking down rocks into smaller pieces, weathering helps form soil. Table 10-2 gives a list of agents and examples of weathering.

Figure 10-3. *Physical weathering caused by water freezing in rock cracks.*

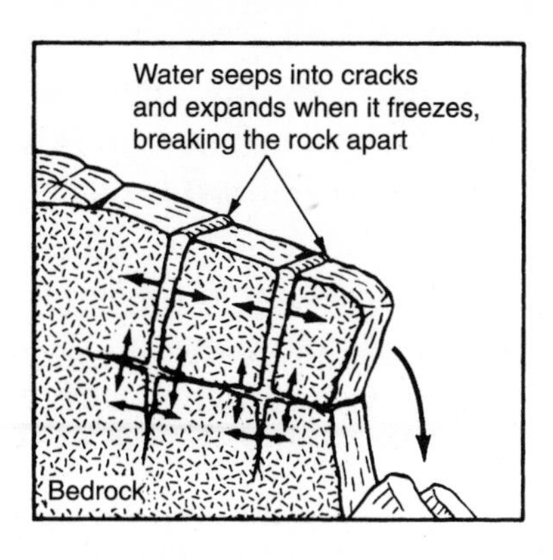

Table 10-2. *Agents and Examples of Weathering*

Agent of Weathering	***Description and Example***
Physical Weathering	
Ice wedging (frost action)	Water freezes and expands in a rock crack, splitting the rock. Finding a large rock surrounded by smaller pieces after the winter thaw is an example of the results of ice wedging, also known as frost action. This can be seen along roadside rock cuts in the spring.
Water abrasion	Sand carried by running water scrapes and polishes rocks in a stream. This process produces smooth and rounded rocks in the stream.
Wind abrasion	Blowing sand scrapes rock surfaces. Wind abrasion forms smooth surfaces on rocks in dry regions.
Exfoliation	Rocks formed underground are under great pressure. When pressure is removed from rock, thick slabs may peel off the rock's surface. This is commonly seen in granite at the top of mountains.
Plant roots	Trees and plants growing out of cracks in large rocks split the rock apart.
Animals	Burrowing animals bring rocks to the surface. The rocks are then exposed to the other agents of weathering.
Chemical Weathering	
Carbonic acid	Carbonic acid forms when carbon dioxide in the atmosphere dissolves in rainwater, producing *acid rain*. Carbonic acid in rainwater dissolves limestone and marble. For example, the lettering on old marble monuments is hard to read.
Oxidation	Oxygen reacts with some materials, changing them chemically. When oxygen reacts with iron in rocks, a red rustlike stain forms on the rock.
Plant acids	Organic matter decays in water, producing acids. Some minerals crumble when they react with the acid.
Water	Some minerals react with water. Water causes feldspar to decompose, producing clay.

Erosion

Erosion is the process that moves rock material at Earth's surface and carries it away. Erosion needs a moving force, such as flowing water, that can carry the rock particles. You can see this after a heavy rain, when streams turn a muddy brown from the rock material in the water.

Gravity and *water* play important roles in erosion. Gravity is the main force that moves water and rock downhill. Flowing water is very powerful; more rock material is eroded by running water than by all other agents of erosion combined. The Grand Canyon in Arizona is a spectacular example of erosion caused by running water. (See Figure 10-4.)

Figure 10-4. *Erosion caused by running water carved the Grand Canyon, which is more than a mile deep.*

The forces of erosion are constantly at work, moving rock material from high places to low places. Table 10-3 lists the agents of erosion and an example of each.

Table 10-3. *Agents and Examples of Erosion*

Agent of Erosion	Example of Erosion
Gravity	Landslides and mudslides carry rock and soil downhill.
Running water	Rock particles are carried in the running water.
Groundwater	Rock material is removed underground to form caverns.
Glaciers	Rock material is carried under and in a glacier.
Winds	Rock particles are carried and rolled along near the ground.
Longshore currents	Sand is moved along a beach by waves washing in and out.

Process Skill 1

Predicting the Result of an Experiment

Rocks carried by a stream constantly bump into and scrape against each other and against the streambed. The longer the rocks are in the stream, the more they tumble and hit one another. To model this action and study its effects, Henry carried out the following experiment.

Henry placed 25 marble chips and 1 liter of water in a large coffee can marked *A*. He covered the can with a lid and shook it for 30 minutes. Then, he placed another 25 marble chips and 1 liter of water in a second can, marked *B*. He covered it and shook it for 120 minutes. The illustration below shows the materials used in Henry's experiment. Keep in mind what you have learned about weathering to help you answer the following questions.

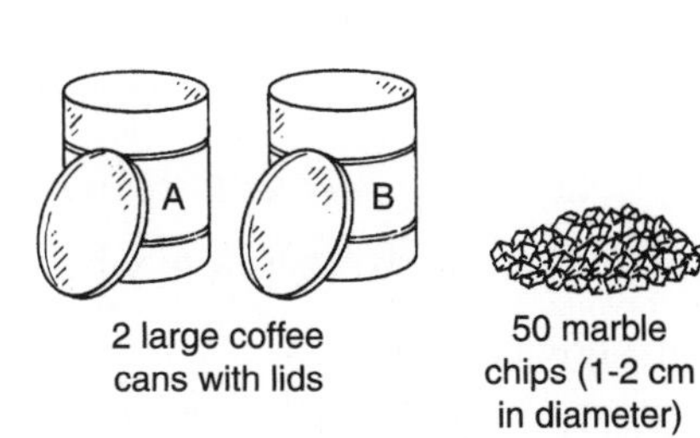

Questions

1. Which is the best prediction of the results of the experiment?
(1) The marble chips in can *A* will be smaller and rounder than the chips in can *B*.
(2) The marble chips in can *B* will be smaller and rounder than the chips in can *A*.
(3) There will be no difference between the marble chips in cans *A* and *B*.

2. Which graph best predicts what would happen over time to rocks in a fast-moving stream?

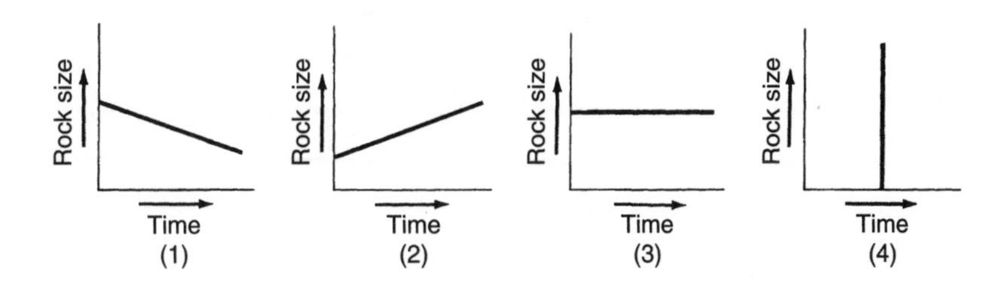

Review Questions

Part I

1. The solid portion of Earth's surface is called the

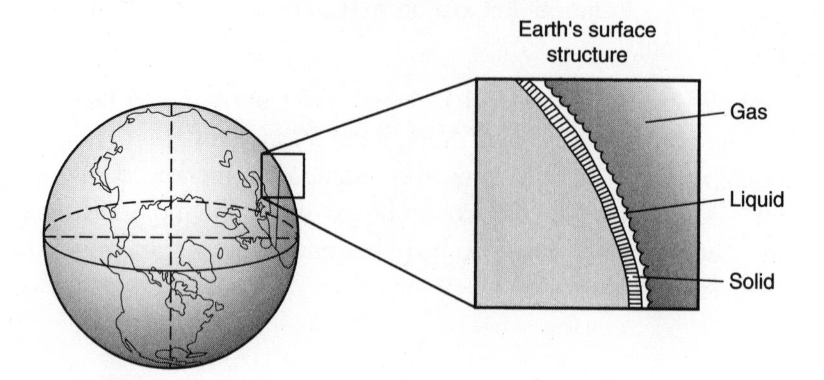

(1) hydrosphere
(2) atmosphere
(3) lithosphere
(4) troposphere

2. The lithosphere interacts with the atmosphere when
   (1) an erupting volcano releases gases
   (2) wind produces waves on the ocean
   (3) water evaporates from the ocean and goes into the air
   (4) rain falls on a slope and washes soil away

3. Which action is an example of physical weathering?
   (1) rainwater dissolving limestone
   (2) particles carried in a stream
   (3) wind blowing sand
   (4) water freezing in a cracked rock

4. The diagram below shows the mineral magnetite, which contains iron, changing into rust particles. This is an example of

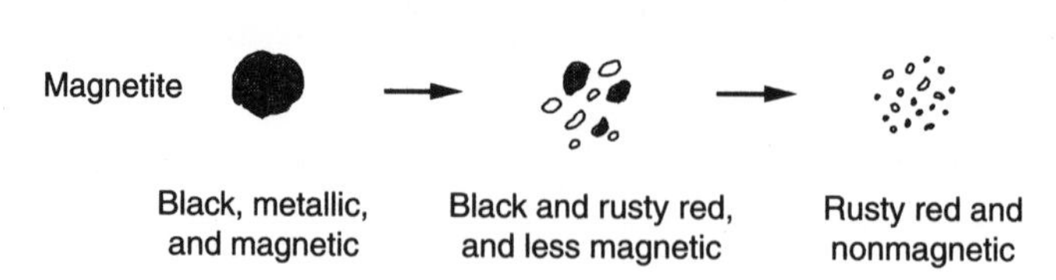

(1) physical weathering
(2) chemical weathering
(3) erosion by running water
(4) the role of gravity in erosion

5. Weathering is the process that causes rocks at Earth's surface to
   (1) be removed and carried away
   (2) crumble and form smaller particles
   (3) melt and form magma
   (4) settle in water

Part II

6. How is weathering different from erosion? Give an example of weathering and an example of erosion.

Base your answers to questions 7 and 8 on the diagram below and your knowledge of science. The series of diagrams shows how a region changed over the past 50 million years.

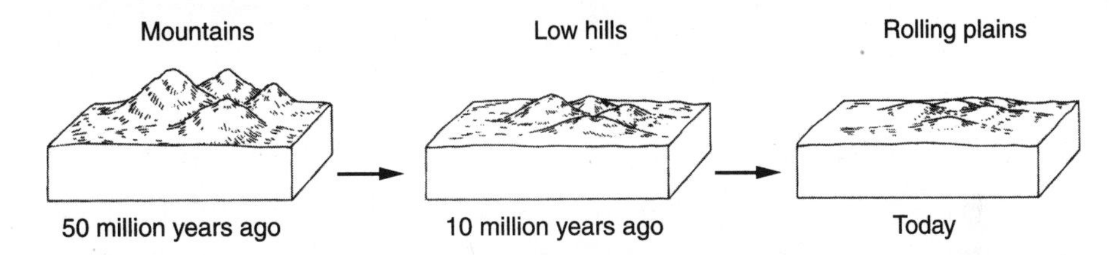

7. What agent of erosion is most responsible for changing the surface of the land?
8. Predict how the land will change if weathering and erosion continue to be the primary processes changing the land during the next 10 million years.

Internal Processes

Earth's internal processes also shape its surface. These processes produce *mountains, earthquakes,* and *volcanoes,* raising the land and building up Earth's surface.

Landforms: Mountains, Plains, and Plateaus

A ***landform*** is a large land feature defined by its height, steepness of slope, and type of bedrock present. The three major landforms are mountains, plains, and plateaus.

A ***mountain*** is a landform that is much higher than the land around it. (See Figure 10-5 on page 280.) Mountains have steep slopes and are usually made up of igneous or metamorphic rock. The processes of folding and faulting build mountains.

Figure 10-5. *The Grand Tetons in Wyoming rise high above the surrounding land.*

Folding occurs when forces in Earth's crust squeeze and bend rock layers. The crust is squeezed into up folds and down folds, forming ridges and valleys (Figure 10-6A). ***Faulting*** occurs when forces in the crust squeeze or pull it until it breaks. When the crust breaks, it slides along a crack or fracture, called a *fault,* relieving the stress in the crust (Figure 10-6B).

Figure 10-6. *(A) Forces in the crust can squeeze rock layers into folds. (B) When forces in Earth's crust reach the breaking point, the crust fractures and produces faults.*

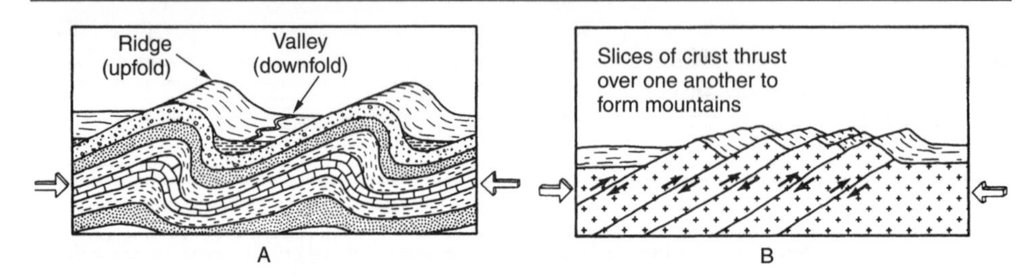

Mountains can also be built by volcanoes. A ***volcano*** is an opening in Earth's crust through which lava flows from underground. During eruptions, lava pours out onto the surface and cools to form solid rock. The rock forms layers that build a mountain, also called a volcano. (See Figure 9-3 on page 246.) Mount St. Helens in Washington State is a volcanic mountain.

A ***plain*** is a broad, flat region found at a relatively low elevation. (See Figure 10-7A.) The land is made up of flat layers of sand or sedimentary rock close to sea level. Long Island and eastern New Jersey are part of the Atlantic Coastal Plain.

A ***plateau*** is a large area of horizontally layered rocks with higher elevation than the plains. (See Figure 10-7B.) Plateaus can form in several ways.

A large block of crust can rise up along faults to form a plateau, or a plateau may be gradually uplifted without faulting. Plateaus can also be built up by lava flows. The Allegheny Plateau in southwestern New York State is an example of a plateau that was cut by streams after it formed.

Figure 10-7. *(A) Plains are broad, flat regions found at low elevations. (B) Plateaus are large areas of horizontally layered rocks with relatively high elevations.*

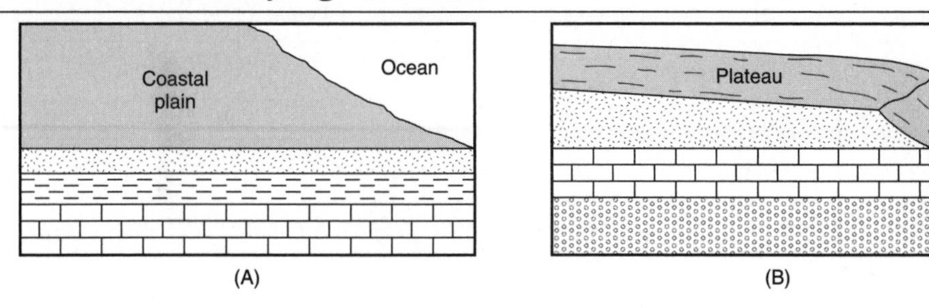

Earthquakes

A sudden movement of rock sliding along a fault in the crust is an ***earthquake***. Most earthquakes are linked with land uplift and mountain building. When an earthquake occurs, it produces strong vibrations called *seismic waves* that travel through Earth. A strong earthquake rocked central China in May 2008.

Earthquakes produce three types of seismic waves. ***Primary waves*** (P-waves) and ***secondary waves*** (S-waves) travel through Earth, while ***longitudinal waves*** (L-waves) travel along Earth's surface. P-waves can travel through liquids and solids; S-waves can travel only through solids. By analyzing how the three seismic waves travel through Earth, scientists find clues to Earth's structure (Figure 10-8).

Figure 10-8. *Paths of seismic waves traveling on the surface (L-waves) and through Earth (P- and S-waves). The study of seismic waves reveals Earth's internal structure.*

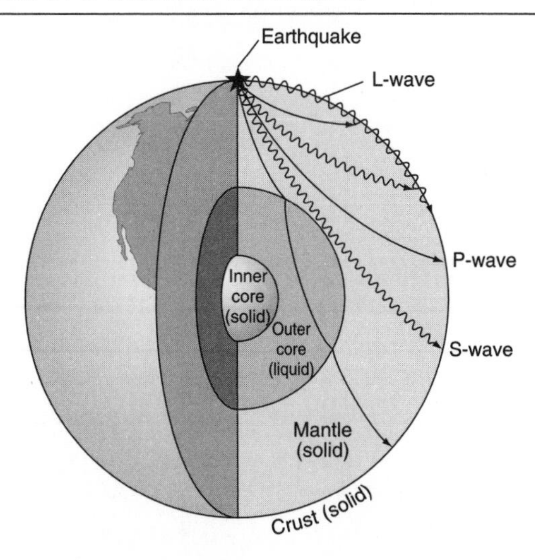

Process Skill 2

Plotting Earthquakes by Latitude and Longitude

New York State has had some moderate earthquakes in the last 300 years. Table 10-4 lists 16 moderate earthquakes that occurred in New York State. Get a copy of a New York State map at *www.lib.utexas.edu/maps/states/new_york.gif.* On the map, plot the position of each earthquake listed in the table.

Table 10-4. Moderate Earthquakes in New York State

Date	*Latitude (° North)*	*Longitude (° West)*	*Magnitude*
Dec. 18, 1737	40.60	73.80	5.0
Mar. 12, 1853	43.70	75.50	4.8
Dec. 18, 1867	44.05	75.15	4.8
Dec. 11, 1874	41.00	73.90	4.8
Aug. 10, 1884	40.59	73.84	5.3
May 28, 1897	44.50	73.50	NA
Mar. 18, 1928	44.50	74.30	4.5
Aug. 12, 1929	42.84	78.24	5.2
Apr. 20, 1931	43.50	73.80	4.5
Apr. 15, 1934	44.70	73.80	4.5
Sep. 5, 1944	45.00	74.85	6.0
Sep. 9, 1944	45.00	74.85	4.0
Jan. 1, 1966	42.84	79.25	4.6
Jun. 13, 1967	42.84	78.23	4.4
Oct. 7, 1983	43.97	74.25	5.1
Apr. 20, 2002	44.50	73.50	5.1

Questions

1. How many of the earthquakes were near New York City?
2. Where did the earthquake with the highest magnitude occur?

Structure of Earth

Earth's outer layer is the *crust.* It is made up of solid rock and forms a layer that covers the whole Earth. Under the oceans the crust is about 5 kilometers (3 miles) thick and contains mostly igneous rocks like basalt. Under the continents, the crust is about 40 kilometers (25 miles) thick and contains mostly igneous rocks like granite. (See Figure 10-9.)

Figure 10-9. *Cross section of Earth showing its internal structure.*

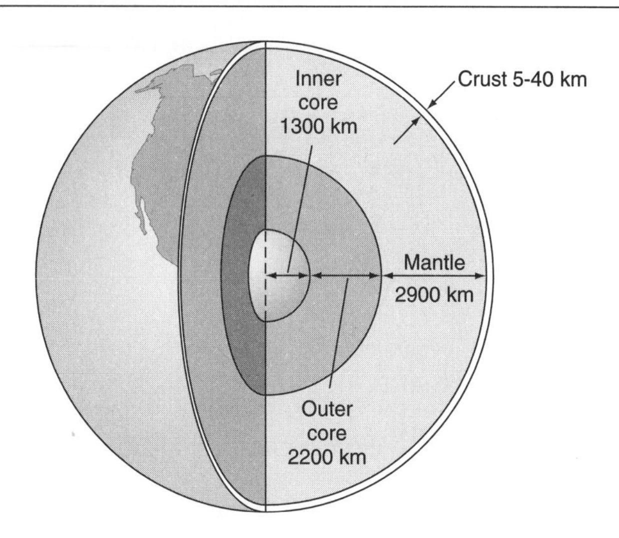

Below the crust is the ***mantle***. It is about 2900 kilometers (1800 miles) thick and probably is made up of dense iron- and magnesium-rich rocks. Scientists know it is solid because S-waves can travel through it. Yet it flows very slowly and causes plates of the crust to move.

At Earth's center is the ***core***, which has an outer and inner zone. The *outer core* is about 2200 kilometers (1400 miles) thick. It is liquid because S-waves cannot travel through it. The *inner core* has a radius of 1300 kilometers (800 miles), and it is solid because P-waves travel faster through it. The core is thought to be mainly an iron and nickel mixture.

Origin of Continental Drift

In 1912, Alfred Wegener, a German meteorologist, proposed a theory that the continents were drifting across Earth's surface. He said that the continents were once connected in a single landmass. The landmass split and the pieces slowly drifted apart, forming the Atlantic Ocean. He based his theory on the way the shapes of the continents fit together like pieces of a jigsaw puzzle. He also supported his theory with evidence of matching fossils, rocks, mountains, and glacial features on both sides of the Atlantic Ocean. (See Figure 10-10 on page 284.)

Figure 10-10. *Continental shapes fit together like pieces of a jigsaw puzzle. Fossils, rock formations, mountains, and glacial features match.*

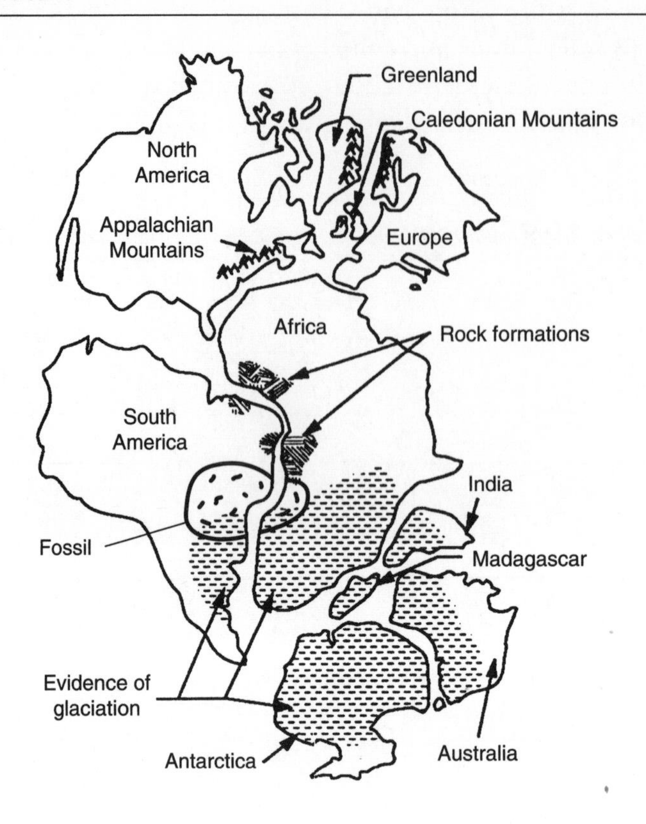

For about 50 years, the idea of drifting continents was rejected. No process or force within Earth could be identified as the cause of this massive movement in the crust. Although the theory and evidence presented by Wegener was intriguing, only when the mechanism for crustal movement was identified was the theory accepted. In the 1960s, the discovery of some ocean floor features supported the *continental drift* theory.

Seafloor Spreading

In the 1960s, oceanographers discovered an underwater mountain ridge running north-south down the middle of the Atlantic Ocean. Along the ridge, there was much volcanic activity. Scientists suggested that new rock material rises along the ridge. This rock moves east and west away from the ridge like two conveyor belts, pushing out in opposite directions. This movement away from the ridge is ***seafloor spreading***.

In 1969, on opposite sides of the Mid-Atlantic Ridge, scientists discovered strips of ocean floor with matching magnetic polarity. This provided evidence that seafloor spreading was occurring. (See Figure 10-11.) Other evidence such as matching the type and age of sediments and fossils on each side of the ridge, determining that the ridge is a high-heat-flow area, and finding trenches where crust subsides helped to support the idea of seafloor spreading.

Figure 10-11. *Support for the theory of seafloor spreading: (1) the age of oceanic crust increases with increasing distance from the mid-ocean ridge, and (2) matching magnetic bands are found on both sides of the mid-ocean ridge.*

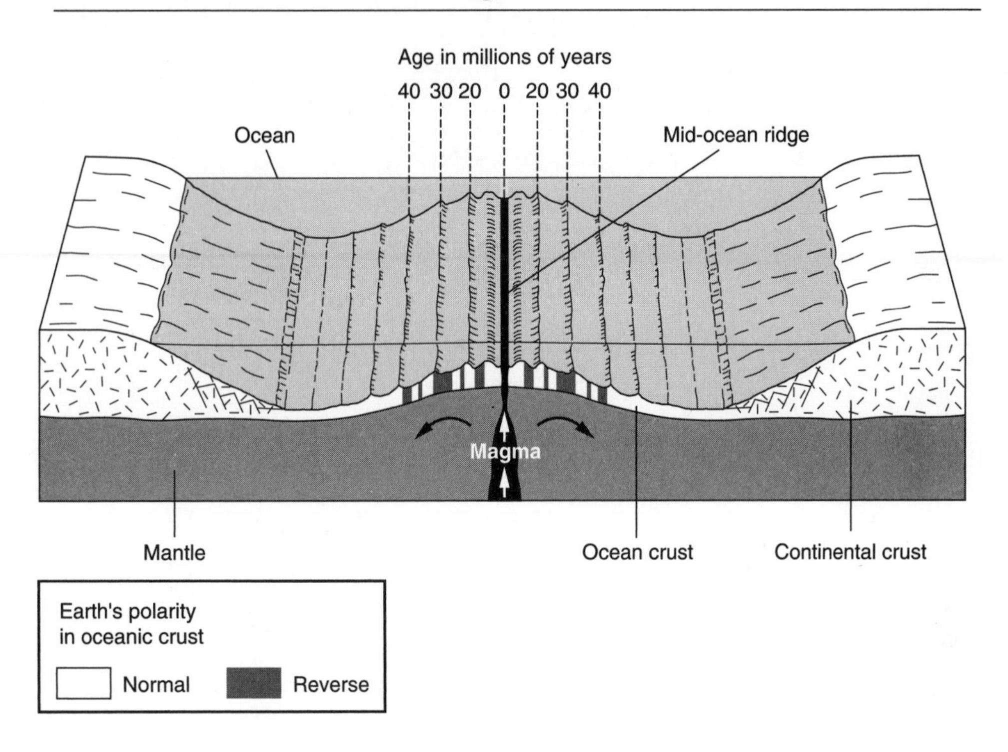

The internal process causing the continents to drift and seafloor to spread appears to be heat flowing upward from inside Earth. The upwelling carries new rock to the surface, slides the continents apart, and in some locations the crust eventually subsides and forms a deep trench in the ocean floor.

Plate Tectonics

There is much evidence that processes at work inside Earth have raised the level of the land. For example, many mountaintops are made of sedimentary rock containing fossils. This rock was formed originally on the ocean floor. Folds and faults seen in many rock outcrops are also signs of crustal movements caused by internal processes. Scientists explain these processes and the movements they produce by the theory of ***plate tectonics***.

According to this theory, Earth's crust is broken up into a number of large pieces called *plates*. These plates slowly move and interact in various ways. Some plates are spreading apart, some are sliding past each other, and some are colliding. These movements cause mountain building, volcanic activity, and earthquakes along the plate edges. Figure 10-12 on page 286 shows Earth's major crustal plates.

Figure 10-12. *Earth's major crustal plates. (Arrows show where plates are spreading apart, triangular "teeth" show where one plate is sliding under another plate.)*

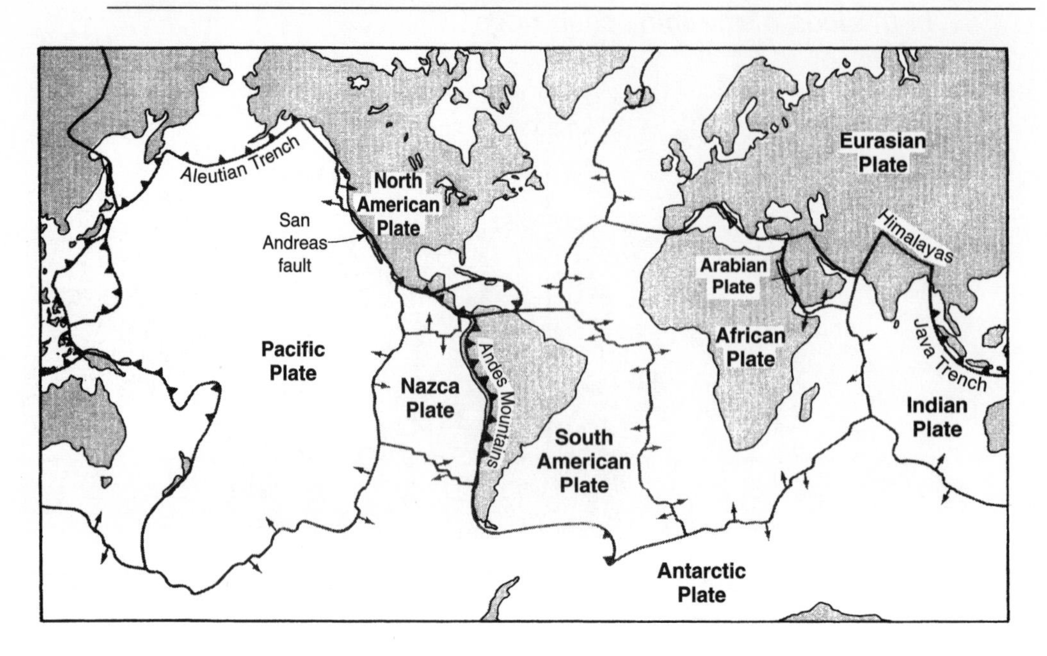

Scientists understand that plate motions are caused by heat circulating in Earth's *mantle,* the thick zone of fluidlike rock beneath the crust. The heat softens mantle rock so that it flows very slowly. At the top of the mantle, the currents spread sideways and carry along overlying pieces of crust. (See Figure 10-13.)

Plate boundaries are areas where plates rub against each other. Most earthquake and volcanic activity occurs along these boundaries. There are four ways plates interact at their boundaries.

1. Plates spread apart along the mid-ocean ridges, producing new ocean basins. (See Figure 10-14A.) As stated in the sea-floor-spreading section above, this is occurring along the Mid-Atlantic Ridge.
2. Oceanic plates subside when they collide with a continental plate. The denser oceanic plate slides down under the continental plate. (See Figure 10-14B.) At the Aleutian Trench off the coast of Alaska, the Pacific Plate is sliding under the North American Plate.
3. Continental plates collide, producing great mountain ranges. (See Figure 10-14C.) The Indian Plate is colliding with the Eurasian Plate, pushing up the Himalaya Mountains in Asia.
4. Plates slide sideways past each other, producing major fault and earthquake zones. (See Figure 10-14D.) In California, the Pacific Plate is sliding past the North American Plate along the San Andreas Fault.

Figure 10-13. *Plate tectonics: Heat currents in the mantle cause movements of Earth's crustal plates, producing many features on the seafloor and the continents.*

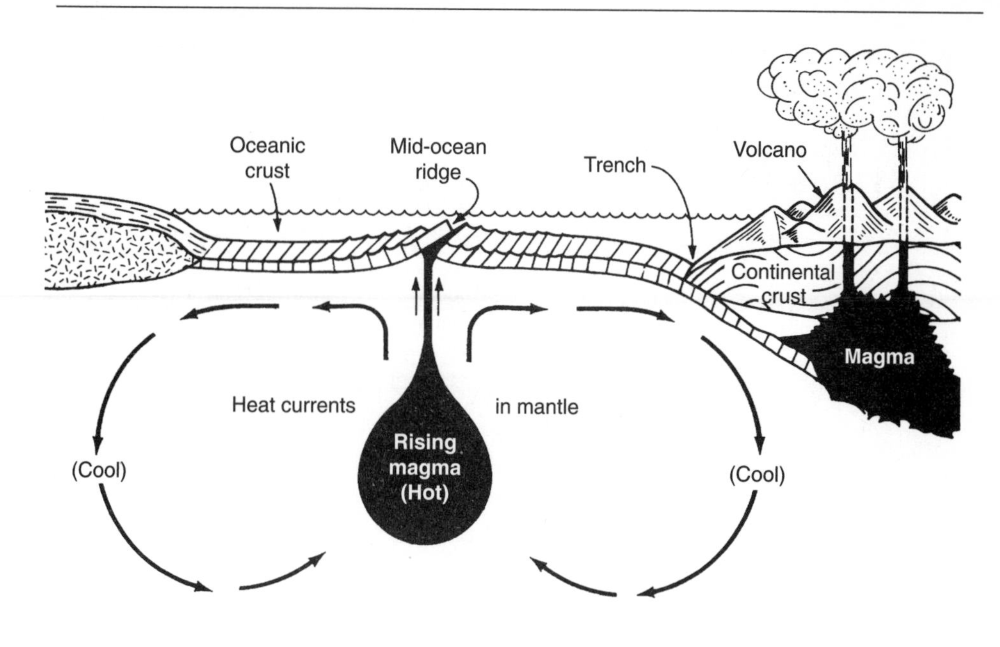

Figure 10-14. *Types of crustal plate boundaries. Most earthquake and volcanic activity is located along plate boundaries.*

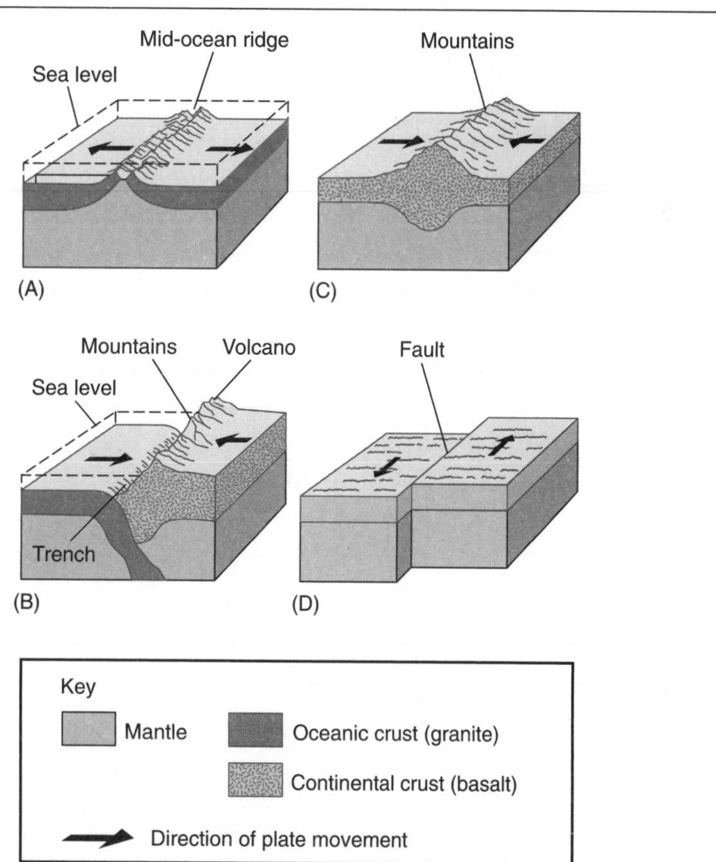

Ocean Floor Features

Almost three-quarters of Earth's surface is covered by ocean water. The floor of the ocean is not flat and featureless. Scientists have found that the ocean floor has mountains, valleys, plains, and plateaus. Many of these features, such as mid-ocean ridges and ocean trenches, are produced by plate tectonics.

Figure 10-15. *Generalized features of the ocean floor.*

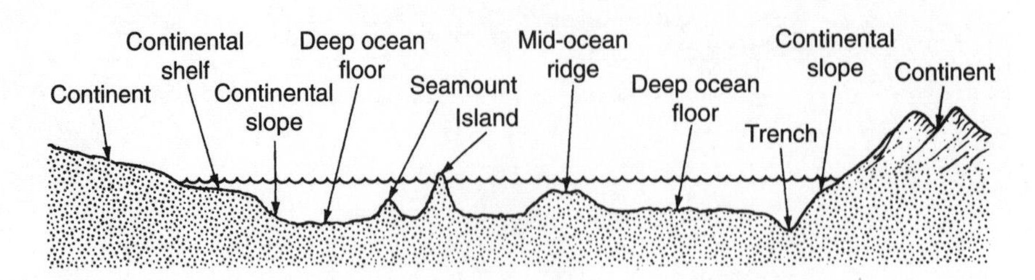

1. A *mid-ocean ridge* is a long underwater mountain chain where rising magma forms new ocean crust. The new crust is added to plates that spread away from the ridge, as shown in Figure 10-11 on page 285. This process is called *seafloor spreading.*
2. *Trenches* are underwater canyons that form the deepest parts of the ocean floor. A trench forms where an oceanic plate collides with a continental plate. The collision forces the oceanic plate to slide under the continental plate, back into Earth's mantle. This produces volcanic activity and mountain building along the edge of the continental plate. (See Figure 10-13 on page 287.)

Other ocean floor features include continental shelves, continental slopes, the deep ocean floor, and seamounts. Continental shelves are shallow features found along some coastlines. Continental slopes level off into the *deep ocean floor*. The deep ocean floor is not a flat plain; it has ridges and valleys. Rising here and there from the ocean floor are tall underwater mountains called *seamounts*. Most seamounts were formed by volcanoes.

When the top of a seamount or mid-ocean ridge rises above the water's surface, it forms an island. The Hawaiian Islands are the tops of a chain of volcanic seamounts. Figure 10-15 shows a generalized profile of an ocean floor that includes many of these features.

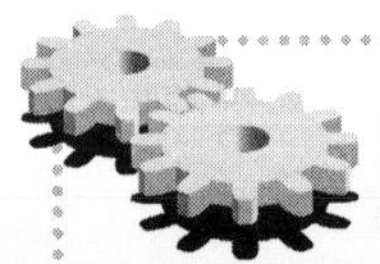

Process Skill 3

Locating Recent Earthquakes and Volcanic Eruptions

Earthquakes and volcanoes usually occur at the edges of crustal plates. At these locations, the plates rub against each other (colliding or sliding) or upwelling (emerging) from under the crust. The map below shows the major earthquake and volcano zones on Earth. Compare the map with Figure 10-12 on page 286.

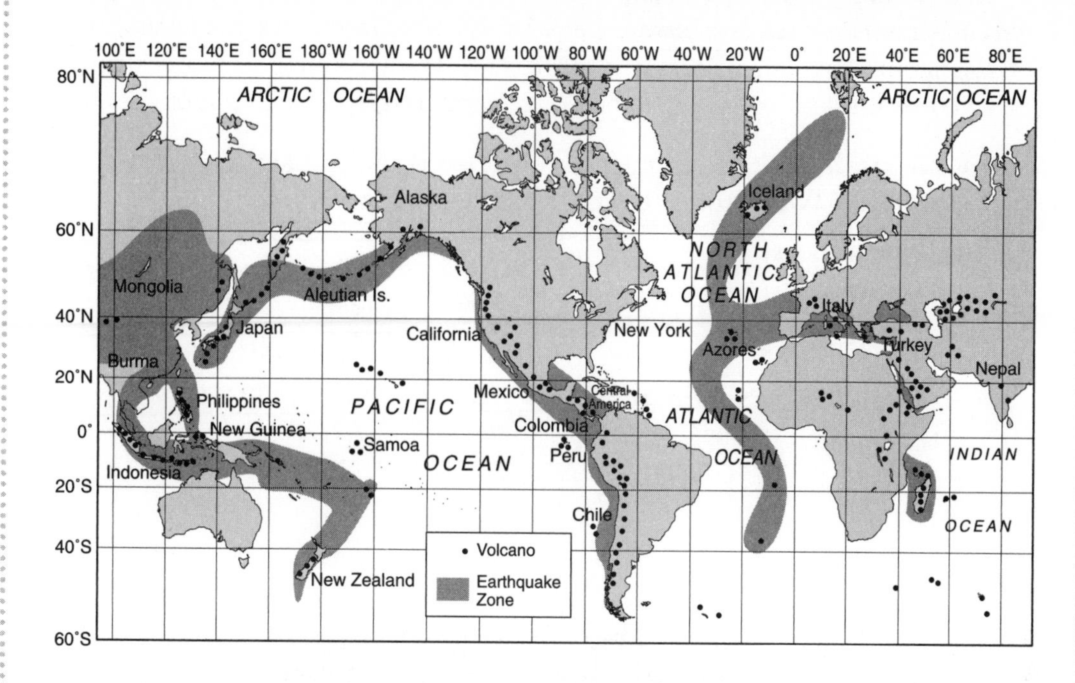

Questions

1. How does the map on page 286 compare with the map above?
2. Use the latitude and longitude of the volcano and earthquake sites listed in the table below to plot the location of each site on a copy of the map. Plot the earthquakes using dots and the volcanoes using triangles.

Latitude and Longitude of 5 Volcanoes and 5 Earthquakes

Volcanoes		*Earthquakes*	
Lat.	*Long.*	*Lat.*	*Long.*
19°N	98°W	31°S	99°W
6°S	105°E	34°N	139°E
38°N	15°E	53°N	168°W
64°N	18°W	36°N	36°W
31°N	131°E	11°N	85°W

3. Determine the location of five recent earthquakes and five recent volcanic events by gathering information from the Internet at the following sites:

Earthquakes: USGS *http://earthquake.usgs.gov/eqcenter/*
Volcanoes: USGS and Smithsonian *www.volcano.si.edu/reports/usgs/*

Make a table listing the latitude, longitude, and location of each earthquake and volcano site. Locate and plot each of the sites on your map.

Part I

9. The position of the sedimentary rock layers in the diagram has changed. What process has changed the rock layers?
   (1) volcanoes
   (2) folding
   (3) weathering
   (4) faulting

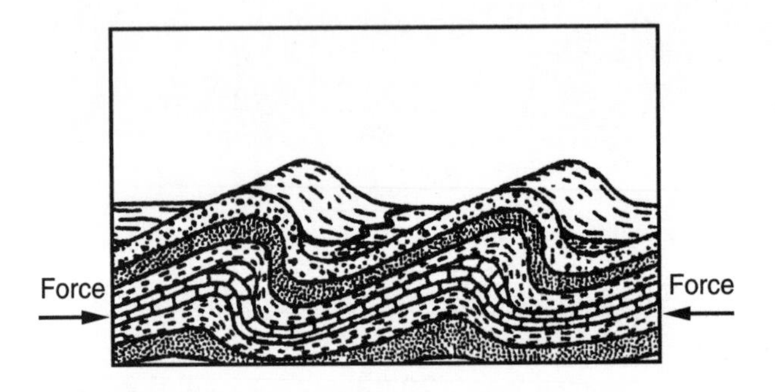

10. Which process can form a mountain?
    (1) erosion
    (2) sediment deposition
    (3) weathering
    (4) faulting

11. The theory of plate tectonics refers to
    (1) movement within Earth's core
    (2) interaction of the lithosphere and atmosphere
    (3) Earth's crust being made up of moving pieces
    (4) Earth's crust providing protection from space rocks

12. Earthquakes, volcanic eruptions, and the building of mountains and ocean basins are events that occur in response to
    (1) heat flow from within Earth
    (2) glacial activity on Earth's surface
    (3) ocean currents
    (4) pressure difference between the lithosphere and atmosphere

13. Most earthquakes are produced by

(1) faulting

(2) weathering and erosion

(3) the formation of plains

(4) deposition of sediment in the ocean

14. Which feature on the ocean floor is formed by new rock material?

(1) continental shelf

(2) deep ocean plain

(3) trench

(4) mid-ocean ridge

15. The table below lists the depth of the ocean at various distances from a continent. What ocean floor feature does the data represent?

Distance from Continent	*Ocean Depth*
50 km	400 m
100 km	9000 m
150 km	1250 m
200 km	1100 m
250 km	1000 m
300 km	950 m

(1) seamount

(2) mid-ocean ridge

(3) trench

(4) continental slope

16. In the profile of the Atlantic Ocean floor, which letter shows the location of the oldest rocks?

(1) A

(2) B

(3) C

(4) D

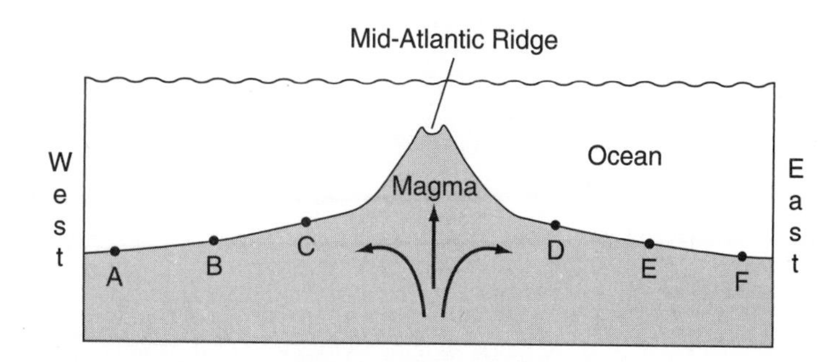

17. In which state is an earthquake least likely to occur?

(1) New York (2) Alaska (3) Oregon (4) California

18. Which block diagram represents the Mid-Atlantic Ridge?

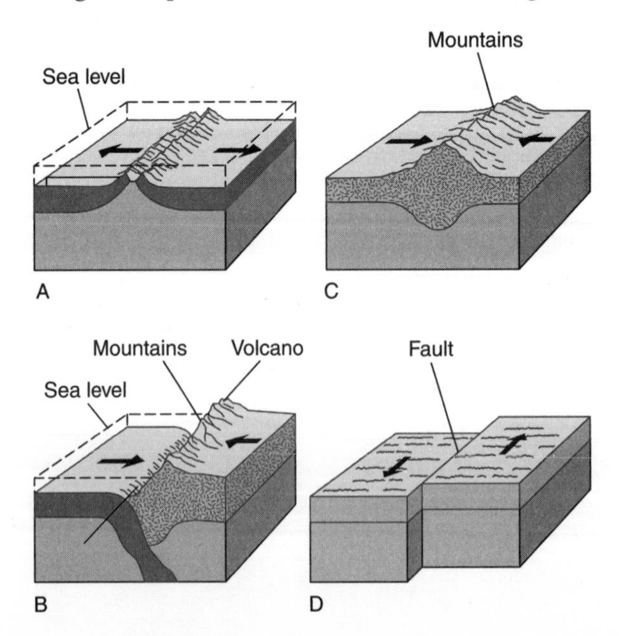

(1) A
(2) B
(3) C
(4) D

19. Earthquakes produce seismic waves that travel through Earth. Seismic waves provide scientists with information about Earth's
(1) structure
(2) size
(3) age
(4) history

20. At first, scientists rejected the theory of continental drift because
(1) Earth's gravity would not allow drifting continents
(2) no evidence existed
(3) the motion of drifting continents could not be observed
(4) the cause of drifting continents could not be identified

Part II

Base your answers to questions 21 and 22 on the passage below, the map on page 289, and your knowledge of science. The passage describes the distribution of earthquakes and the map shows the distribution of earthquakes (shaded area).

Distribution of Earthquakes

Earthquakes can occur anywhere on Earth's surface; however, they occur much more often on the edge of crustal plates. A very high percentage of earthquakes occurs in three zones:

1—Around the edge of the Pacific Ocean
2—Across Europe and Asia
3—Down the center of the Atlantic Ocean

Two recent earthquakes occurred at the following locations.

Earthquake	*Latitude*	*Longitude*
A	36°N	40°E
B	34°N	140°E

21. Based on the latitude and longitude of the two earthquakes in the table, identify the location of each using the world map on page 289. In what earthquake zone is each earthquake located?

22. How are earthquakes and volcanoes distributed on Earth?

Base your answers to questions 23 through 25 on the world map below and your knowledge of science. The world map has a latitude and longitude grid and shows the location of a recent volcanic eruption.

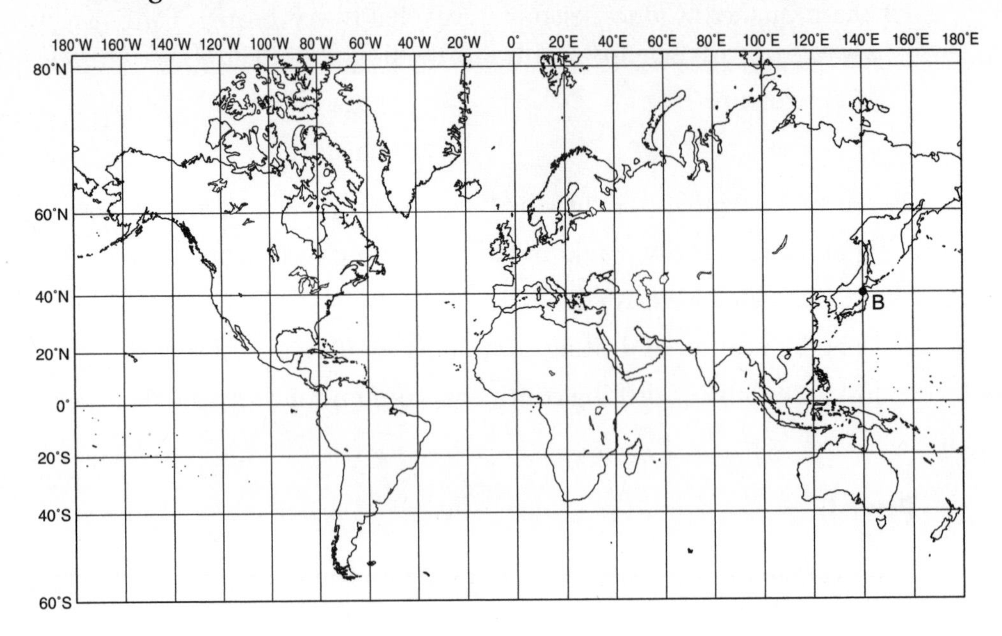

23. A volcanic eruption occurred at point B on the map. What is the latitude and longitude of the volcano? (Your answer must include a value, unit, and direction for each.)
24. A volcanic eruption occurred at 63°N latitude and 20°W longitude. With what ocean floor feature was this earthquake associated?
25. Why is New York State unlikely to have a volcanic eruption?

Base your answers to questions 26 through 28 on the diagram below and your knowledge of science. The diagram shows the Mid-Atlantic Ridge located in the center of the Atlantic Ocean.

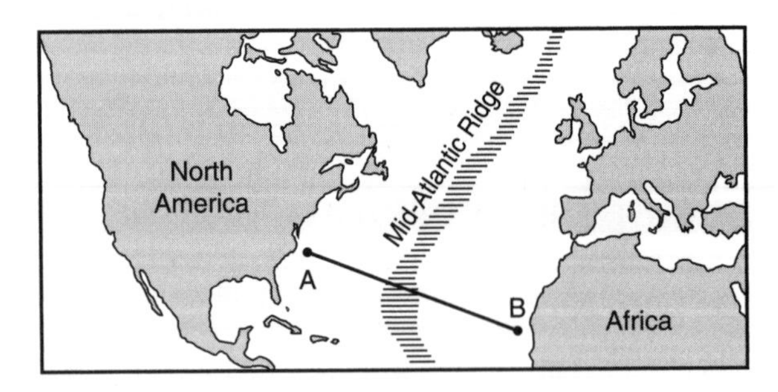

26. How is the distance between points A and B changing?
27. Describe how Earth's crust is changing along the Mid-Atlantic Ridge.
28. List two differences between continental crust and oceanic crust.

Chapter 11 Weather and Climate

What some consider bad weather can look beautiful to others, as shown in this wintery scene.

Major Concepts

- Weather is the short-term condition of the atmosphere over a small area.
- Climate is the long-term average weather over a large area.
- The atmosphere is mostly nitrogen and oxygen. Scientists divide it into layers by temperature. Almost all weather happens in the lowest layer, the troposphere.
- The sun is Earth's primary source of energy. Unequal heating of Earth's surface causes the weather to change.
- An air mass is a large body of air that has similar temperature and humidity throughout. Global winds move air masses across the United States from west to east.
- A high-air-pressure system brings fair weather; a low-air-pressure system brings stormy weather.
- Weather maps show the position of air masses, fronts, and weather conditions.

Vocabulary at a Glance

acid rain
air mass
air pressure
air temperature
altitude
climate
cloud
cold front
front
greenhouse effect
high-pressure system
humidity
hurricane
low-pressure system
pollutant
precipitation
prevailing winds
smog
storm
thunderstorm
tornado
troposphere
warm front
water cycle
weather
weather forecasting
wind
wind direction
winter storm

Weather Variables and Causes

Defining Weather

A layer of gases called the atmosphere surrounds Earth. These gases make up air. ***Weather*** is the short-term condition of the atmosphere over a small area. Weather is described by its conditions, or variables—temperature, air pressure, winds, humidity (moisture content), and clouds. These conditions change every day. The unequal heating of Earth's surface by the sun causes these weather variables to change.

Structure of the Atmosphere

The atmosphere is a layer of gases held to Earth's surface by gravity. About 97 percent of the gases are in the 30 kilometers closest to Earth's surface. The gases become thinner and thinner the higher they rise. The upper atmosphere has no sharp boundary. The top of the atmosphere is set at about 150 kilometers above Earth's surface.

Scientists use the temperature changes that occur with increasing altitude to identify layers in the atmosphere. The lowest layer, the ***troposphere***, is about 10 kilometers thick. Nearly all weather takes place in the troposphere. The troposphere supports life with oxygen, water vapor, and carbon dioxide. As altitude increases in the troposphere, the temperature decreases. The next layer is the stratosphere. The ozone layer, which protects us from ultraviolet light, is part of the stratosphere. Figure 11-1 shows the layers of the atmosphere.

Figure 11-1. *Layers of Earth's atmosphere as described by temperature changes.*

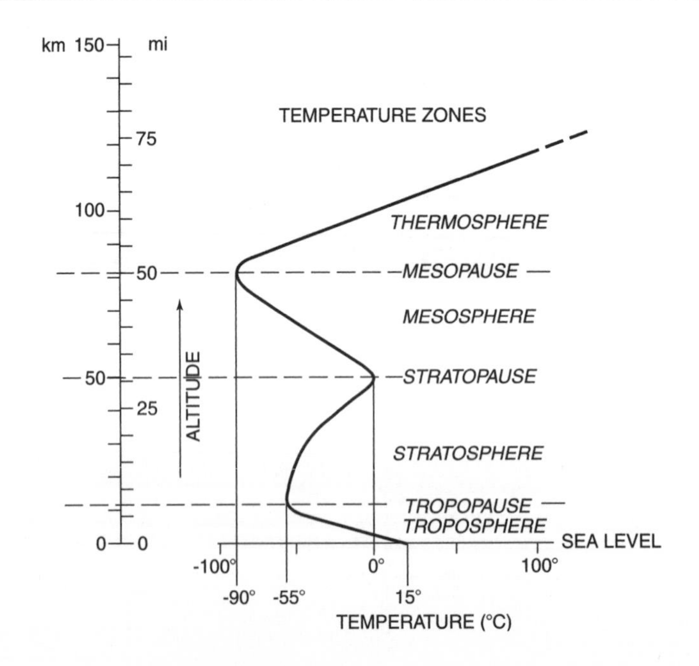

Composition

Table 11-1 lists the average composition of Earth's atmosphere. The atmosphere's composition is relatively constant. The lithosphere, hydrosphere, and atmosphere replace these gases through the water cycle, the nitrogen cycle, and the oxygen–carbon dioxide cycle.

Table 11-1. *Composition of the Atmosphere*

Nitrogen	78%
Oxygen	21%
Carbon dioxide	0.03%
Other gases	0.17%
Water vapor (gas)	1–3%

Weather Variables

Weather is made up of a number of variables, including air temperature, air pressure, humidity, wind speed and direction, clouds, and precipitation. The weather at any location is described in terms of its variables.

1. ***Air temperature*** indicates the amount of heat in the atmosphere. It is measured with a *thermometer*. The two temperature scales commonly used for thermometers are the Celsius scale and the Fahrenheit scale, shown and compared in Figure 11-2. Both scales are divided into units called degrees.

Figure 11-2. *Celsius and Fahrenheit temperature scales compared.*

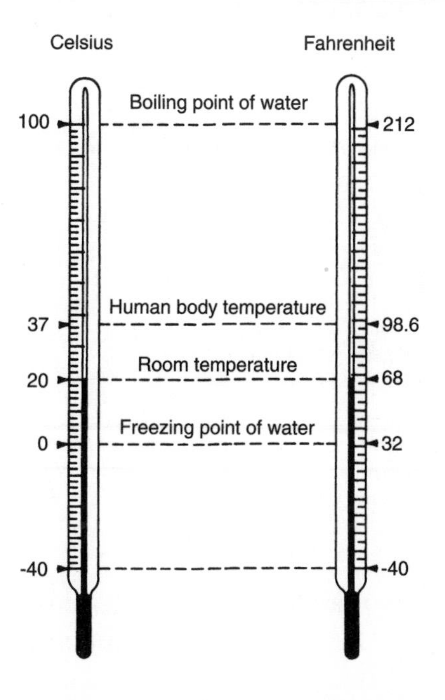

2. Air has mass; it presses down on Earth's surface with a force called ***air pressure***. The force is measured with a barometer, in either inches of mercury or millibars. Figure 11-3 shows some instruments used to measure weather variables. Changes in air pressure are caused mainly by changes in temperature and altitude. Temperature affects air pressure because a volume of cool air is denser and weighs more than the same volume of warm air. Altitude affects air pressure because places at high elevations, being higher in the atmosphere, have less air pressing down on them. Therefore, as altitude increases, air pressure decreases.
3. ***Humidity*** is the amount of water vapor (water in the form of a gas) in the air. Warm air can absorb more water vapor than cool air. Therefore, the temperature of the air determines how much water vapor can be absorbed by the air. Relative humidity is a comparison of the actual amount of water vapor in the air with the maximum amount the air could absorb at that temperature. It is measured with an instrument called a *psychrometer*. A wet-and-dry-bulb thermometer is one type of psychrometer.
4. ***Wind*** is the horizontal movement of air over Earth's surface. Wind speed is a measure of how fast the air is moving, in miles or kilometers per hour. Wind speed is measured with an *anemometer*. ***Wind direction*** is the direction from which the wind is coming. For instance, a wind blowing from north to south is called a north wind. Wind direction is determined with a *wind vane*, which points in the direction from which the wind is blowing.

Figure 11-3. *Weather instruments: A barometer measures air pressure. A psychrometer (wet- and dry-bulb thermometers) is used to determine humidity. A wind vane shows wind direction. The anemometer measures wind speed.*

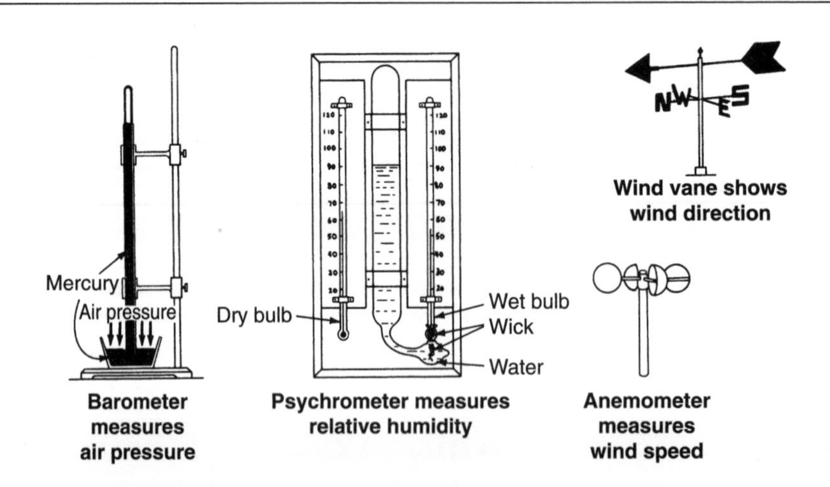

5. A ***cloud*** is a mass of tiny water droplets or ice crystals floating high in the atmosphere. Cloudiness, the amount of sky covered

by clouds, is described by phrases like "partly cloudy" or "mostly cloudy." A sky completely covered by clouds is described as "overcast."

6. *Precipitation* is water, in any form, that falls from the atmosphere. Temperature determines the type of precipitation produced. Rain is a liquid form of precipitation. Snow, sleet, and hail are solid forms of precipitation. A *rain gauge* measures the amount of precipitation in inches.

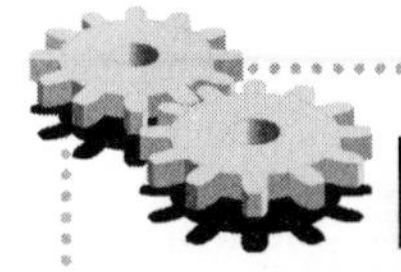

Laboratory Skill

Reading Weather Instruments

Scientists use instruments to measure weather variables such as temperature and air pressure. Each of these variables is measured in more than one unit. The following text describes the units used for each instrument.

Thermometers measure air temperature in degrees Celsius or degrees Fahrenheit. If the scale is calibrated for Celsius, the number scale starts at about −40°C and goes to 50°C. For Fahrenheit, the number scale starts at about −40°F and goes to 120°F. Weather data in the United States is usually given in degrees Fahrenheit; however, some weather forecasts also give it in degrees Celsius.

Barometers measure air pressure in inches of mercury or millibars. If the scale is in inches of mercury, weather readings generally fall between 29.00 and 31.00 inches of mercury. When the scale is in millibars, readings fall between 980 and 1050 millibars. Inches of mercury are still used in weather forecasting; however, weather maps use millibars.

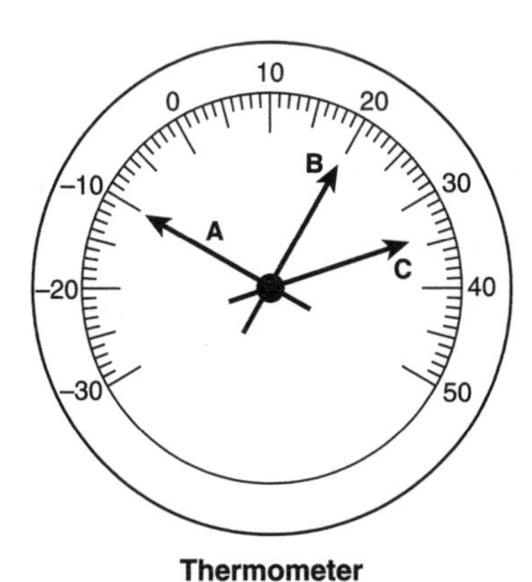

Thermometer

Barometer

Questions

1. What are the units on the thermometer scale?
2. What are the units on the barometer scale?

3. The letters *A*, *B*, and *C* indicate pointer positions on the thermometer. What value is each pointer indicating on the thermometer scale? (Label the units for each of your answers.)
4. The letters *D*, *E*, and *F* indicate pointer positions on the barometer. What value is each pointer indicating on the barometer scale? (Label the units for each of your answers.)
5. Copy the thermometer and barometer into your notebook. On your copy, draw a pointer to indicate each temperature and air pressure reading listed below. (DO NOT WRITE IN YOUR BOOK.)

Temperature: (G) 0°C
(H) −18°C
(I) 27°C

Air Pressure: (J) 29.40 inches of mercury
(K) 30.35 inches of mercury
(L) 30.08 inches of mercury

The Sun's Energy

Earth gets almost all its energy from the sun in the form of radiant energy. Light from the sun travels through space and strikes Earth. Much of the energy that reaches Earth's surface radiates back into the atmosphere. The sun does not heat Earth's surface evenly; therefore, the atmosphere is not heated evenly. This uneven heating of the atmosphere causes weather changes.

Different surfaces have different abilities to absorb heat. For example, pavement and sand absorb more heat than do grass and water. The surfaces of oceans, forests, and deserts each absorb heat from the sun differently. These surfaces, in turn, heat the air above them differently, producing variations in air temperature.

When air is heated, it becomes less dense (lighter) than the surrounding air. Therefore, warm air rises. Cool air is more dense (heavier) so it tends to sink. As air rises or sinks, the surrounding air rushes in to replace it, causing air to circulate. This produces convection currents in the air. (See Figure 11-4.) You learned about the transfer of heat by convection in Chapter 2. Convection can take place over a few kilometers or over thousands of kilometers.

The amount of heat Earth's surface absorbs also depends on the angle of the sun's rays as they strike the surface. (See Figure 12-7 on page 335.) Near the equator, the sun's rays strike Earth vertically or nearly so, since the sun rises high in the sky every day. This concentrates the sun's energy within a small area, heating the surface very effectively. However, since Earth's surface is curved, the sun's rays strike areas away from the equator at an angle. This spreads energy over a wider area, heating Earth's surface less effectively. The changing angle of the sun's rays plays a major role in seasonal temperature changes.

Figure 11-4. *Warm air rises and cool air sinks, creating a convection circulation pattern.*

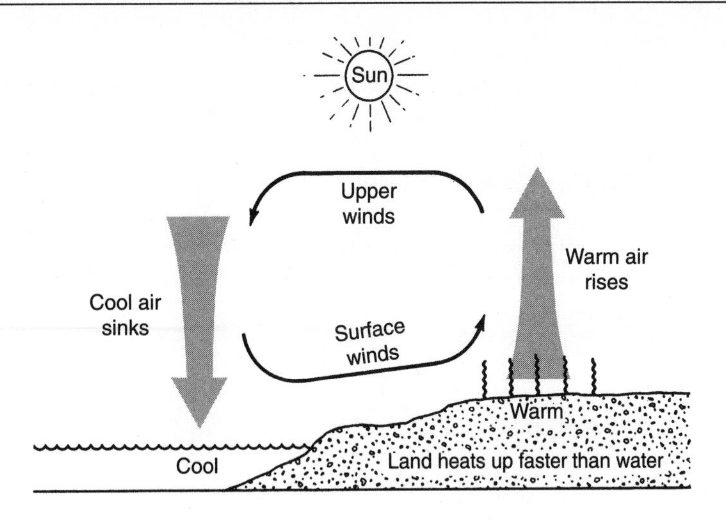

The uneven heating of Earth's curved surface causes hotter air close to the equator to rise and spread to the north and south. At the same time, cooler air near Earth's poles moves toward the equator to replace the rising air. Earth's rotation breaks up this simple circulation into complex global wind belts, shown in Figure 11-5, in which winds blow in six broad belts around Earth.

Figure 11-5. *Global winds are separated into six broad wind belts around Earth.*

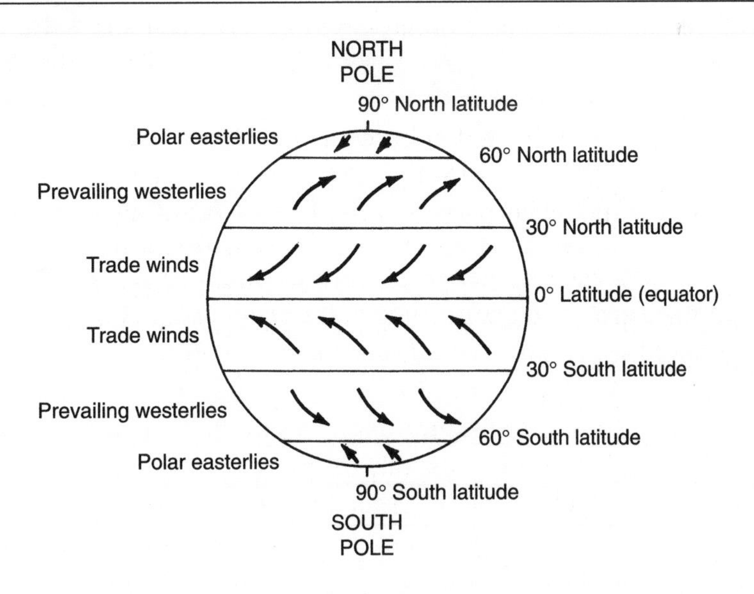

North of the equator, between 0° and 30°, the ***prevailing winds*** generally blow from the northeast, and are called the *trade winds*. Between 30° and 60°

north latitude, the winds generally blow from the southwest and are called the *prevailing westerlies*. Most of the United States lies in the prevailing westerlies. These winds cause most weather systems to move from west to east across the United States. Between 60° and 90° north latitude, the North Pole, there are the *polar easterlies*. South of the equator the wind belts follow a similar pattern.

The Water Cycle

The sun's energy also powers the ***water cycle***. The water cycle moves water between Earth's surface and the atmosphere by evaporation, condensation, and precipitation. (See Figure 11-6.) Heat from the sun converts liquid water into water vapor. This is *evaporation*. For example, evaporation causes a puddle of rainwater to shrink and dry up on a hot, sunny day. Water enters the atmosphere by evaporation from oceans, lakes, and rivers, and by plants releasing water vapor through their leaves.

Figure 11-6. *In the water cycle, water circulates from the clouds to the ground and back to the clouds.*

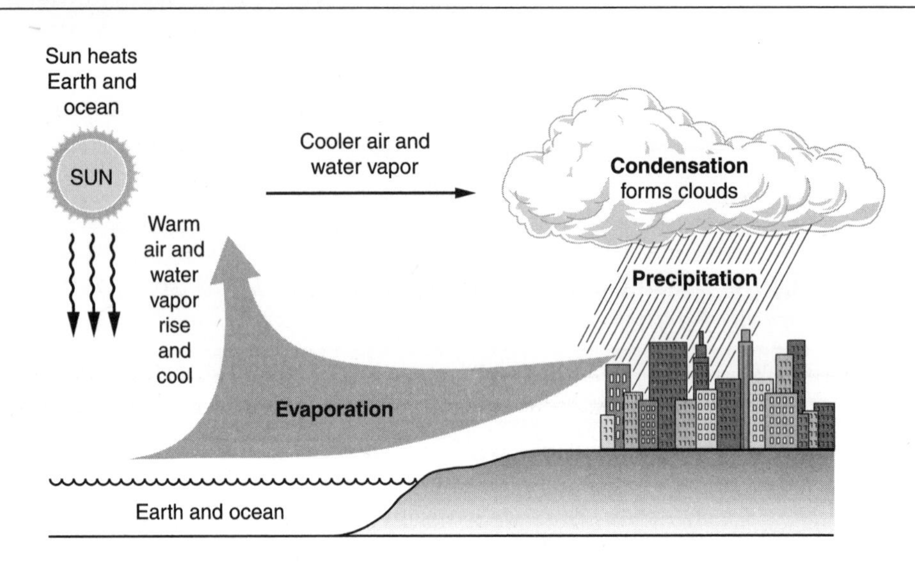

Warm air can absorb more water than cool air can. Therefore, when warm air rises and cools, some of the water vapor changes back into droplets of liquid water. This is *condensation*, which produces clouds, fog, and dew.

1. When warm-humid air rises in the atmosphere and cools, condensation forms tiny droplets of water. The water droplets form clouds. If enough water vapor condenses, the tiny water droplets join, forming larger, heavier drops that fall to Earth as *rain*. Rain is a liquid form of precipitation. Under colder conditions the precipitation may fall in solid form, such as *snow, sleet,* or *hail*.
2. Condensation may also form tiny water droplets that remain suspended in the air close to the ground. This is called *fog*. For

example, when you breathe warm-humid air out from your mouth on a cold day, a little fog forms.

3. *Dew* forms when water vapor condenses on cool surfaces. When cool nights follow warm days in the fall, dew forms on the surface of grass. A similar process occurs when water droplets form on the outside of a glass that holds a cold drink. (See Figure 11-7.)

Figure 11-7. *Condensation forms water droplets on the outside surface of a glass holding a cold drink.*

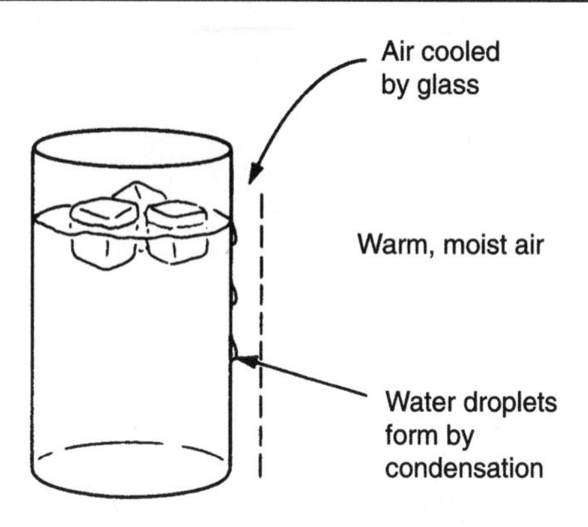

Climate

Climate is the long-term average weather over a large area. A number of factors combine to produce different climates.

1. *Latitude. Latitude* refers to distance from the equator. Places at high latitudes, far from the equator, tend to have colder climates than places at lower latitudes, close to the equator. For instance, Canada is at a higher latitude than Mexico, so it has a colder climate.
2. *Altitude.* ***Altitude*** is the height (elevation) of a place above sea level. Higher elevations are cooler than lower elevations. Just as a mountaintop is colder than its base, a city at a high altitude will have a colder climate than a nearby city at a lower altitude.
3. *Large Bodies of Water.* Land areas close to oceans or large lakes usually have more moderate climates (cooler summers and warmer winters) than areas far from water. Water absorbs and gives off heat more slowly than land does. Therefore, as the land heats up during summer, the water stays relatively cool. This keeps coastal areas cooler in summer than places farther inland. In winter, the situation is reversed. The water loses heat that built up during summer more slowly than the land does. This keeps coastal areas warmer in winter than areas farther inland.
4. *Mountain Barriers.* The side of a mountain range facing the global winds (windward side) tends to have a cool, humid climate,

while the opposite side (leeward side) of the mountains has a warmer, drier climate. This is illustrated in Figure 11-8.

Figure 11-8. *A mountain range can affect climate. A cool, humid climate occurs on the side of the mountain where the winds force air up the mountain, and a warm, dry climate occurs on the side of the mountain where air descends.*

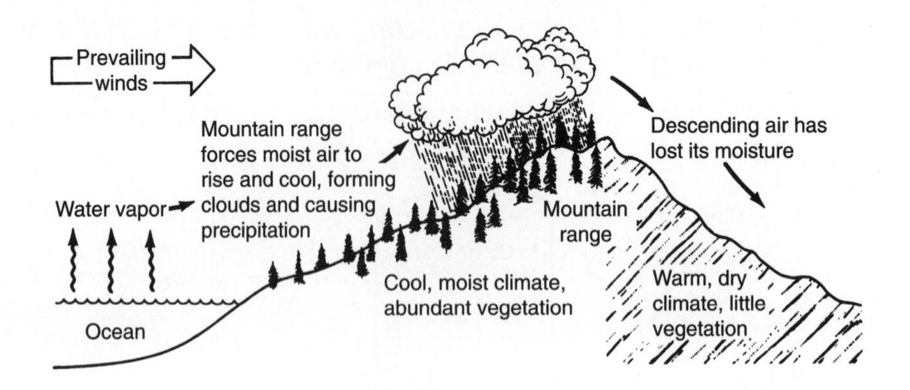

5. *Ocean Currents.* Ocean currents carry warm water away from the equator and cool water toward the equator, affecting land temperatures. Great Britain and Canada are at approximately the same latitude and same altitude. However, due to the warm ocean current that flows past Great Britain, the winters are much milder than those in Canada.

Review Questions

Part I

1. What temperature is indicated by the thermometer in the diagram?

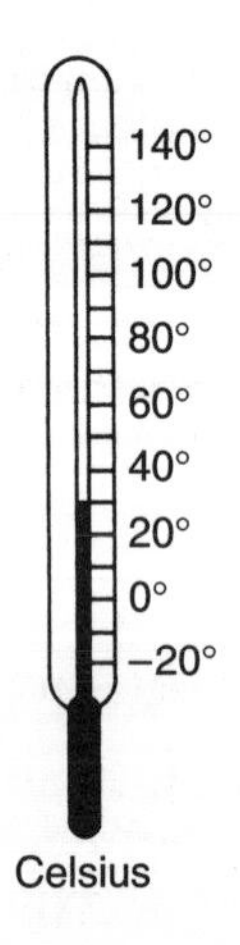

(1) 10°C (2) 20°C (3) 25°C (4) 30°C

2. In the diagram, the most likely reason that City A has a warmer climate than City B is that

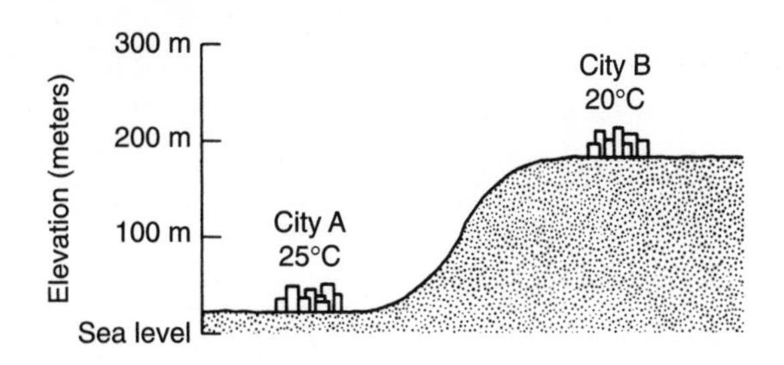

(1) City B is at a different latitude.
(2) City A is at a lower altitude.
(3) City B is closer to the ocean.
(4) City B is closer to the sun.

3. The diagram below shows two cities, X and Y, and their positions on a continent. How will the climates of the cities compare?

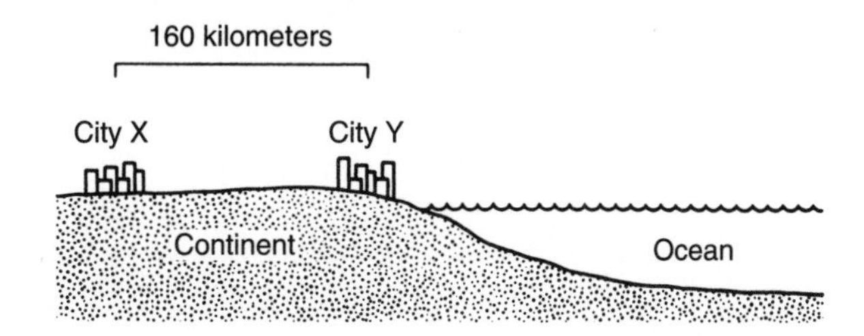

(1) City Y will have warmer summers and colder winters than City X.
(2) City Y will have cooler summers and warmer winters than City X.
(3) City Y will have cooler summers and colder winters than City X.
(4) City Y will have warmer summers and warmer winters than City X.

4. The most likely reason that Miami has a warmer climate than New York City is

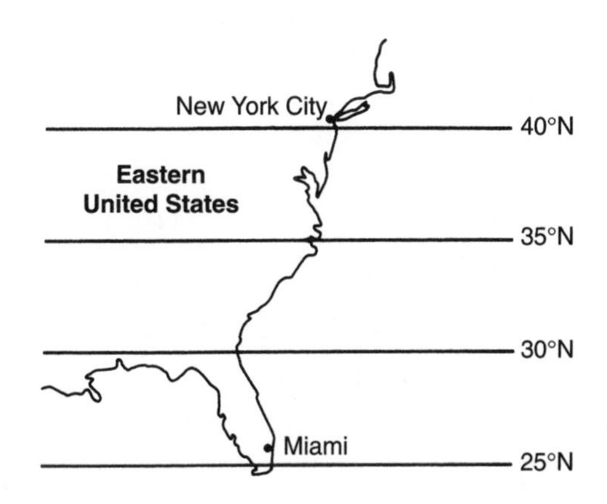

(1) a difference in distance from the ocean
(2) a difference in altitude
(3) a difference in air pressure
(4) a difference in latitude

5. Which graph indicates the general temperature change as you travel from the North Pole (NP) to the equator?

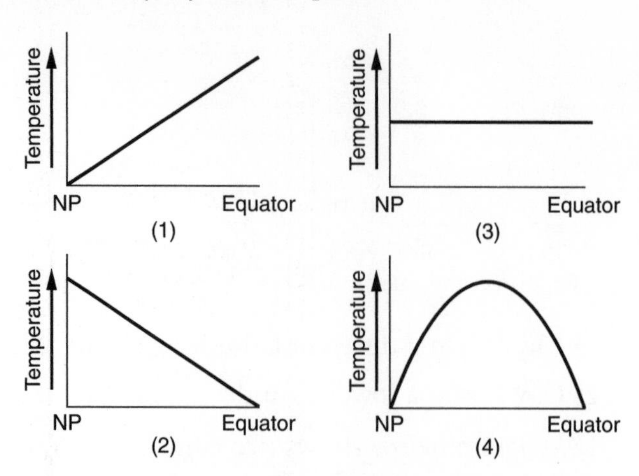

Base your answers to questions 6 and 7 on the diagram below, which shows the water cycle.

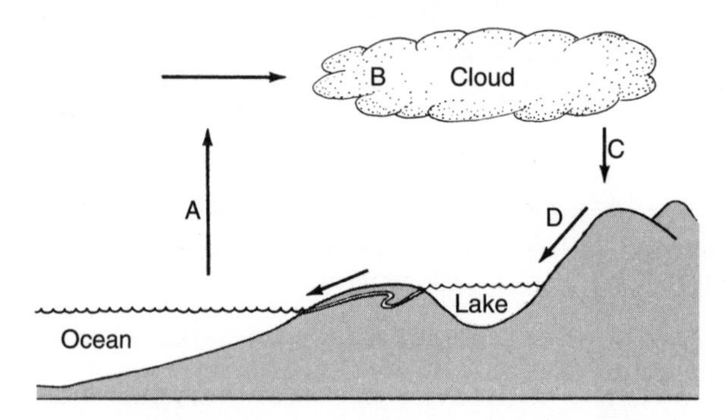

6. Which letter in the diagram indicates falling precipitation?

   (1) A (3) C

   (2) B (4) D

7. Which process is taking place at position B?

   (1) evaporation

   (2) condensation

   (3) precipitation

   (4) transpiration

8. The source of energy that sets Earth's atmosphere in motion and causes weather changes is

   (1) volcanism

   (2) gravity

   (3) the ocean

   (4) the sun

Base your answers to questions 9 and 10 on the diagram on page 307, which shows a temperature comparison of a Celsius and Fahrenheit thermometers.

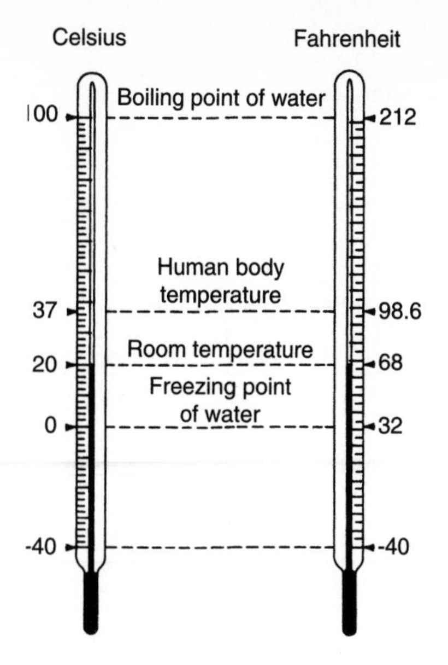

9. What temperature on the Fahrenheit scale is equivalent to 25°C?

   (1) 170°F (2) 100°F (3) 77°F (4) −4°F

10. What temperature on the Celsius scale is equivalent to 5°F?

   (1) −15°C (2) 0°C (3) 20°C (4) 68°C

Base your answer to question 11 on the diagram below, which shows the temperature distribution in the atmosphere.

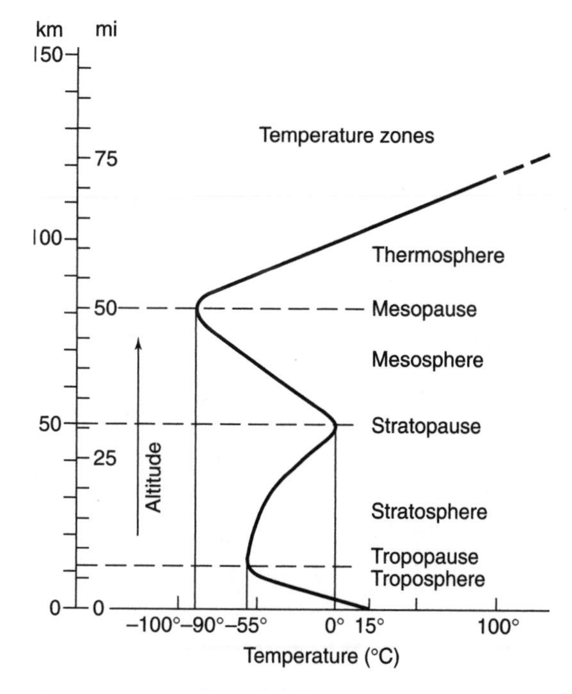

11. As you travel from an altitude of 0 km to an altitude of 80 km, the temperature

   (1) increases, decreases, and increases

   (2) decreases

(3) remains the same

(4) decreases, increases, and decreases

Base your answer to question 12 on the table below, which shows weather station information from Binghamton, New York.

Temperature	−10°C
Air pressure	29.82 inches of mercury
Wind	Southeast 5–10 miles per hour
Relative humidity	100%
Cloud cover	100%

12. What type of precipitation is most likely occurring?

(1) heavy rain

(2) snow

(3) drizzle

(4) light rain

Part II

Base your answers to questions 13 through 15 on the diagram below and your knowledge of science. The diagram shows a beach on a hot summer day. Positions A and C are 1 meter above the surface and positions B and D are 1500 meters above the surface.

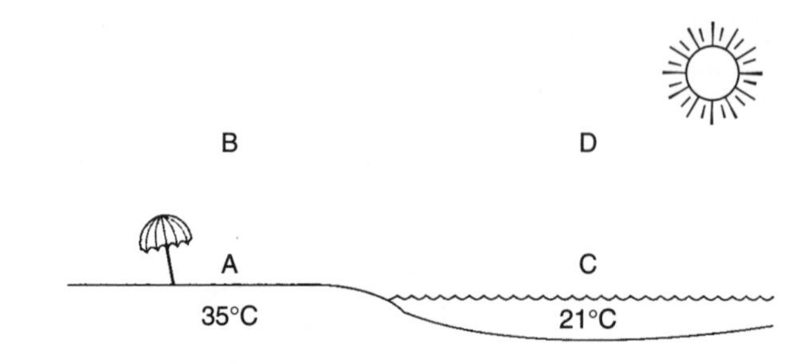

13. Air is sinking from position ___ to position ___.

14. At what position are clouds most likely forming?

15. At position A, what heat-transfer process heats the air?

Base your answers to questions 16 and 17 on the diagram on page 309 and your knowledge of science. The diagram shows a car driving up Whiteface Mountain in New York State. The car carries a thermometer and a barometer.

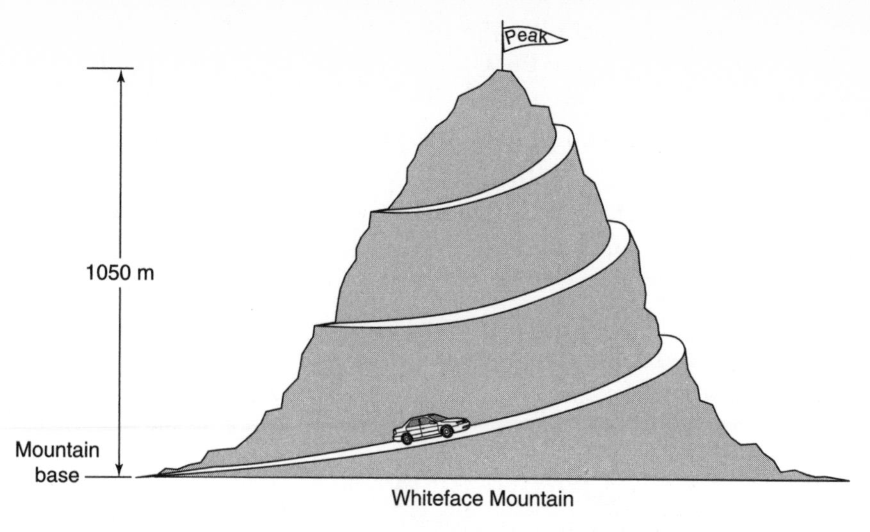

16. How will the thermometer reading change as the car goes up the mountain?

17. How will the barometer reading change as the car goes up the mountain?

Base your answers to questions 18 through 21 on the following information and your knowledge of science.

After a rain shower on a summer day, the sun came out and the puddle in front of Aldo's house evaporated quickly. A few days later another rain shower occurred, and the sky remained cloudy. This time, the puddle took much longer to evaporate. Aldo thinks the sun caused the puddle to evaporate quickly. Design an experiment to test Aldo's hypothesis. Include the following information:

18. Clearly state the hypothesis.

19. State the factor to be changed.

20. State two constant factors that will not be changed.

21. Describe the procedure.

Large-Scale Weather Systems

Air Masses

An ***air mass*** is a large body of air that has about the same conditions of temperature and humidity throughout. Air masses develop when air stays over a large area of Earth's surface and takes on the temperature and humidity characteristics of the land or water under it. Air masses that develop over land are dry. Air masses that develop near the poles are cold. For example, air masses that form over Canada are usually cool and dry because they formed over a cool land surface. Warm air masses develop near the equator. Humid air masses form over water. For example, an air mass that forms over a warm body of water, like the Gulf of Mexico, will be warm and humid.

Air masses that affect the continental United States, shown in Figure 11-9, enter the country from the north, west, and south. They move from west to east, pushed by the global winds. As an air mass moves, it changes the local weather conditions below it at Earth's surface. The weather may become warmer or cooler, wetter or drier, depending on the type of air mass passing.

Figure 11-9. *Six major air masses affect the weather across the United States.*

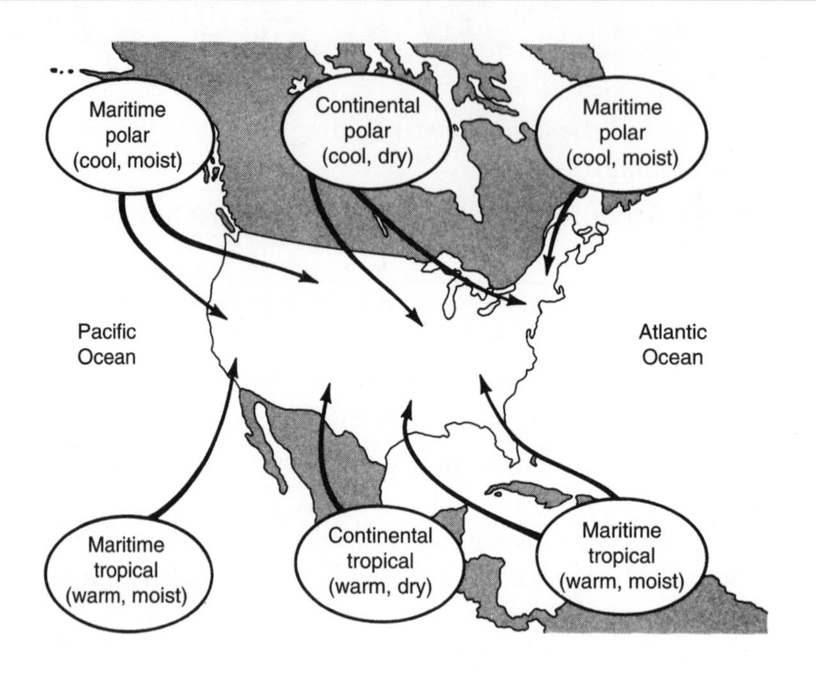

Process Skill 1

Creating a Table

Air masses are named for the temperature and humidity found in the region over which they form. Continental air masses form over land and are dry. Maritime air masses form over water and are humid. Tropical air masses form south of the United States and are warm. Polar air masses form north of the United States and are cool.

An air mass that forms over Canada is dry because it forms over land. It is also cool because it forms north of the United States. This air mass is called continental polar. A maritime tropical air mass would be humid and warm.

Information can sometimes be put into an orderly form by creating a table. For instance, the table below represents information about air masses. Copy the table in your notebook and fill in the information missing from each column.

Name of Air Mass	*Temperature (Cool or Warm)*	*Moisture (Humid or Dry)*	*Direction It Enters the U.S.*
Continental Polar			
Maritime Polar			
Continental Tropical			
Maritime Tropical			

High- and Low-Pressure Systems

Surface air pressure is highest in the center of an air mass, called a ***high-pressure system***, or high. The air in a high-pressure system sinks, and winds blow outward from the center, turning in a clockwise direction in the Northern Hemisphere, as shown in Figure 11-10(a). High-pressure systems usually bring clear skies, dry weather, and gentle winds.

Figure 11-10. *Movement of air currents in (a) a high-pressure system and (b) a low-pressure system (mb stands for millibars).*

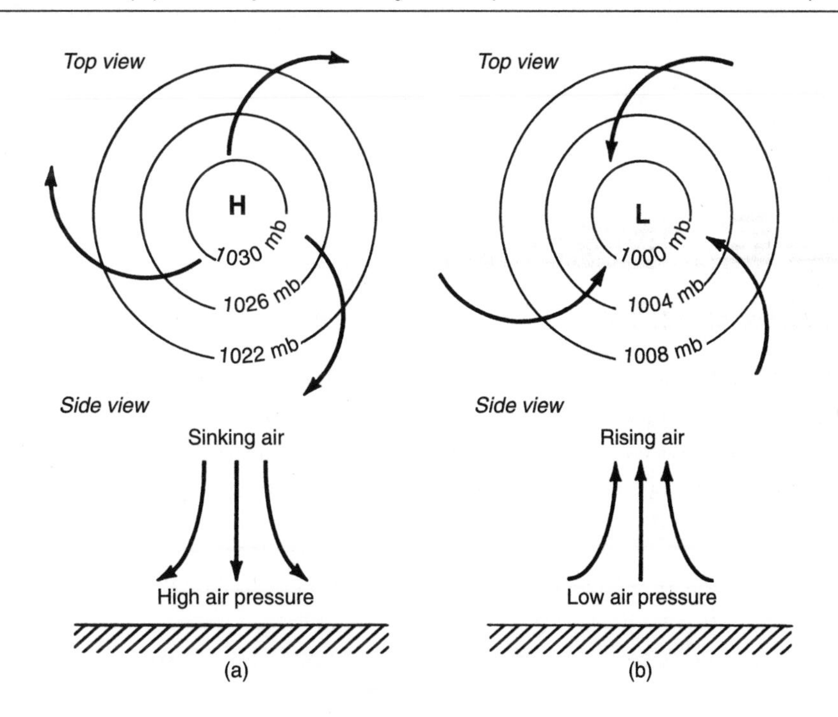

Surface air pressure is lowest along the edge of an air mass. Along the edge of an air mass a ***low-pressure system***, or low, may form. The air in a low-pressure system rises, and winds spiral in toward the center in a counter-clockwise direction in the Northern Hemisphere, as shown in Figure 11-10(b). (Note that in both highs and lows, winds always blow from areas of high air pressure toward areas of low air pressure.) Low-pressure systems usually bring cloudy, wet weather, often with strong, gusty winds.

Highs and lows are shown on weather maps by the letters **H** and **L**. (See Figure 11-15 on page 315.) The **H**'s on weather maps are the centers of air masses. A low forms between highs, much like a valley between two mountains. Changes in air pressure readings usually mean that a high or low is passing by.

Fronts

When air masses meet, a boundary, called a ***front***, forms between them. Sudden changes in weather conditions can occur as a front passes. The air masses that meet often have different temperatures, humidities, and densities. These differences prevent the air masses from mixing. The cooler, drier air is denser and remains close to the ground. The warmer, more humid air is less dense and rises upward. The rising air forms a low-pressure area along the front, often producing stormy weather conditions.

There are four types of fronts. The type of front formed depends on the types and motion of the air masses that meet.

1. *Stationary Front.* When a cool air mass and a warm air mass meet, but neither air mass moves the other, a stationary front forms. The map symbol for a stationary front is a line with arrowheads on the warm side and half circles on the cold side. (See Figure 11-11.)

Figure 11-11. *A stationary front forms when a cold air mass and a warm air mass meet and neither air mass pushes the other.*

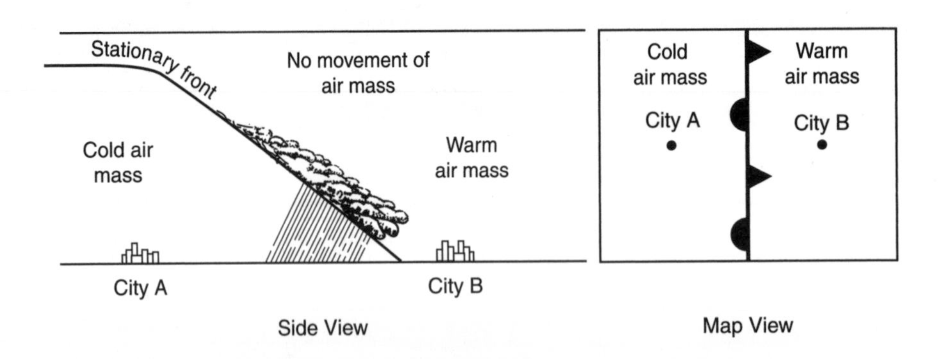

2. *Cold Front.* A cold air mass pushing into and under a warm air mass forms a ***cold front***. Cold fronts usually bring stormy

weather and cooler temperatures. The map symbol for a cold front is a line with arrowheads on one side, pointing in the direction the front is moving. (See Figure 11-12.)

Figure 11-12. *A cold front forms along the leading edge of a forward-moving cold air mass.*

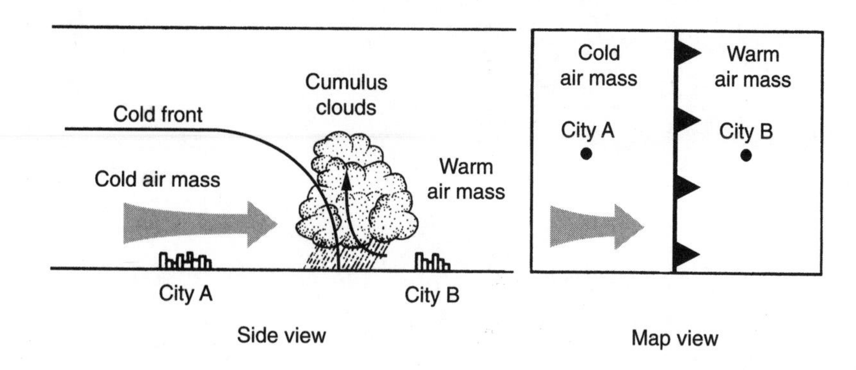

3. *Warm Front.* A warm air mass pushing into and over a cold air mass forms a ***warm front***. The warm air slides up and over the cooler air. Warm fronts usually bring light precipitation and warmer temperatures. The map symbol for a warm front is a line with half circles on one side, showing the direction the front is moving. (See Figure 11-13.)

Figure 11-13. *A warm front forms along the leading edge of a forward-moving warm air mass.*

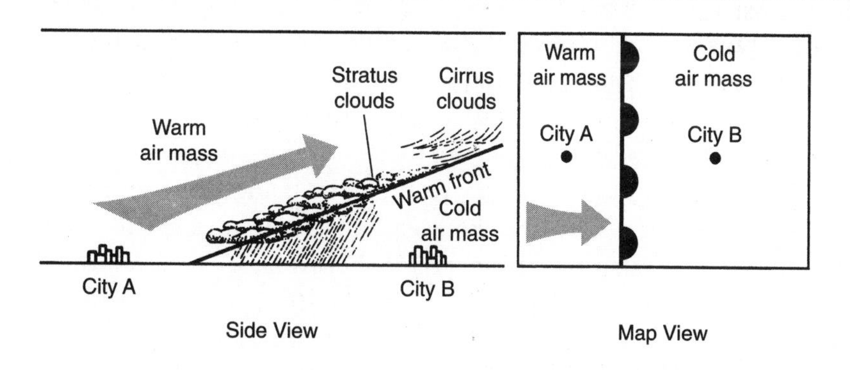

4. *Occluded Front.* An occluded front forms when cool air is trapped and squeezed upward between a fast-moving cold front and a slow-moving warm front. This produces stormy weather conditions. The map symbol for an occluded front is a line with arrowheads and half circles on the same side, showing the direction the front is moving. (See Figure 11-14.)

Figure 11-14. *An occluded front forms when a cool air mass is squeezed upward between a fast-moving cold front and a slow-moving warm front.*

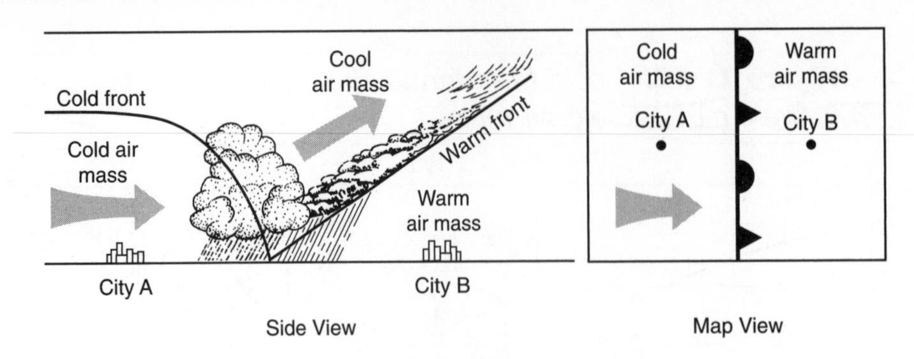

The way air rises determines the type of clouds that form. If air is pushed up vertically, as it is along a cold front, puffy *cumulus* clouds form. If air rises at a low angle, as it does along a warm front, flat layers of *stratus* clouds form. At higher altitudes, wispy *cirrus* clouds made up of ice crystals form. They may look like feathers, or tufts of hair. *Nimbus* clouds are dark clouds that usually produce rain or snow.

Weather Forecasting

Weather forecasting is an attempt to make accurate predictions of future weather. The accuracy of weather forecasting improves as technology improves. In addition to older technology such as weather balloons, thermometers, and barometers, weather forecasters now have weather satellites, Doppler radar, and computers to provide data.

Short-range local forecasts are relatively easy. They are based mostly on air pressure, wind direction, and cloud types. Changes in these weather variables are usually good predictors of the weather for the next 24 hours. (1) Decreasing air pressure suggests that stormy weather is likely. Rising air pressure suggests that fair weather is coming. (2) In New York State, winds blowing from the north or west usually bring fair weather. Winds from the south or east often bring wet weather. (3) An increase in cloudiness is a sign that a front is approaching, probably bringing precipitation.

Weather forecasters use information from weather satellites and Doppler radar to produce up-to-date weather maps like the one in Figure 11-15. These maps can help us predict coming changes. Weather systems, pushed by the prevailing westerlies, generally move from west to east across the United States. Therefore, if a weather map shows a high immediately to our west, we can forecast fair weather for the next day. On the other hand, if the map shows a low to our west, we can expect stormy weather.

Long-range weather forecasting is difficult. Computers, satellite photographs, and Doppler radar images allow forecasters to track the movements of weather systems, like air masses and fronts, over great distances.

Figure 11-15. *A weather map provides the weather data used to make weather forecasts.*

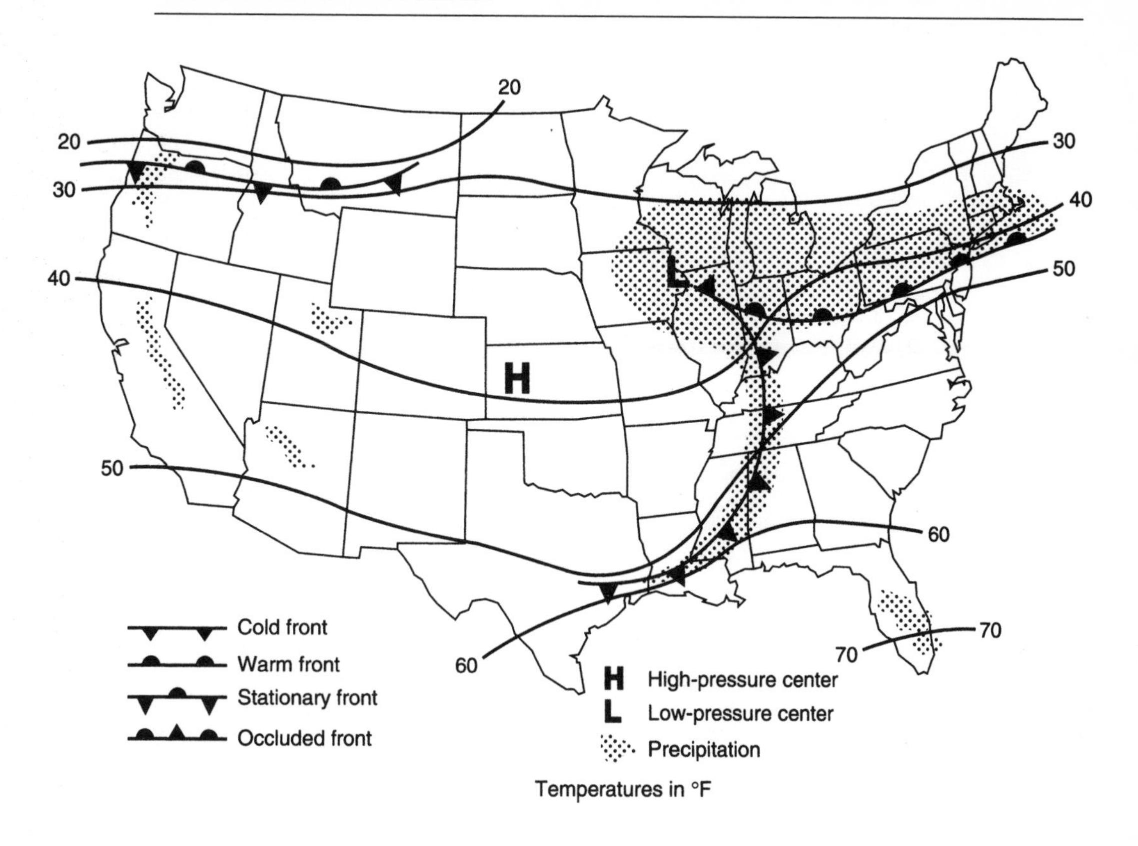

With this information, they can predict the weather several days ahead. However, such forecasts are not always accurate. Generally, weather forecasts for the next day or two are reliable, while long-range forecasts usually have to be revised.

Process Skill 2

Predicting Weather Changes

You can make fairly good 6-to-12-hour weather predictions using the table "Weather Forecasting by Wind and Barometer." You will need three pieces of information to use the table: (1) the wind direction, (2) the barometer reading, and (3) the tendency of the barometer to rise or fall. Measure local wind conditions and barometer readings or get the readings from a weather report given on television, radio, or the Internet. To use the table, locate the present wind direction in the column on the left. Then find the best-fit barometer reading in the center column. The weather forecast is found in the column on the right.

Weather Forecasting by Wind and Barometer

Wind Direction	*Barometer Reading*	*Forecast*
SW to NW	30.20 and above; stationary	Continued fair, with no temperature change
SW to NW	30.20 and above; falling slowly	Slowly rising temperature and fair for 2 days
SW to NW	30.10 to 30.20; steady	Fair, with slight temperature changes for 1 to 2 days
SW to NW	30.10 to 30.20; rising rapidly	Fair, followed within 2 days by rain
S to SE	30.10 to 30.20; falling slowly	Rain within 24 hours
S to SE	30.10 to 30.20; falling rapidly	Wind increasing, with rain within 12 to 24 hours
SE to NE	30.10 to 30.20; falling slowly	Rain in 12 to 18 hours
SE to NE	30.10 to 30.20; falling rapidly	Increasing wind, rain within 12 hours
E to NE	30.10 and above; falling slowly	In summer, with light winds, rain may not fall for several days. In winter, rain within 24 hours.
E to NE	30.10 and above; falling rapidly	In summer, rain likely within 12 to 24 hours. In winter, rain or snow, with increasing winds, often starts when barometer begins to fall and the wind sets in from the NE.
SE to NE	30.00 or below; falling slowly	Rain will continue 1 to 2 days
SE to NE	30.00 or below; falling rapidly	Rain, with high wind, followed within 36 hours by clearing; in winter by colder temperatures
S to SW	30.00 or below; rising slowly	Clearing within a few hours; fair for several days
S to E	29.80 or below; falling rapidly	Severe storm imminent, followed within 24 hours by clearing; in winter by colder temperatures
E to N	29.80 or below; falling rapidly	Severe northeast gale and heavy precipitation; in winter, heavy snow followed by a cold wave
Changing to W	29.80 or below; rising rapidly	Clearing and colder

Using the the table "Weather Forecasting by Wind and Barometer," forecast the next day's weather from the four sets of weather conditions below.

Winds	*Barometer Reading*	*Tendency*
(1) Southeast	30.10	Falling rapidly
(2) Northwest	30.15	Steady
(3) Southeast	29.75	Falling rapidly
(4) West	30.30	Stationary

Part I

22. During summer, air masses that form over the Gulf of Mexico often affect the weather in New York State. Such an air mass would be
    (1) dry and warm
    (2) dry and cool
    (3) humid and warm
    (4) humid and cool

Base your answers to questions 23 and 24 on the weather map.

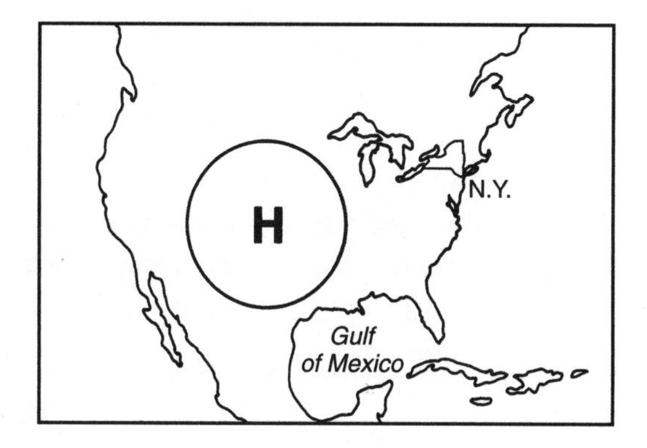

23. What does the **H** on the weather map indicate?
    (1) high temperature
    (2) high air pressure
    (3) high amount of precipitation
    (4) high clouds

24. As the air mass moves, what section of the United States will it most likely affect in the next few days?
    (1) North
    (2) South
    (3) East
    (4) West

25. Stormy weather conditions are associated with
    (1) high-pressure systems
    (2) low-pressure systems
    (3) centers of air masses
    (4) cold, dry air

26. At which city in the diagram would the air pressure most likely be the highest?

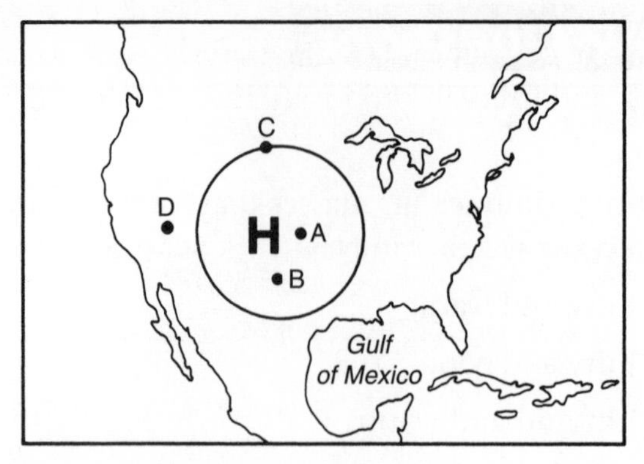

(1) city A (2) city B (3) city C (4) city D

27. A rising barometer indicates

(1) a high-pressure system is coming

(2) a low-pressure system is coming

(3) a high-temperature system is coming

(4) stormy weather is coming

28. Andrew observed the weather from Monday night to Tuesday night. First it was warm and humid, then there were thunderstorms, and finally the air became cooler and drier. These changes were most likely due to the passing of a

(1) warm front
(2) cold front
(3) hurricane
(4) wind belt

29. On a warm summer day, Tony noticed drops of water appear on the outside of a glass of ice water. Where did the water on the outside of the glass come from?

(1) the ice

(2) the water in the glass

(3) the water vapor in the air

(4) a crack in the glass

30. Which factor has the greatest influence on the movement of an air mass?

(1) ocean currents
(2) global winds
(3) sea breeze
(4) mountain barriers

31. If on January 15 the people of New York City are having a sunny but very cold day, they are probably in an air mass that formed in

(1) Canada

(2) the North Atlantic Ocean

(3) the Gulf of Mexico

(4) Mexico

Part II

Base your answers to questions 32 through 34 on the diagram and your knowledge of science. The diagram shows a warm front between two air masses, and four labeled locations—A, B, C, and D.

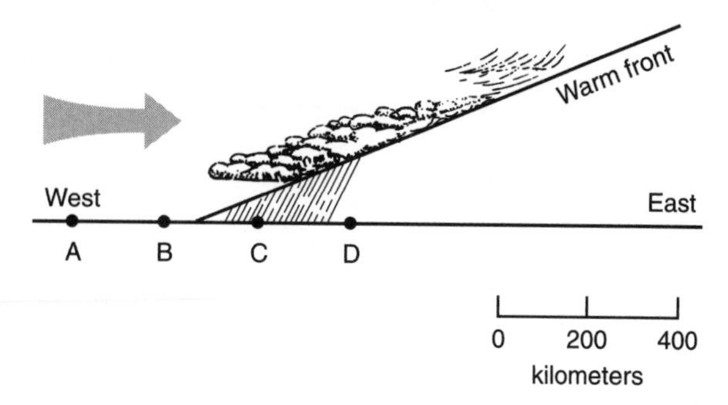

32. Describe the present weather at location C.

33. Compare the temperature at locations A and D.

34. If the front moves east at 400 km per day, what would the weather be like at location D the next day?

Base your answers to questions 35 and 36 on the weather map of New York State.

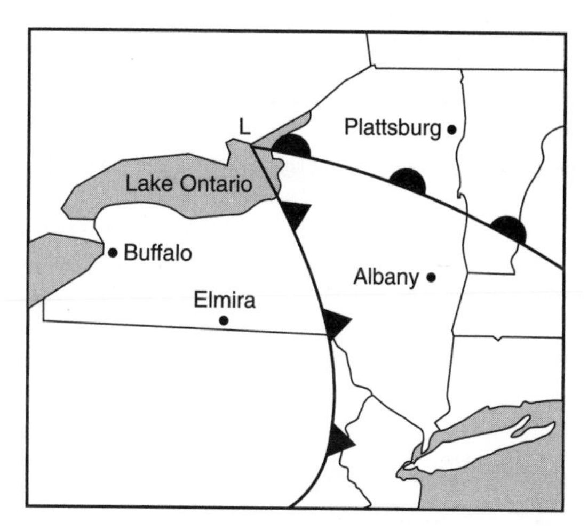

35. Describe how the weather has changed over the past six hours in Elmira, New York.

36. Predict the weather most likely to occur in Plattsburg within the next six hours.

37. Describe the typical weather conditions associated with a low-pressure system. Make sure your description includes the following:

(a) vertical air motion

(b) circular air motion

(c) possibility of precipitation

(d) amount of cloud cover

Weather Hazards

Weather affects our lives. Lack of rain can ruin crops and force emergency conditions such as water rationing. Too much rain can cause destructive floods. Storms are weather hazards that can make travel dangerous, force schools and businesses to close, and cause property damage, injuries, and deaths. With enough warning, people can prepare for hazardous weather and reduce damage to life and property.

Storms

Storms are natural disturbances in the atmosphere that include low air pressure, clouds, precipitation, and strong winds. The major storms are thunderstorms, hurricanes, tornadoes, and winter storms. Each type of storm has its own characteristics and dangers.

1. ***Thunderstorms*** are brief (few hours), intense storms that affect a relatively small area and are usually about 10–20 km in size. They occur when rapidly rising warm air builds cumulus clouds upward, forming a *cumulonimbus cloud,* commonly called a *thunderhead* (Figure 11-16). Thunderstorms produce lightning and thunder, strong gusty winds, rain, and sometimes hail.

Figure 11-16. *Thunderstorm development occurs in stages.*

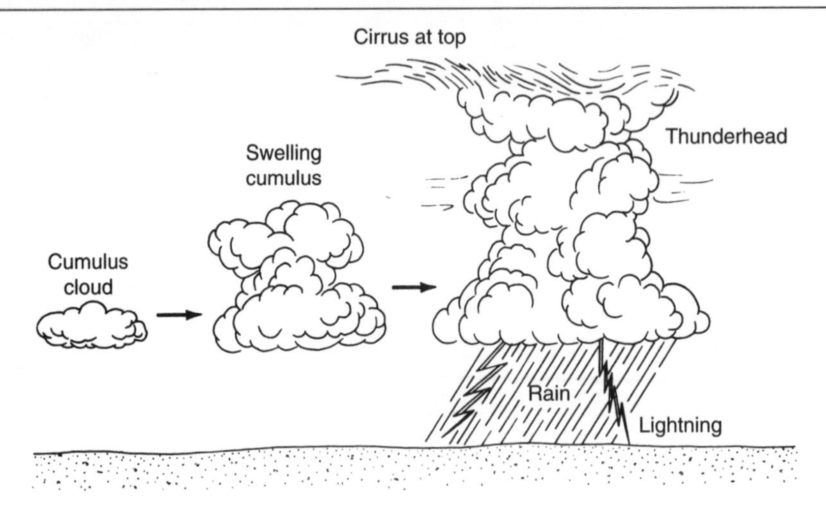

Lightning is a huge electrical discharge, like a giant spark. Lightning strikes are very dangerous. Large hailstones that sometimes fall during thunderstorms are also dangerous. During a thunderstorm, you should stay indoors or get into a car. If you cannot get indoors, avoid hilltops, trees, open fields, beaches, and bodies of water.

2. *Hurricanes* are huge rotating storms that form over the ocean near the equator. Usually, they are 700–800 km in diameter and take about a day to pass. (See Figure 11-17.) They produce very strong winds, heavy rains, and large, powerful ocean waves. The eye of the hurricane is a calm region in the storm's center.

Figure 11-17. *A cross section of a hurricane showing cloud structure and the eye of the hurricane.*

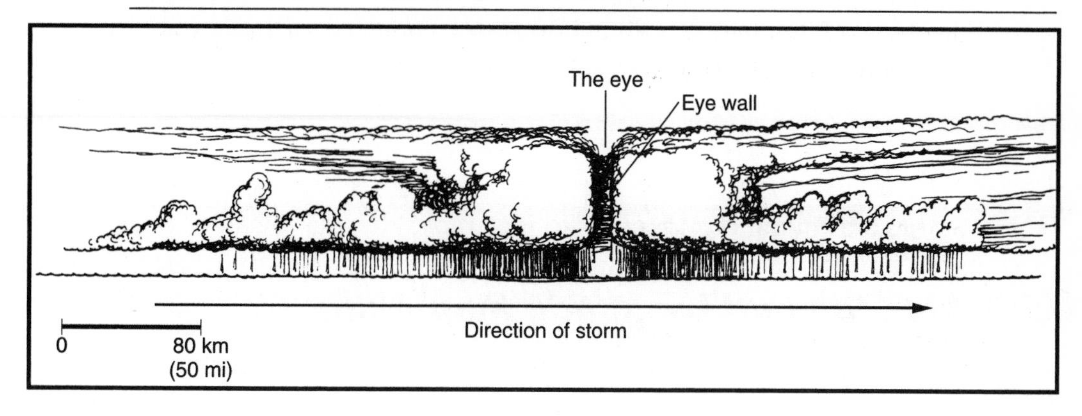

The typical hurricane path in the United States is north along the East Coast and eventually east into the North Atlantic Ocean. (See Figure 11-18.)

Figure 11-18. *Typical path of a hurricane along the East Coast of the United States.*

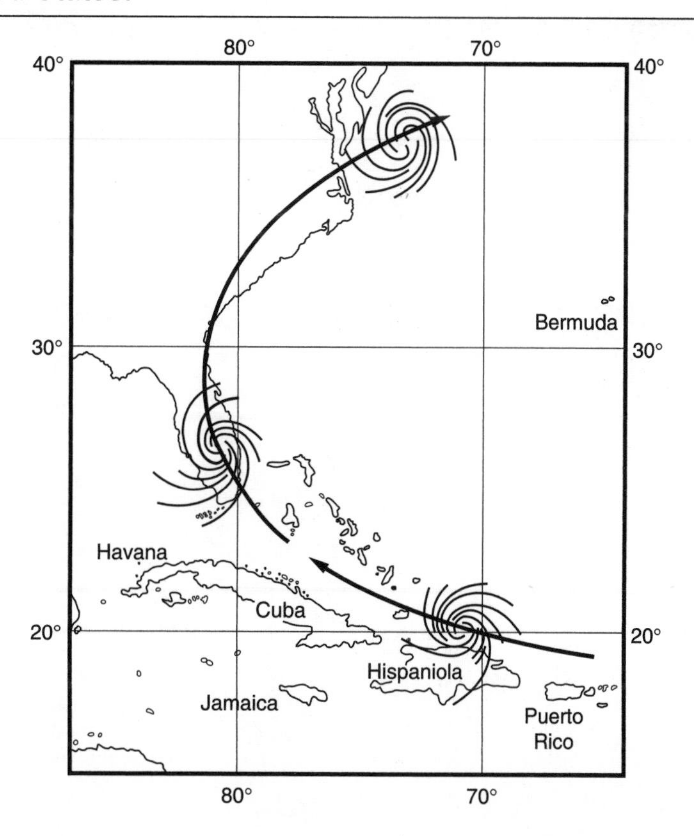

Hurricanes can cause severe flooding and damage from their high winds. When hurricanes strike land, the wind speed decreases, but large amounts of rain may fall. People living along the coast and in flood-prone areas should leave their homes and move to higher ground when a hurricane threatens. In the Pacific Ocean, these large rotating storms are called typhoons. In the Indian Ocean, these storms are called cyclones. In May 2008, cyclone Nargis caused much property damage and the deaths of more than 75,000 people in Myanmar.

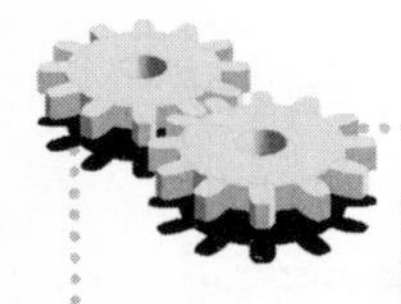

Process Skill 3

Using Latitude and Longitude Map Skills

The Atlantic Basin Hurricane Tracking Chart shows the path of Hurricane Floyd (1999) as a dashed line, and Hurricane Katrina (2005) as a solid line. Each dot or hurricane symbol § along the hurricane path represents a 1-day change in position of the hurricane.

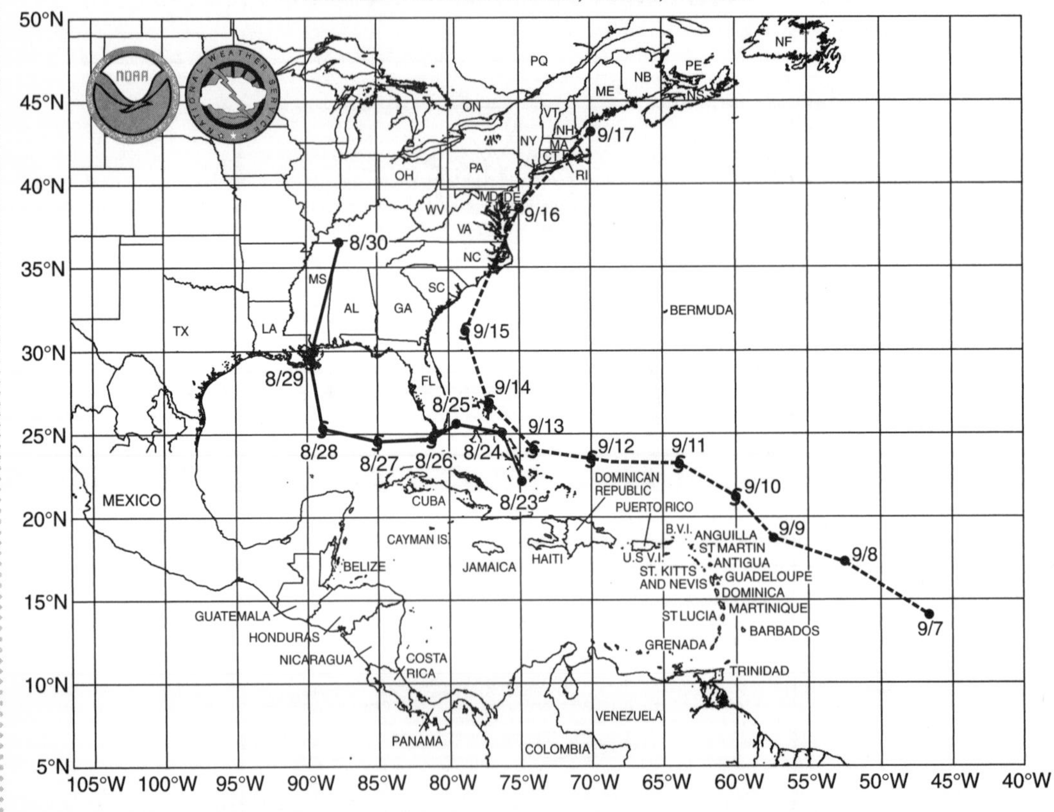

QUESTIONS

1. Did Hurricane Floyd take a typical hurricane path? Explain.
2. What was Hurricane Floyd's latitude and longitude on September 7 (9/7)?
3. On August 29, Hurricane Katrina had a wind speed of 125 mph. On August 30, the wind speed decreased to 30 mph. What caused the decrease in wind speed?
4. What is the most likely location of Hurricane Katrina on August 31 (8/31)?
   (1) latitude 35°N, longitude 75°W
   (2) latitude 40°N, longitude 95°W
   (3) latitude 35°N, longitude 65°W
   (4) latitude 40°N, longitude 80°W

3. *Tornadoes* are violently whirling winds, usually visible as a funnel-shaped cloud extending down to the ground. They develop in severe thunderstorms. Figure 11-19 shows the development of a tornado. Tornadoes usually appear suddenly, carve a narrow path of destruction, and disappear as suddenly as they came. Spiraling, high-speed winds and extremely low air pressure are the identifying features of tornadoes.

Figure 11-19. *A tornado funnel developing from cloud to ground.*

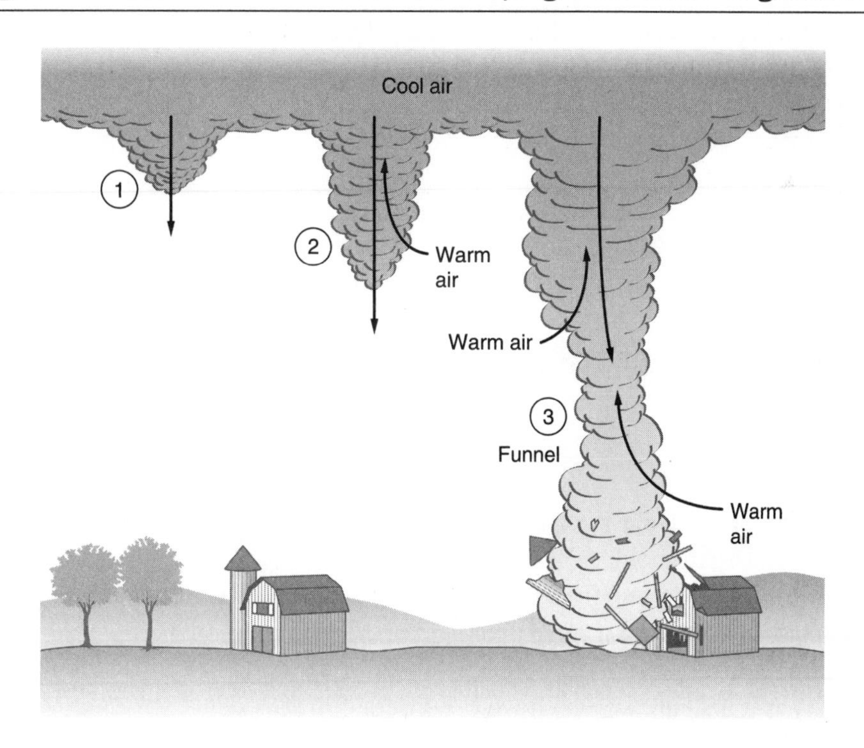

A tornado can lift and toss large objects, including cars, into the air. It can destroy houses in its path in a matter of seconds.

An underground cellar or basement is the safest place to be during a tornado.

4. ***Winter storms*** include blizzards and ice storms. Blizzards are fierce storms with strong winds, blowing snow, and very cold temperatures. Ice storms occur when falling rain freezes at Earth's surface, coating everything with ice. Under these conditions, you should remain indoors and not try to travel.

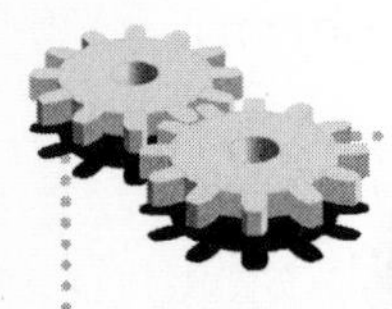

Process Skill 4

Interpreting Data in a Table

Storms form when the temperature, air pressure, and humidity conditions necessary for their development exist. At certain times of the year, weather conditions are more likely to produce one of these storms—thunderstorms, tornadoes, and hurricanes—than at other times. Therefore, each storm type should have a "storm season."

You can determine if there are "storm seasons" by recording the number of storms that occur each month for a period of time. The table below lists the monthly number of thunderstorm days, the monthly number of tornadoes in 2007, and the monthly number of hurricanes from 2005 to 2007.

Storm Data

Storm	*Jan*	*Feb*	*Mar*	*Apr*	*May*	*Jun*	*Jul*	*Aug*	*Sep*	*Oct*	*Nov*	*Dec*
Thunderstorm	0	1	3	6	9	15	18	16	5	4	2	0
Tornado	21	52	171	165	251	128	69	73	51	87	6	20
Hurricane	0	0	0	0	1	4	9	10	18	8	4	1

Questions

1. Use the Storm Data to draw a graph of the number of thunderstorms that occurred each month.
2. During what three-month period of time are thunderstorms most likely to occur? Are tornadoes most likely to occur? Are hurricanes most likely to occur?
3. Why is it unlikely we will have a hurricane in January, February, March, or April?

Pollution

Human activities can affect the atmosphere and the weather. Factories, power plants, cars, and airplanes produce harmful substances called ***pollutants***. The buildup of pollutants in the atmosphere can cause a number of weather problems.

Smog is a haze in the atmosphere. Smog forms when sunlight reacts with chemicals in the air. The chemicals come from automobile exhaust and factory smoke. The smog hanging over large cities gives the air a hazy, dirty look. Inhaling smog is very dangerous for people with breathing problems like asthma. It is even harmful to the lungs of healthy people.

Chemicals in the air from factory smoke and car and truck exhaust can increase the acidity of the tiny water droplets in clouds. When the ***acid rain*** falls to Earth, it harms lakes and forests and the creatures that live in them. Because air pollutants are often carried along by the global winds, acid rain usually falls far from the source of pollution. (See Figure 11-20.)

Figure 11-20. *Acid rain may affect areas far downwind from the source of pollution.*

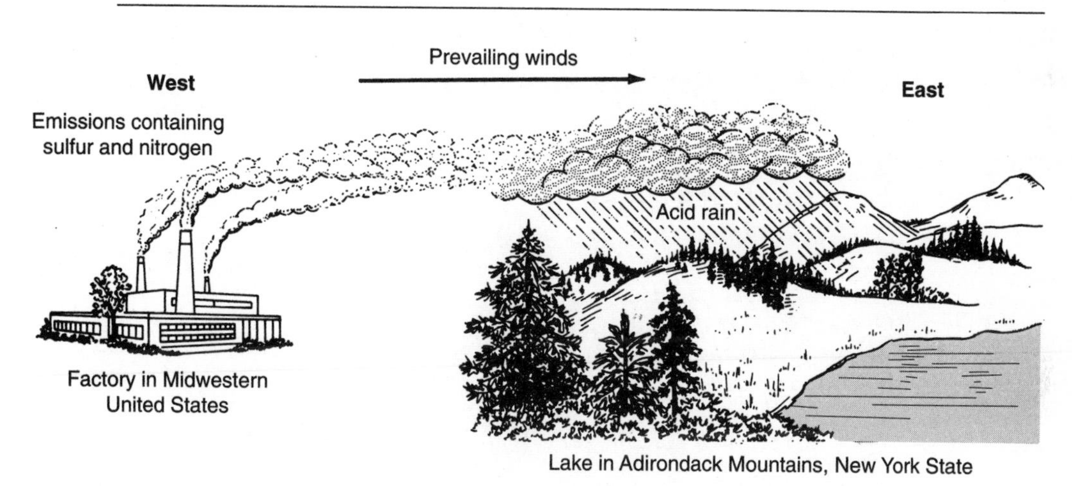

Many industries and most forms of transportation produce carbon dioxide. In the atmosphere, carbon dioxide acts like the glass of a greenhouse. It traps heat close to Earth instead of letting it radiate back into space. This is called the ***greenhouse effect***. The greenhouse effect helps keep Earth warm. Without it, Earth would be too cold to support life as we know it. However, many scientists fear that the buildup of atmospheric carbon dioxide caused by human activities may lead to global warming. Global warming is a rise in worldwide average temperatures. This condition, often called a runaway greenhouse effect, could have disastrous results. It could make some climates hotter and drier, and interfere with agriculture. Polar ice caps could melt, raising sea level and flooding coastal areas and many major cities.

Some natural events also release pollutants. Forest fires and volcanic eruptions produce huge amounts of dust and ash. These particles collect high in the atmosphere, where they block sunlight. This causes cooler temperatures on Earth. Plants release irritating pollen into the air, causing health problems for people with hay fever and asthma. There is little that we can do to control natural pollutants. If we are to avoid harming our planet and the plants and animals that live on it, we can and must control our own activities that cause pollution.

Review Questions

Part I

38. A dangerous characteristic of an average thunderstorm is
    (1) thunder
    (2) 100 mph winds
    (3) funnel-shaped winds
    (4) lightning

39. The violent storm visible in this picture is called a

    (1) hurricane
    (2) thunderstorm
    (3) tornado
    (4) blizzard

40. During what months do most hurricanes form?
    (1) January to June
    (2) June to November
    (3) May to September
    (4) November to January

41. Hurricanes are about the size of
    (1) New York City
    (3) Texas

(2) New York State (4) the Atlantic Ocean

42. The map shows the relative size and track of a certain weather event. Which weather event is shown on the map?

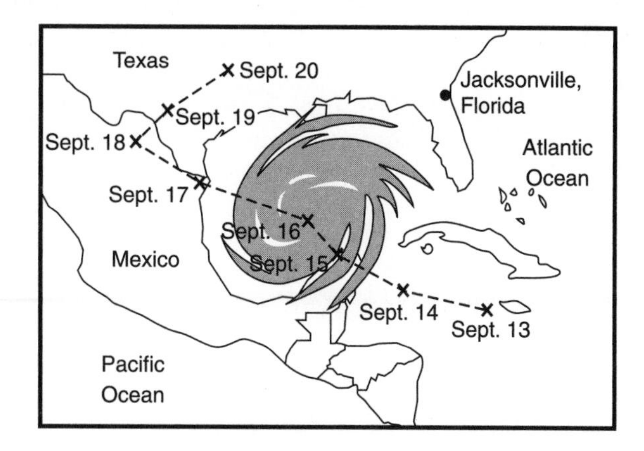

(1) tornado (3) hurricane

(2) blizzard (4) thunderstorm

43. What are some hazards associated with a hurricane?

(1) cold temperatures, blowing snow, and poor visibility

(2) funnel-shaped winds that can lift large objects

(3) lightning and hailstones

(4) flooding, large waves, and strong winds

44. John visited New York City on a warm summer day. He noticed that the air was hazy, and that his eyes and throat burned. This was probably caused by

(1) an approaching storm (3) low clouds

(2) smog (4) global warming

Base your answer to question 45 on the table below listing the sources and percentages of air pollutants.

Sources and Percentages of Air Pollutants

Percentage of Pollutant Source	*Total Pollutants*
Transportation	42%
Heating fuel	21%
Solid waste disposal	5%
Forest fires	8%
Miscellaneous	10%
Industrial processes	14%

45. Which two processes produce the greatest percentage of air pollution?

(1) transportation and miscellaneous

(2) transportation and forest fires

(3) transportation and solid waste disposal

(4) transportation and heating fuel

46. Which atmospheric gas causes the greenhouse effect?

(1) nitrogen (2) oxygen (3) carbon dioxide (4) argon

Part II

Base your answers to questions 47 through 49 on the diagram (A) and graph (B) below and on your knowledge of science. The diagram shows an aerial view of a hurricane with a cross section line X–Y, and the graph shows the wind speed within the hurricane along the X–Y line.

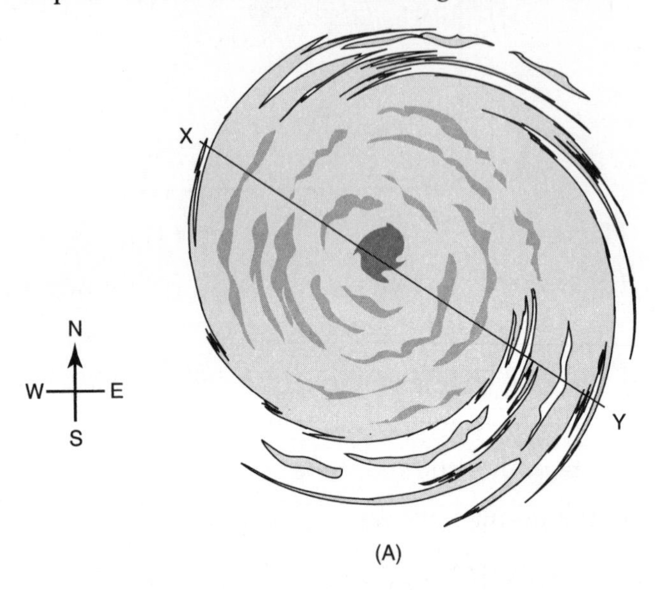

(A)

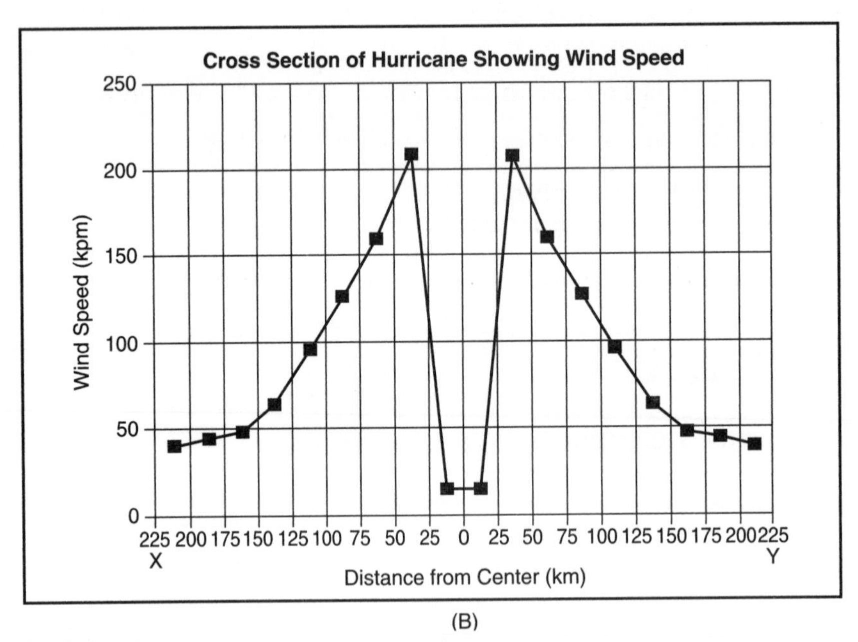

(B)

47. Describe the relationship of the winds to the structure of the hurricane.

48. What feature is in the center of the hurricane? Describe the size and wind conditions of this feature.

49. What is the diameter of this hurricane?

Base your answers to questions 50 and 51 on the map and your knowledge of science. That map shows a major industrial area and three lake regions in the United States.

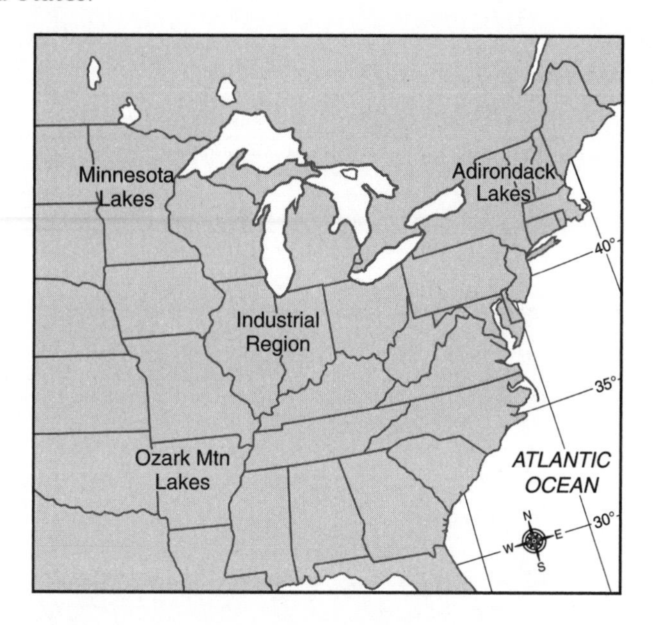

50. Which lake region is most likely affected by acid rain caused by pollution from the industrial area? Why?

51. How does acid rain form?

Base your answers to questions 52 through 54 on the table below and your knowledge of science. The table lists the number of tornadoes in a 25-year period and the number of tornadoes per 10,000 square miles in five states.

State	*Tornadoes in 25-Year Period*	*Tornadoes per 10,000 sq Miles*
New York	93	0.7
Oklahoma	1401	7.7
Texas	3035	4.4
Florida	1006	6.6
California	84	0.2

52. In which state is a house least likely to be hit by a tornado?

53. What might explain the reason Texas has many more tornadoes than Oklahoma, but a house in Oklahoma is more likely to be hit by a tornado?

54. About how many tornadoes per year occur in New York State?

Chapter 12 Astronomy

Vocabulary at a Glance

asteroid
astronomical unit (AU)
axis of rotation
comet
constellation
dwarf planet
eclipse
full moon
galaxy
light-year
meteor
meteorite
meteoroid
moon
moon phase
new moon
orbit
planet
plutoid
revolution
rotation
solar system
tide
Tropic of Cancer
Tropic of Capricorn

Our solar system is part of the Milky Way galaxy, which is a spiral galaxy similar to the one shown here, called M101.

Major Concepts

- Rotation is the spinning of Earth around its axis, which causes day and night. One rotation takes Earth 24 hours (one day) to complete.
- Revolution is Earth's motion around the sun; it takes Earth 365.25 days (one year) to complete one revolution around the sun.
- Earth's axis is tilted 23.5° to the plane of its orbit. The northern end of Earth's axis (the North Pole) points toward the North Star.
- The seasons are caused by the tilt of Earth's axis as it revolves around the sun. This tilt causes the sun's vertical rays to strike Earth farthest north on June 21 and farthest south on December 21.
- Our solar system consists of the sun, eight planets, dwarf planets, satellites, asteroids, meteoroids, and comets. The sun is at the center; its gravity keeps all other members of the solar system in orbits around the sun.
- Our sun is an average star, one of hundreds of billion stars clustered in a spiral galaxy called the Milky Way.
- The universe contains trillions of stars, most clustered in galaxies. Yet most of the universe is thought to be empty space.

Earth Motions, Time, and Seasons

Earth's Rotation

People once believed that Earth stood still while the sun, moon, stars, and planets revolved around it each day. This seemed to make sense for several reasons:

- We do not feel Earth moving.
- The sun does appear to move across the sky during the day.
- The moon and stars appear to move through the sky at night.

Today we know that these apparent motions of the sun, moon, and stars are caused by Earth's rotation.

Earth spins like a top. This spinning motion is called ***rotation***. Earth rotates from west to east, or, put another way, if we could look down at the North Pole from space, we would see Earth spinning in a counterclockwise direction.

The ***axis of rotation*** is an imaginary line on which Earth spins (see Figure 12-1). It extends through Earth between the North and South Poles. A basketball spinning on a fingertip is a good model of how Earth spins on its axis. The imaginary line from the fingertip through the basketball to the top of the ball is the axis of rotation.

Figure 12-1. *Rotation of Earth: (a) Earth rotates from west to east around its axis. (b) Viewed from above the North Pole (NP), Earth rotates in a counterclockwise direction.*

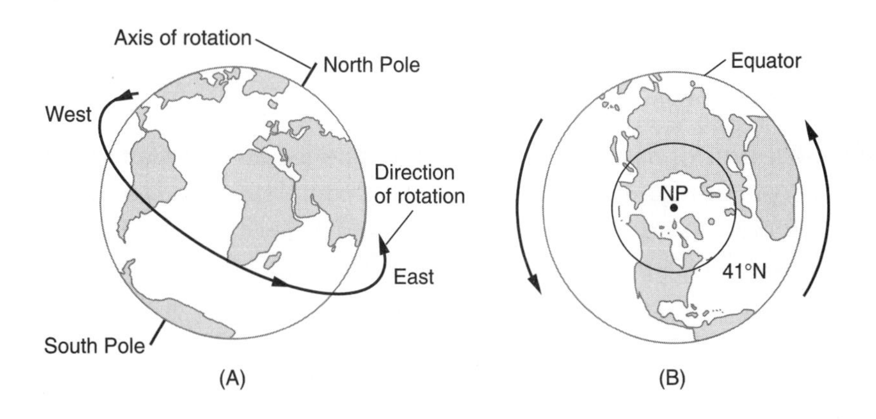

Earth's rotation produces several effects:

1. Earth's rotation causes the daily change from day to night. At any given time, half of Earth is in daylight, facing the sun, while half is in darkness, facing away from the sun. This is shown in Figure

12-2. Each day, all places on Earth, except the areas near the poles, experience this change from daylight to darkness. (Areas within the Arctic and Antarctic circles experience several weeks of continuous daylight or darkness at certain times of the year.)

Figure 12-2. *Earth's rotation causes the change from day to night.*

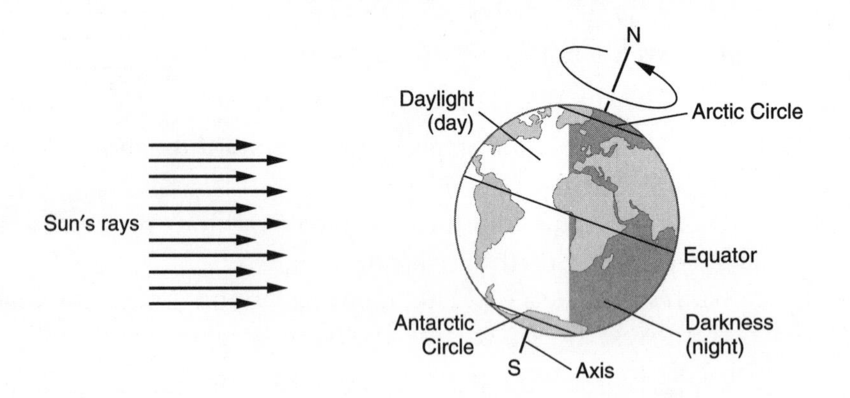

2. The speed of Earth's rotation causes the length of one day to be about 24 hours. This is the amount of time Earth takes to rotate once on its axis.
3. The apparent daily motions of the sun, moon, planets, and stars across the sky are also caused by Earth's rotation. These objects appear to rise in the eastern sky and set in the western sky because Earth rotates in the opposite direction, from west to east.

Daily Time

The time of day at any location on Earth is based on Earth's rotation and the position of the sun in the sky. Earth completes one rotation each day. That is, every location on Earth's surface rotates 360° around Earth's axis each day. There are 24 hours in a day. If we divide 360° by 24 hours, we find that Earth rotates 15° per hour.

Figure 12-3 shows how the location of the sun in the sky is related to the time of day. At any location along longitude line NP–A_1 the sun is highest in the southern sky and it is 12 o'clock noon. In one hour the Earth rotates 15° and the longitude line is at a new position: NP–A_2. The time changes by 1 hour; therefore, it is 1 P.M. In 3 hours, Earth rotates 45° and the time is 3 P.M. Each hour the sun appears to move westward until after 6 hours (6 P.M.) the sun is setting along the western horizon. In 24 hours the Earth rotates 360° and the time is 12 o'clock noon again. Each 15° of longitude represents a time difference of one hour.

Figure 12-3. *The time of day and the sun's location in the sky are related to Earth's rotating at 15° per hour.*

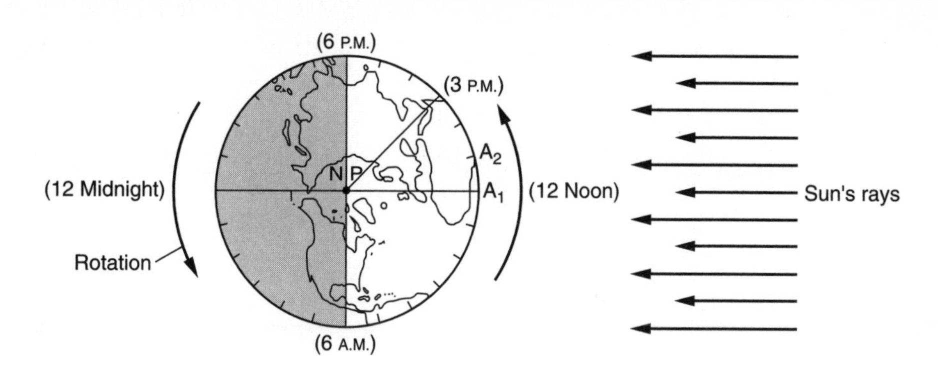

Earth's Revolution

Earth's motion around the sun is ***revolution***. The path Earth travels around the sun is called an ***orbit***. Earth's orbit is an ellipse. An ellipse is an oval. (See Figure 12-4.)

Figure 12-4. *Earth's orbit around the sun is an ellipse, not a circle. Earth is closest to the sun on January 3, and farthest from the sun on July 4.*

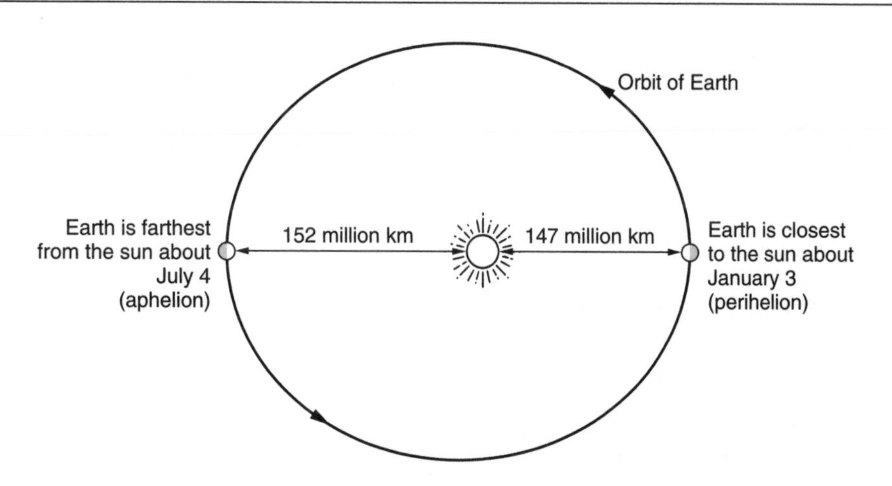

Earth's revolution has two major effects.

1. The time Earth takes to revolve once around the sun defines the length of a year. During one year, Earth rotates on its axis 365.25 times. For convenience, the calendar year is 365 days long, and an extra day is added every fourth year (called a leap year) to make up for the four extra 0.25 days.

2. Earth's revolution around the sun, combined with the *tilt* of Earth's axis, causes the changing seasons on the planet. Earth's axis of rotation is not perpendicular (at a right angle) to the plane of its orbit; rather, it is tilted 23.5°. (See Figure 12-5.) No matter where Earth is in its orbit, its axis is always tilted in the same direction in space, pointing toward the North Star. While all the other stars seem to move across the night sky, the North Star remains motionless because Earth's axis points toward it.

Figure 12-5. *Earth's axis of rotation is tilted 23.5°.*

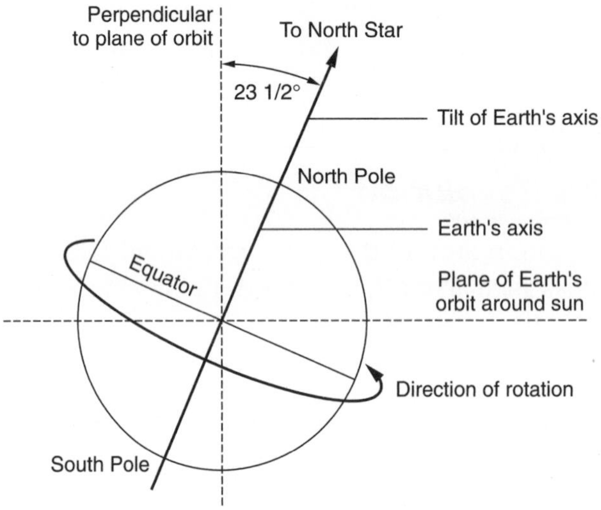

Seasonal Changes

Earth is closest to the sun (*perihelion*) in early January and farthest from the sun (*aphelion*) in early July. (See Figure 12-4). This means that Earth's changing distance from the sun is not the cause of the changing seasons. The seasons change because Earth's axis is tilted. Earth's axis always points in the same direction while Earth orbits the sun. This tilt causes the Northern Hemisphere to point toward the sun for half the year and away from the sun the other half of the year. (See Figure 12-6.) This changes the angle at which the sun's rays strike Earth and the number of hours of daylight each day.

When the sun is directly overhead, its rays strike Earth's surface vertically. Vertical rays concentrate the sun's energy over the smallest area. These are the strongest and hottest rays that strike Earth. Due to the tilt of Earth's axis, the sun's vertical rays move north and south of the equator, bringing about the change of seasons. The sun's rays that strike the surface on a slant spread their energy over a larger area. The greater the slant, the larger the area the energy spreads over, and the less effective the rays are in heating Earth. (See Figure 12-7.)

Figure 12-6. *The seasons are caused by the tilt of Earth's axis and Earth's revolution around the sun.*

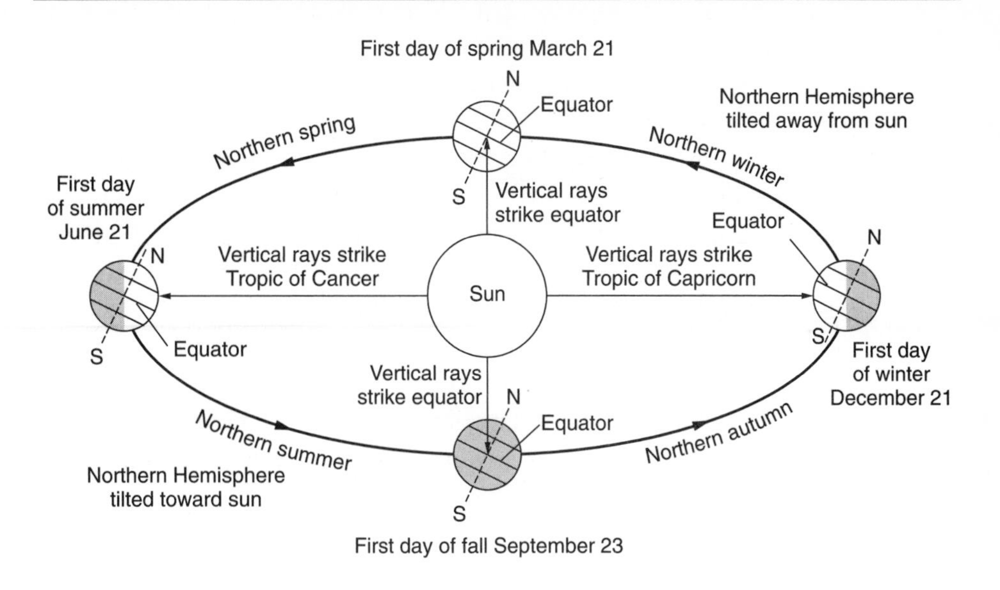

Figure 12-7. *Vertical rays of the sun heat Earth more effectively than slanted rays.*

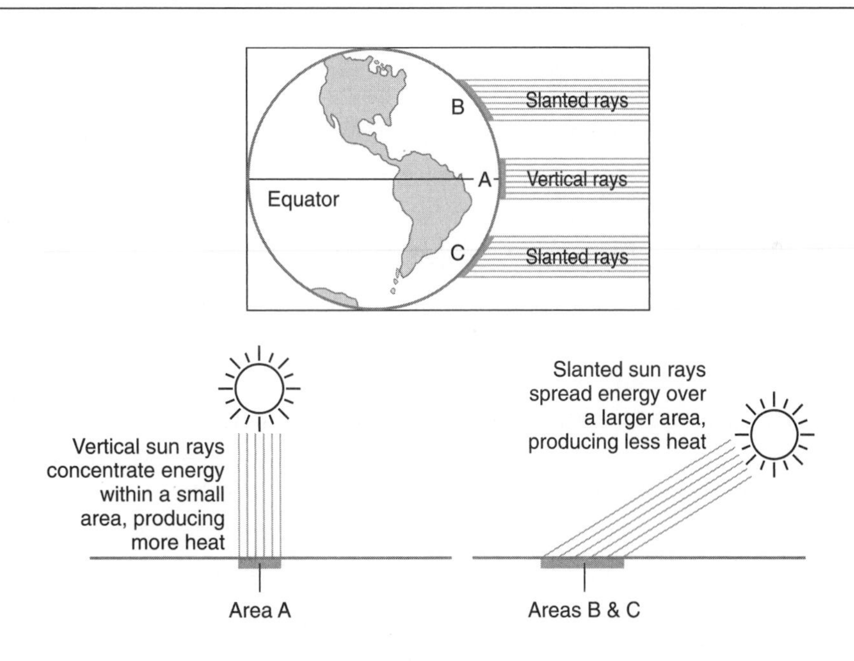

For a place to receive the sun's vertical rays, the sun must be directly overhead at some time during the year. The only places on Earth where the sun is directly overhead lie between 23.5°N latitude and 23.5°S latitude, in the tropics. Because New York City is about 41°N latitude, the sun is *never* directly overhead. Therefore, the vertical rays of the sun *never* strike as far north as New York City.

On or about June 21, the Northern Hemisphere is tilted toward the sun. The sun's vertical rays strike 23.5° north of the equator at the ***Tropic of Cancer.*** (Look again at Figure 12-6.) The sun's rays heat the Northern Hemisphere more effectively, and summer begins in the Northern Hemisphere. At the same time, winter begins in the Southern Hemisphere, which is tilted away from the sun.

On or about December 21, the Northern Hemisphere is tilted away from the sun. The sun's vertical rays strike 23.5° south of the equator at the ***Tropic of Capricorn.*** (Look again at Figure 12-6.) As a result, the Northern Hemisphere is not heated effectively and winter begins. At the same time, summer begins in the Southern Hemisphere, which is tilted toward the sun.

On or about March 21 and September 23, neither hemisphere is tilted toward the sun. The sun's vertical rays strike the equator, and all areas on Earth have equal periods of daylight and darkness. (Look again at Figure 12-6.) Table 12-1 summarizes the information about seasonal dates in the Northern Hemisphere.

Table 12-1. *Seasonal Information for the Northern Hemisphere*

Date (On or About)	Northern Hemisphere Season	View from Earth of Sun's Path Across the Sky	Location of Vertical Rays	Length of Day and Night
June 21	First day of summer	Sun follows highest path across sky	Tropic of Cancer	Longest day; shortest night
Sept. 23	First day of autumn	Sun follows path midway between summer and winter paths	Equator	Day and night of equal length (12 hours each)
Dec. 21	First day of winter	Sun follows lowest path across sky	Tropic of Capricorn	Shortest day; longest night
March 21	First day of spring	Sun follows path midway between summer and winter paths	Equator	Day and night of equal length (12 hours each)

During any season, the sun is highest in the sky each day at noon. In the continental United States, the noon sun approaches but never reaches the point directly overhead; it is always in the southern half of the sky. As a result, shadows at noon always point to the north.

Process Skill 1

Observing the Changing Position of the Noon Sun

In New York State, the noon sun is never directly overhead; it is always in the southern sky. However, the noon sun is much higher in the sky during the summer than it is during the winter. The sun's noon position changes a little each day, moving higher in the sky from December 21 to June 21, and lower in the sky from June 21 to December 21. As the height of the noon sun changes, so does the length of an object's shadow.

You can determine the changing position of the noon sun by measuring the length of the shadow cast by a vertical pole. Because the noon sun is always in the southern sky, the pole's shadow is always cast to the north. As the noon sun moves higher in the sky, the shadow gets shorter. As the noon sun moves lower in the sky, the shadow gets longer. This is shown in diagram 1.

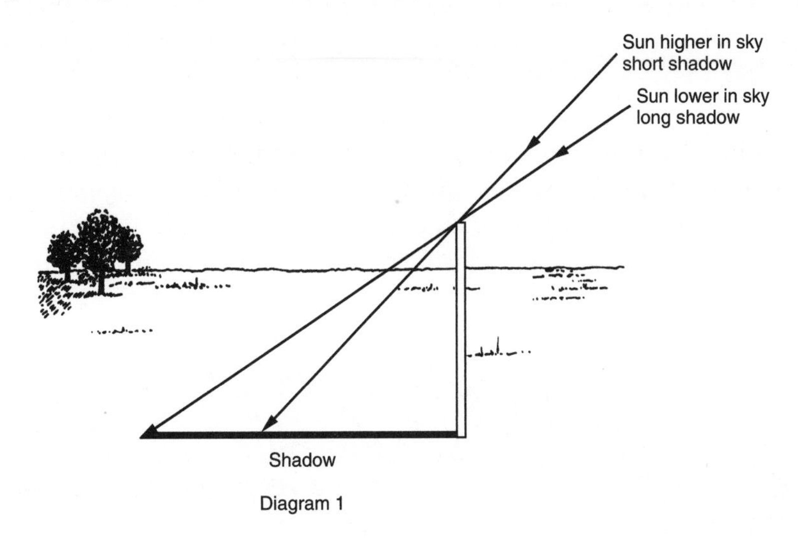

Diagram 1

Kent wanted to observe the changing position of the noon sun near Syracuse, New York. He set up an experiment using a vertical pole and a tape measure that ran north along the ground from the base of the pole, as shown in Diagram 2. For eight weeks, Kent measured the length of the pole's shadow every Friday at 12:00 noon. Answer the questions about Kent's observations of the changing position of the noon sun.

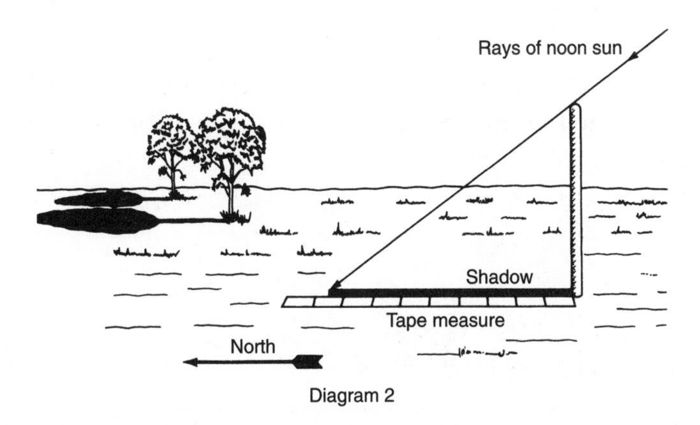

Diagram 2

Questions

1. If the length of the shadow increased each week during the experiment, how did the position of the sun change?

2. If Kent started the experiment on December 21 and recorded a shadow length of 51 centimeters, how would the length of the shadow have changed by January 15? Explain why this would occur.
3. If Kent started the experiment on May 21 and continued his experiment until July 21, describe how the shadow would change, and explain why.

Proof of Earth's Revolution

Proof that Earth revolves around the sun comes from observations of stars. Ancient people saw patterns in the stars that reminded them of animals or characters in their myths. These patterns are called ***constellations***. Two easily recognized constellations are Ursa Major (the Great Bear, which contains the Big Dipper) and Orion, the Hunter.

During the year, we are able to see different constellations at night. This suggests that Earth's night side faces different directions in space (Figure 12-8), so Earth's position in relation to the sun must be changing. In other words, Earth must be moving around the sun.

Figure 12-8. *As Earth orbits the sun, different constellations become visible at night.*

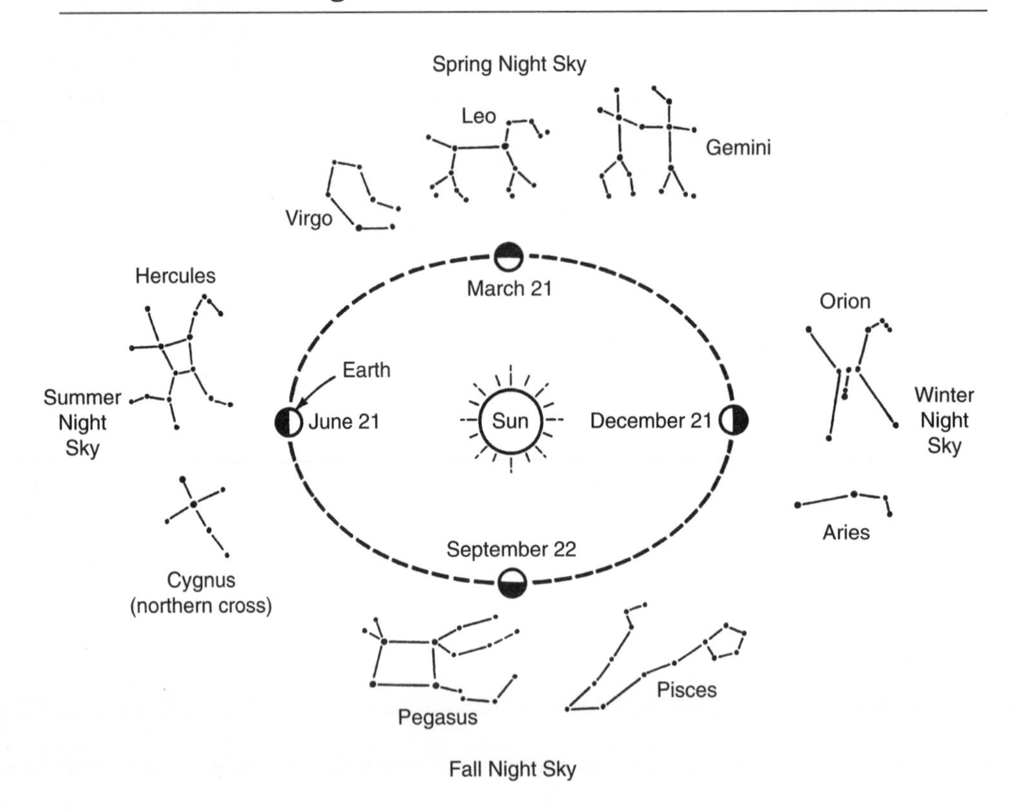

Review Questions

Part I

Base your answers to questions 1 and 2 on the map below, which shows part of New York State in daylight and part in darkness (night).

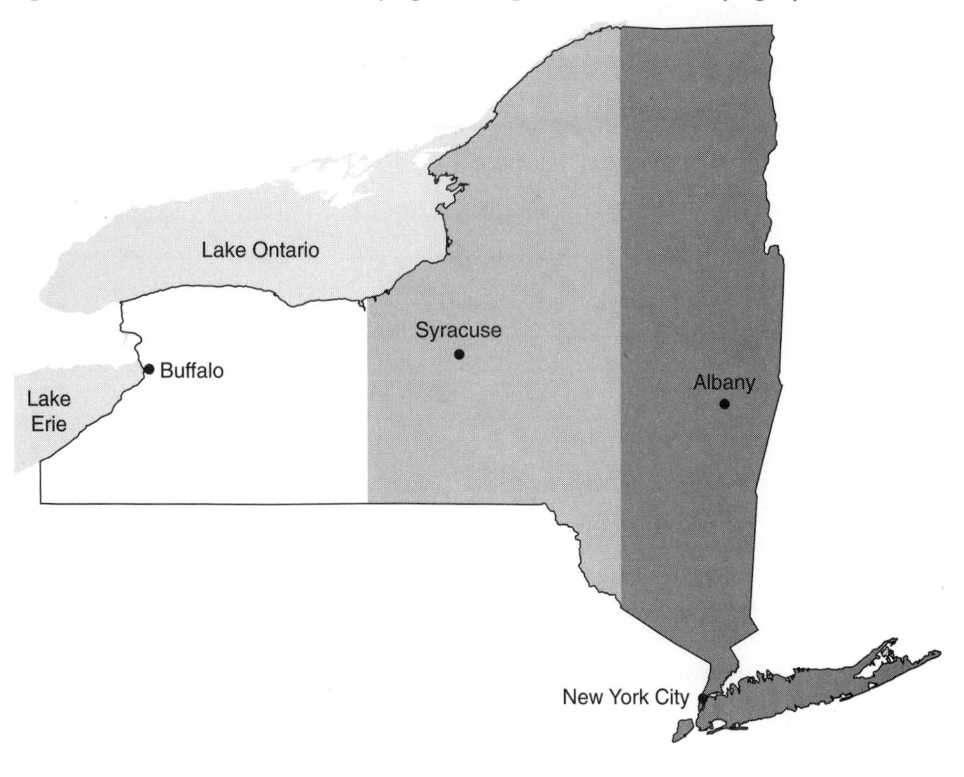

1. The light-gray area on the map represents
   (1) sunrise (dawn) (3) daylight
   (2) sunset (dusk) (4) night
2. During the next hour, New York State will
   (1) be completely into night
   (2) be completely into daylight
   (3) remain unchanged
   (4) become daylight in the east and night in the west
3. Earth rotates from west to east causing the stars to
   (1) rise in the west and set in the east
   (2) rise in the north and set in the south
   (3) rise in the east and set in the west
   (4) rise in the south and set in the north

4. The daily change from day to night is caused by the
   (1) rotation of the sun (3) rotation of Earth
   (2) revolution of the sun (4) revolution of Earth
5. On which date is the sun highest in the sky at noon in Newburgh, New York?
   (1) December 21 (3) March 21
   (2) September 23 (4) June 21

Base your answers to questions 6 and 7 on the graph below, which shows the location of the sun's vertical rays at noon during the year.

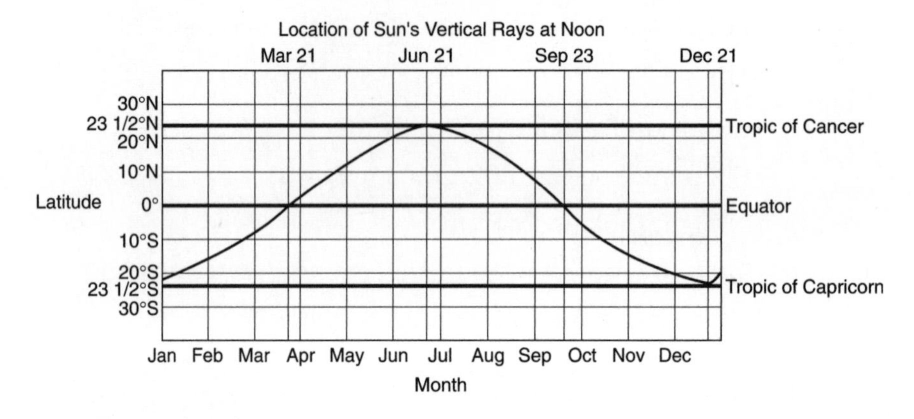

6. On what day are the sun's vertical rays farthest north?
   (1) June 21 (3) September 23
   (2) December 21 (4) March 21
7. New York City is about 41° north latitude. The sun's vertical rays strike New York City
   (1) between March 21 and June 21
   (2) between June 21 and September 23
   (3) June 21
   (4) never
8. Use the diagram to determine what change would occur if Earth's axis were tilted 35° from the vertical rather than 23.5°.

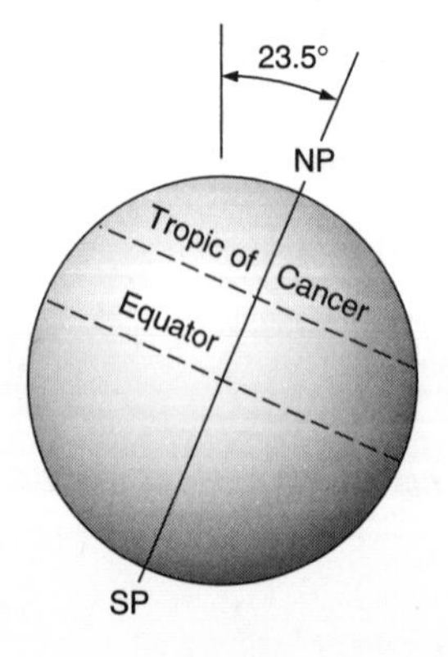

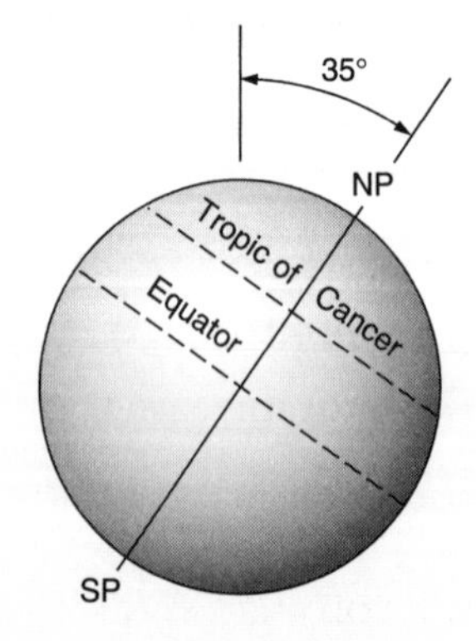

(1) The Tropic of Cancer would be farther north.

(2) The Tropic of Cancer would be farther south.

(3) The equator would be farther north.

(4) The equator would be farther south.

9. How much time does it take Earth to revolve once around the sun?

(1) 1 day
(2) 1 week
(3) 1 month
(4) 1 year

10. Aaron read in the newspaper that the next day would have the shortest amount of daylight. That next day's date most likely is

(1) October 21
(2) June 21
(3) December 21
(4) March 21

11. Seasons on Earth are caused by the

(1) distance between the sun and Earth

(2) Earth's rotation and tilt

(3) Earth's revolution and tilt

(4) distance to the sun and tilt of Earth

Part II

Base your answers to questions 12 and 13 on the diagram below and your knowledge of science. The diagram shows the position of the moon over a tree at 9:00 P.M. (position A in the diagram). An hour later, the moon has moved to position B.

12. What caused this apparent change in position of the moon?

13. What will be the moon's position in another hour?

Base your answers to questions 14 through 16 on the diagram on page 342 and your knowledge of science. The diagram shows Earth at different positions in its orbit around the sun.

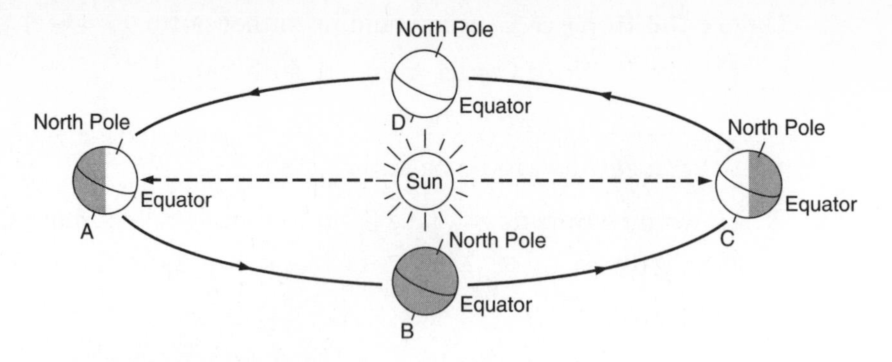

14. On June 21, Earth is at position A in the diagram. At what point in its orbit will Earth be in three months? In nine months?

15. Earth is closer to the sun on December 21 than on June 21. What conclusion about seasons can you make from this information?

16. Earth revolves around the sun (360°) in 365 days. About how many degrees per day does the Earth revolve around the sun?

The Solar System

The *solar system* consists of the sun and all the objects that revolve around it. The sun's gravity holds these objects in their orbits. The major members of the solar system are the sun and the eight planets. (See Figure 12-9.) A number of other objects also come under the influence of the sun's gravity and so belong to the solar system. These include dwarf planets, satellites (moons) that revolve around planets, asteroids, comets, and meteoroids. Most objects in the solar system show regular and predictable motion.

The Sun

The sun is a hot, bright ball of gases. Nuclear reactions in the sun's interior release huge amounts of energy, mostly as light and heat. The sun is by far the largest object in the solar system. It is many times larger than Earth. If the sun were hollow, about one million Earths could fit inside. (See Figure 12-10.)

The sun is a star, like the stars we see in the night sky. Although the sun is an average-size star, it seems much larger than other stars because it is much closer to Earth. The sun is about 150,000,000 km (93 million miles) from Earth. The average distance from Earth to the sun is called an ***astronomical unit*** (AU). So, Earth is 1 AU from the sun. In Chapter 2 you learned light travels at 300,000 km per second. Even at that great speed it takes light from the sun about 8 minutes and 20 seconds to reach Earth.

Figure 12-9. *The relative positions of the sun, the eight major planets, and the asteroid belt.*

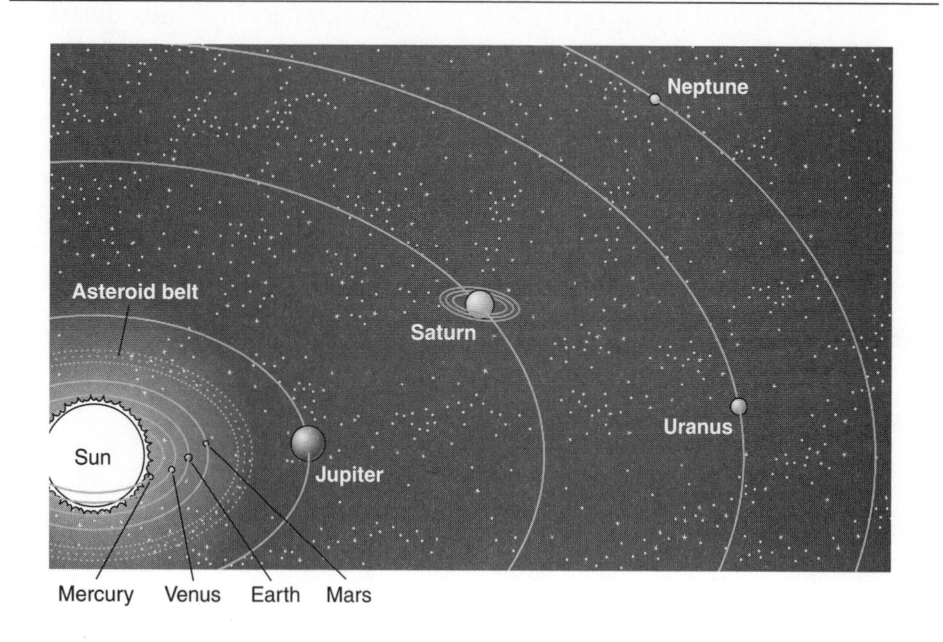

Figure 12-10. *If the sun were hollow, more than a million Earths could fit in it.*

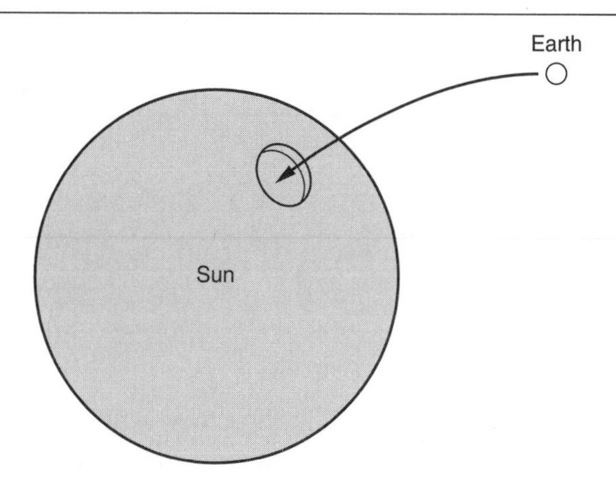

The Planets and Dwarf Planets

In 2006, the International Astronomical Union defined planets and dwarf planets. A ***planet*** is a celestial body that orbits around a star, is spherically shaped, and has no rocky objects in its orbit. A ***dwarf planet*** is defined similarly, but it had rocky objects in its orbit. In 2008, they created a new subclass of dwarf planets called plutoids. ***Plutoids*** are dwarf planets that orbit the sun beyond Neptune. Pluto Eris, and Makemake (MAH-keh MAH-keh) are plutoids. Our solar system now contains eight planets and four dwarf planets (three of the dwarf planets are plutoids). Other plutoids are expected to be discovered in the future. Table 12-2 lists some of the physical characteristics of the eight planets and four dwarf planets in our solar system.

Table 12-2. *Physical Characteristics of Planets and Plutoids in Our Solar System*

Planet	Average distance from sun (AU)	Diameter (km)	Time to Revolve Around Sun (Earth Time)	Time to Rotate on Axis (Earth Time)
Mercury	0.4	4,878	88 days	59 days
Venus	0.7	12,104	225 days	−243 days*
Earth	1.0	12,756	365.25 days	24 hr
Mars	1.5	6,787	1.88 yr	24.6 hr
Jupiter	5.2	142,800	11.86 yr	9.9 hr
Saturn	9.5	120,660	29.63 yr	10.6 hr
Uranus	19.2	51,118	83.97 yr	17 hr
Neptune	30.1	49,528	165 yr	16 hr
Dwarf Planets				
Ceres	2.8	950	4.6 yr	9.1 hr
Pluto	39.5	2,300	248 yr	6.4 days
Makemake	46.0	1,600	310 yr	?
Eris	67.7	3,000	560 yr	8.0? hr

*A negative number indicates it rotates east to west, rather than west to ease.

In the night sky, the planets look much like stars. However, as days and weeks go by, planets change their positions against the background of motionless stars. (See Figure 12-11.) Also, planets do not usually "twinkle" the way stars do. Unlike stars, planets do not give off their own light. Planets are visible because they reflect sunlight.

Figure 12-11. *A planet changes position compared to the background stars, which appear motionless.*

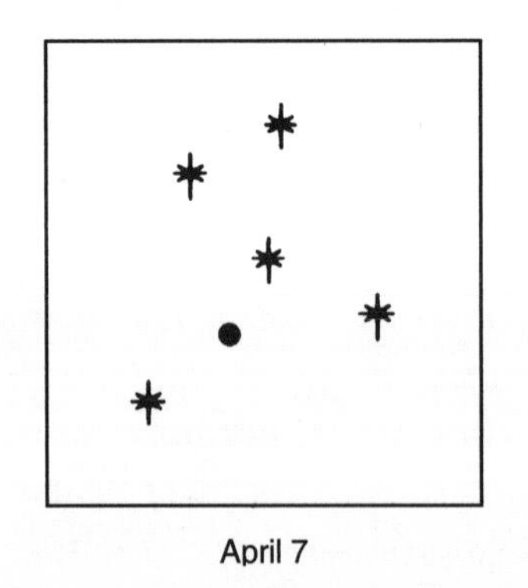
April 7

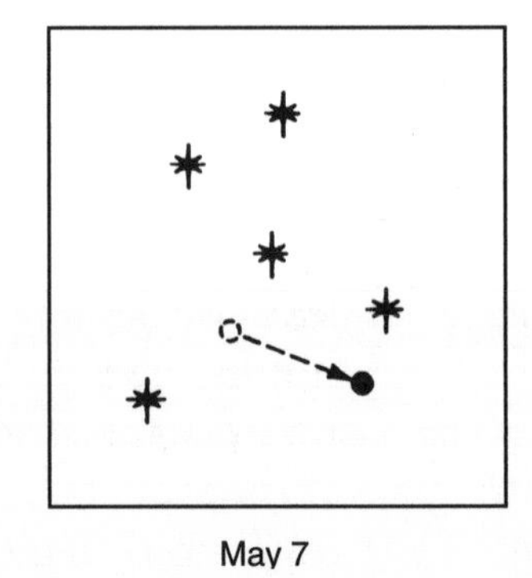
May 7

Laboratory Skill

Determining a Quantitative Relationship

For an object to remain in orbit around the sun, the gravitational pull of the sun on the object must be in balance with the object's forward speed, or orbital speed. Without this balancing, the object would not stay in orbit.

The closer an object is to the sun, the stronger the sun's gravity pulls on the object. To avoid being pulled into the sun, objects near the sun move faster than objects farther away. This suggests that there is a relationship between an object's orbital speed and its distance from the sun.

Planet and Dwarf Planet Data Table

Object	*Average Orbital Speed (km/s)*	*Average Distance from the Sun (km)*
Mercury	47.60	57,900,000
Venus	34.82	108,200,000
Earth	29.62	149,600,000
Mars	23.98	227,900,000
Jupiter	12.99	778,000,000
Saturn	9.58	1,427,000,000
Uranus	6.77	2,871,000,000
Neptune	5.41	4,497,000,000
Pluto*	4.72	5,913,000,000
Makemake*	4.42	6,900,000,000
Eris*	3.44	10,000,000,000

*Dwarf planet

Base your answers to the questions on the information presented above and the data table.

Questions

1. If Earth moved closer to the sun, to remain in orbit its orbital speed would have to ____________ .
2. If a dwarf planet were discovered farther from the sun than Eris, how would its orbital speed compare to that of Eris?
3. If the orbital speed of a newly discovered asteroid were found to be 7.61 km/s, between the orbits of what two planets would the asteroid's orbit be located?
4. Describe the relationship between an object's average orbital speed and its average distance from the sun.

The Moon

The ***moon*** is a ball of rock that revolves around Earth. It is Earth's only natural satellite and our nearest neighbor in space. There is no water or air on the moon, so it cannot support life. Because the moon is much smaller than Earth, the moon's gravity is too weak to hold moisture or an atmosphere at its surface.

As shown in Figure 12-12, the moon has many surface features. The dark areas of the moon are low, flat plains. The light areas are mountainous highlands. Circular features are *craters.* Scientists think that most craters were formed by the impacts of space rocks.

Figure 12-12. *In this photograph of the moon's surface, darker areas are plains, and brighter areas are highlands. (Courtesy of NASA.)*

The shape of the moon's orbit around Earth is an ellipse. Remember, an ellipse is shaped like an oval. The average distance between Earth and the moon is about 382,000 km (237,000 mi). The motions of the moon and Earth are known with great accuracy. These motions cause the sun, Earth, and the moon to change their positions in a regular, repeating pattern that produces predictable events. These events include the apparent changing shape of the visible moon (phases), the eclipses of the sun and moon (shadows cast by the moon and Earth on each other), and the rising and falling of ocean water (tides).

Motions and Phases of the Moon

The moon takes 29.5 days, about one month, to revolve around Earth from one full moon to the next. (See Figure 12-13.) As the moon orbits Earth, it also rotates on its axis. Because the moon completes one rotation in the time it takes to orbit Earth once, the same side of the moon always faces Earth. Not until 1959, when a Russian spacecraft circled the moon and took photographs, did we learn what the moon's far side looks like.

Figure 12-13. *Phases of the moon: The moon takes 29 ½ days to complete the cycle of its phases. The dot represents a crater that always faces Earth.*

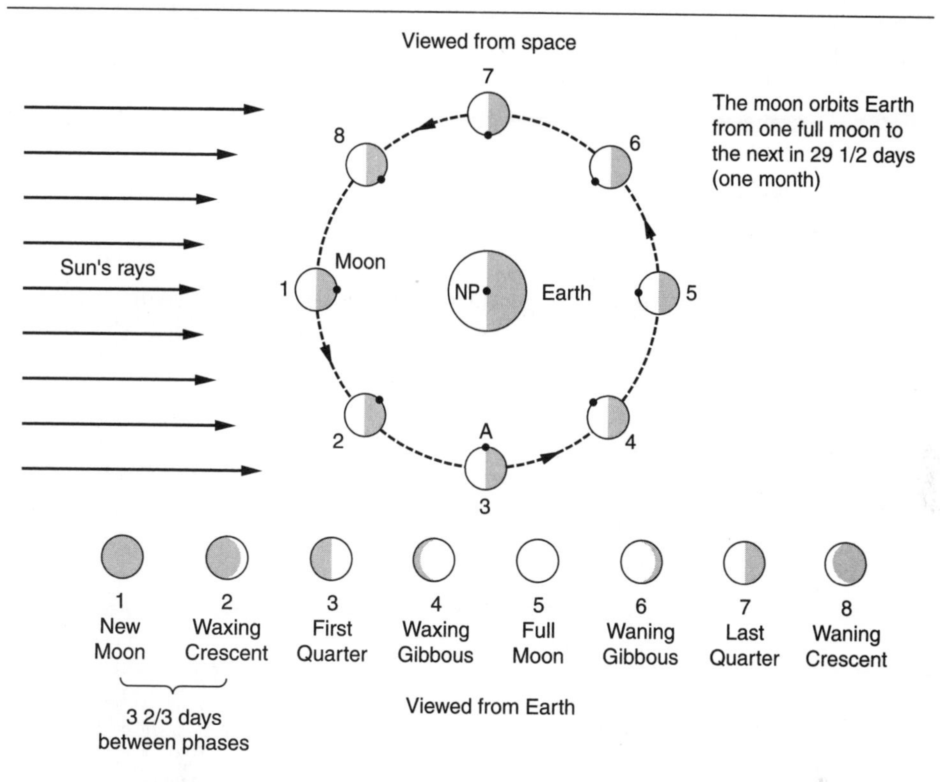

The moon does not give off its own light. We see it because it reflects light from the sun. One half of the moon's surface is always illuminated by the sun. Because the moon is moving in its orbit, the amount of sunlit side that faces toward Earth changes from night to night. As the moon revolves around Earth, we see different amounts of its lighted surface, so the moon appears to change shape. The *apparent* changes in the shape of the moon are called ***moon phases***. (See Figure 12-13.)

During the ***new moon*** phase, the moon is between Earth and the sun. The side facing Earth is dark, so we cannot see the moon. (However, the other side is lighted.) In a little more than one week, we see the moon's lighted surface slowly increase in size (referred to as a *waxing moon*), until the right half of the moon facing Earth is visible. This phase is called *first quarter*, because the moon has completed one-quarter of its journey around Earth. During the next week, the moon continues to *wax* until the entire side of the moon facing Earth is visible (lighted). This is called the ***full moon***. During the next week, we see the moon's lighted surface slowly decrease in size (referred to as a *waning moon*), until only the left half of the moon is visible. This phase is called *last quarter*. Slowly, the moon continues to wane, until once again the moon is at the new moon phase and is not visible.

Eclipses

An ***eclipse*** can occur when the sun, Earth, and moon are lined up. A lunar eclipse (eclipse of the moon) takes place when the moon passes through Earth's shadow. This can happen only when the moon is full, as shown in Figure 12-14A. A lunar eclipse is visible at night over a large area on Earth and takes a few hours from start to finish.

A solar eclipse (eclipse of the sun) takes place when the moon casts its shadow on Earth, which can occur only during new moon phase, as shown in Figure 12-14B. A solar eclipse is visible in daytime over a small area on Earth. A total solar eclipse, in which the sun is completely blocked by the moon, lasts only a few minutes. We do not have an eclipse every time the moon is in its new or full phase because the moon's orbit is tilted 5° either above or below Earth's orbital path around the sun.

Figure 12-14. *Eclipses: (a) A lunar eclipse occurs when the moon passes into Earth's shadow. (b) A solar eclipse occurs when the moon casts its shadow on Earth.*

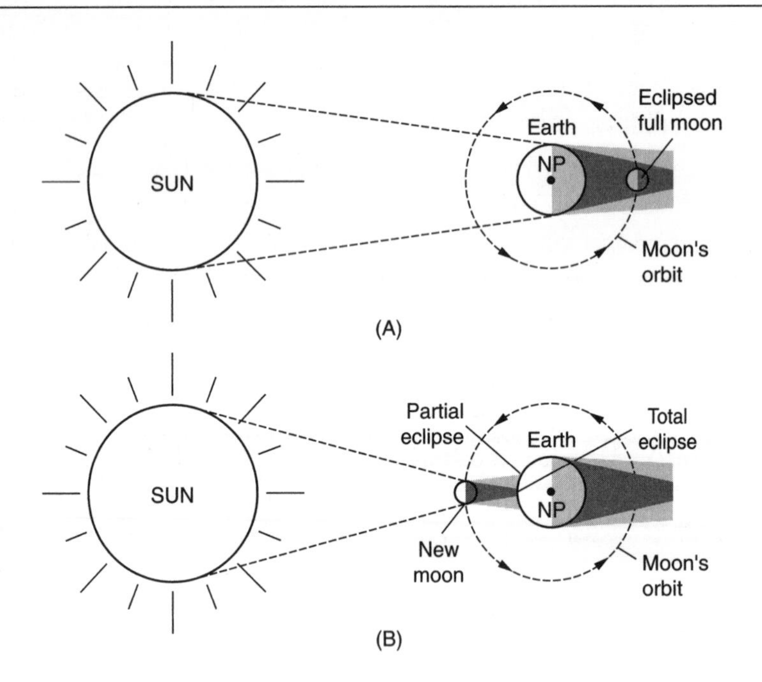

The Moon and Earth's Tides

The ***tides*** are the regular rise and fall in the level of the ocean. These changes in sea level are caused by the gravitational pull of the moon and the gravitational pull of the sun. The moon's gravity affects Earth's tides more strongly than the sun's gravity, because the moon is much closer to Earth.

The moon's gravity pulls the ocean waters into two large bulges. One bulge is on the side of Earth facing the moon. The other bulge is on the opposite side of Earth, the side not facing the moon. High tides occur at the same time at both locations. (See Figure 12-15.) Halfway between the tidal bulges, the ocean level falls, producing low tide. As Earth rotates, the oceans go through two cycles of high tide and low tide in a little more than 24 hours.

Figure 12-15. *High tides occur at positions on Earth facing directly toward or away from the moon. Low tides occur halfway between high tides.*

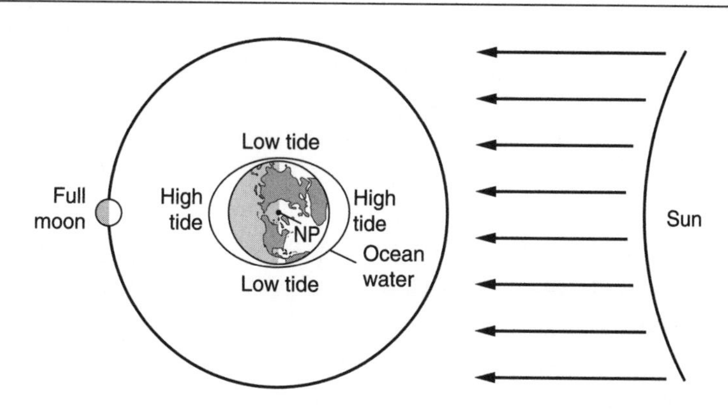

When the sun, Earth, and moon are in a straight line, during full moon and new moon, the high and low tides are greater than normal. The moon and sun's gravity are working in the same line. During first quarter and last quarter moon phases, the sun, Earth, and moon form a right angle with Earth at the vertex. At this time, the high and low tides are less than normal. The gravity of the moon and sun are working at right angles to each other.

Asteroids, Comets, and Meteoroids

Asteroids are pieces of rock of many different sizes and shapes that revolve around the sun. Many are found in the asteroid belt, between the orbits of Mars and Jupiter. (See Figure 12-9 on page 343.) Some scientists think that the asteroids are material left over from the birth of our solar system. This material never combined to form a planet. Other scientists think asteroids are the remains of a planet that broke apart.

A ***comet*** is a small object made up of rock, ice, and dust that orbits the sun. Comets vary in size, but usually are 10 to 30 km (6 to 19 miles) in diameter. They travel in long stretched-out elliptical orbits. (See Figure 12-16 on page 350.) As a comet approaches the sun, energy from the sun makes the comet glow and produce a "tail." The comet's tail always points away from the sun, whether the comet is moving toward or away from the sun.

Figure 12-16. *A comet has a stretched-out elliptical orbit around the sun. A comet's tail forms as it comes near the sun and always points away from the sun.*

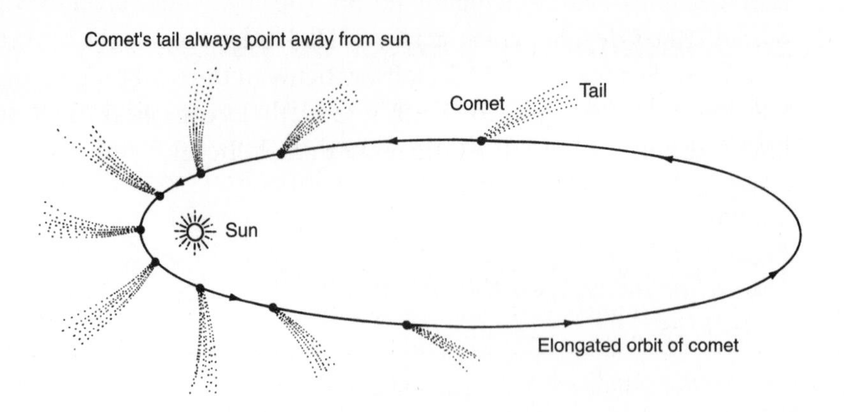

A ***meteoroid*** is a tiny piece of rock, some as small as dust, traveling through space. Scientists think the particles come from comets. Sometimes these rocks enter Earth's atmosphere at high speed. Friction between the rock and the atmosphere causes the meteoroid to heat up and burn. We see the burning meteoroid as a bright streak across the night sky, called a ***meteor*** or a "shooting star." Occasionally, a large meteoroid does not burn up completely and a chunk of rock, called a ***meteorite***, reaches Earth's surface.

Most meteoroids travel in orbits associated with comets. Particles fall off comets and remain in the same orbit around the sun, producing a *stream* of meteoroids. If Earth's orbit crosses the path of the meteoroid stream each year, a predictable meteor shower will occur. (See Figure 12-17.) On these nights, many meteors appear in a specific area of the night sky. The Perseid meteor shower occurs about August 12 when Earth passes through the orbit of a large stream of meteoroids from comet 109P/Swift-Tuttle.

Figure 12-17. *A stream of meteoroids is scattered in the orbit of a comet. If Earth passes through the meteoroids, we experience a meteor shower.*

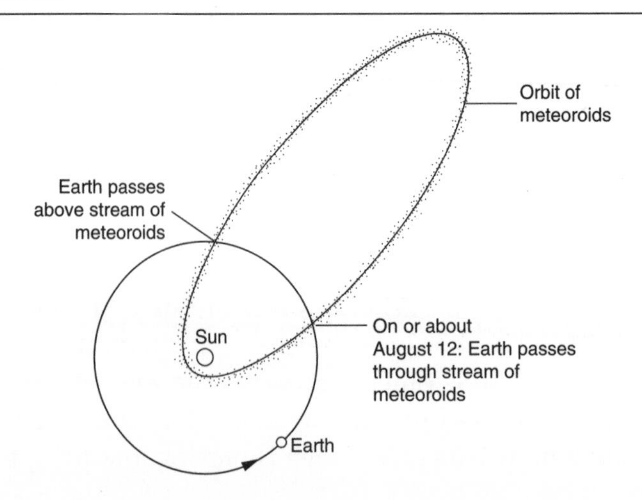

Process Skill 2

Analyzing Technology's Effect on Astronomy

In 1610, Galileo, using his telescope to observe the night sky, discovered four satellites (moons) revolving around Jupiter. During the next few centuries other astronomers using bigger and better telescopes discovered planets, plutoids, asteroids, and moons.

Improvements in telescope technology were slow until the mid-1900s. As a result, discovery of moons around other planet was slow. Five of Saturn's moons were discovered in the 1600s, two more in the 1700s, and one in the 1800s. Phobos and Deimos, the two satellites of Mars, were discovered in 1877. In the mid- to late 1900s new technology helped astronomers to look deeper into space. Telescope improvements, space probes, and computer-assisted imagery have helped astronomers to discover many new objects in space. The number of known satellites in the solar system has increased greatly through the use of this new technology.

The table below lists our solar system's planets and the number of known satellites of each at different dates. Use the information presented above and in the table to answer the questions.

Number of Known Solar System Satellites

Planet	*1609*	*1892*	*1962*	*1982*	*2000*	*2007*
Mercury	0	0	0	0	0	0
Venus	0	0	0	0	0	0
Earth	1	1	1	1	1	1
Mars	0	2	2	2	2	2
Jupiter	3	4	12	16	17	63
Saturn	0	8	9	16	18	56
Uranus	ND*	4	5	5	20	27
Neptune	ND*	1	2	2	8	13

*ND: Planet not yet discovered

Questions

1. Explain why the number of satellites in the solar system increased during the late 1900s.

2. Why has the number of known satellites for the inner planets (Mercury, Venus, Earth, and Mars) remained relatively constant, while the number of satellites for the outer planets has increased during the last four centuries?

3. Astronomers will most likely continue to find additional satellites around the outer planets because
(1) technology keeps improving
(2) new satellites are being produced
(3) their predictions are improving
(4) the laws of motion are changing

Review Questions

Part I

17. Andrea saw a streak of light flash across the night sky. She most likely saw a moving

(1) planet (2) star (3) comet (4) meteor

18. The largest object in the solar system is

(1) Earth (2) the sun (3) the moon (4) Jupiter

19. Most asteroids travel

(1) in highly elongated orbits

(2) in the rings of Saturn

(3) in orbits between Mars and Jupiter

(4) in orbits between Earth and Mars

20. Claude wanted to photograph the full moon on March 15 but the night was cloudy. His next opportunity to take a picture of a full moon will be on

(1) March 20 (2) April 1 (3) April 14 (4) April 30

21. What makes it possible for an astronaut to see Earth from space?

(1) Earth emits light from its surface.

(2) The moon emits light, and this light is reflected from Earth

(3) The sun emits light, and this light is reflected from Earth.

(4) The sun emits light, and this light is reflected from the moon.

22. The diagram shows how the moon looked on three nights in August. Which figure shows how the moon would most likely appear on the night of August 11?

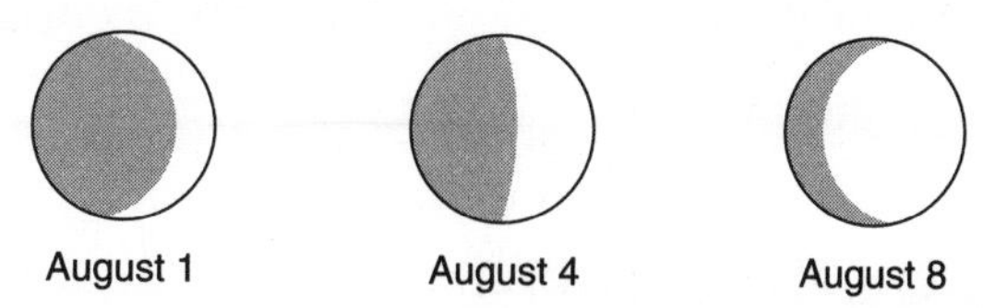

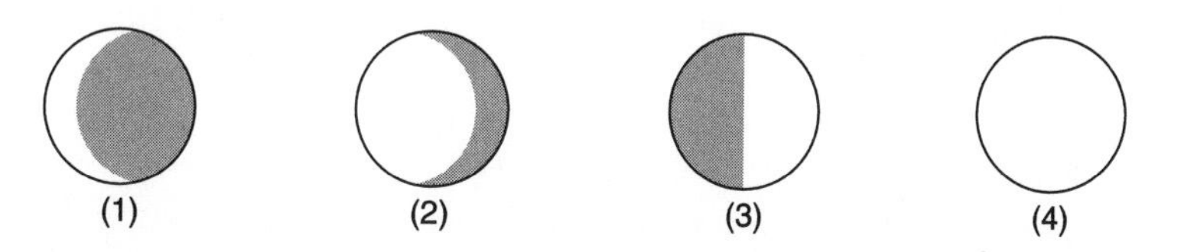

23. How long does it take the moon to revolve once around Earth?

(1) about 1 day

(2) about 1 week

(3) about 1 month

(4) about 1 year

24. We *cannot observe* the moon's surface from Earth when the moon is in which position in its orbit?

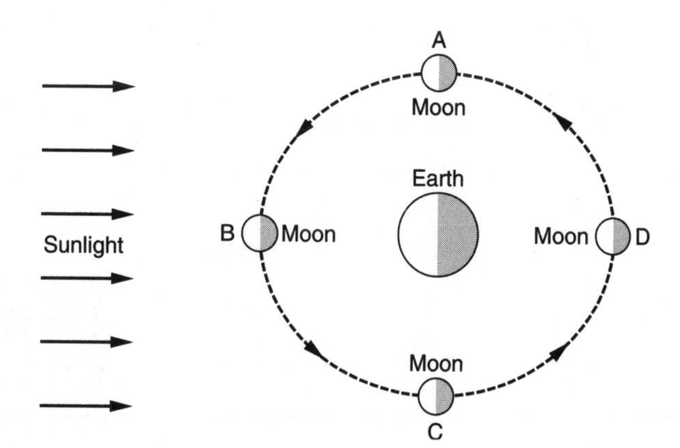

(1) A (2) B (3) C (4) D

25. A lunar eclipse can take place only during the

(1) full moon phase

(2) last quarter phase

(3) new moon phase

(4) first quarter phase

26. How many tides do we have each day?

(1) 2 high tides and 2 low tides
(2) 1 high tide and 1 low tide
(3) 2 high tides and 1 low tide
(4) 4 high tides

27. The most extreme high and low tides occur when

(1) the sun, Earth, and moon are at right angles to one another
(2) the first and last quarter moon phases occur
(3) the moon appears at night
(4) the sun, Earth, and moon form a straight line

Part II

28. The table below shows the times of high and low tides over two days.

Tidal Data Table

Day	*Time*	*Tide*
1	4:20 A.M.	Low
	10:35 A.M.	High
	4:45 P.M.	Low
	11:00 P.M.	High
2	5:10 A.M.	Low
	11:25 A.M.	High
	5:35 P.M.	Low
	11:50 P.M.	High

Using the pattern in the table, at about what time will the next low tide occur?

Base your answers to questions 29 through 31 on the diagram below and your knowledge of science. The diagram shows the orbit and position of the four planets closest to the sun on a particular day.

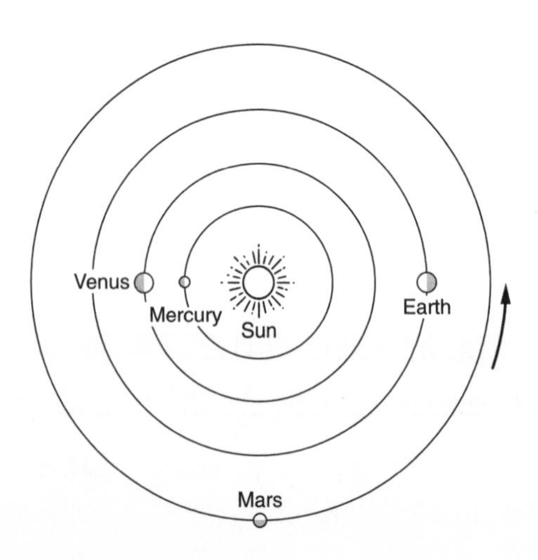

29. Why it is impossible to observe Mercury from Earth.

30. If you were on Mercury, which planet would be in your night sky?

31. Where is the asteroid belt in this diagram?

Base your answers to questions 32 through 34 on the table below and your knowledge of science. The table gives the amount of time it takes for each of the first five planets to make one complete rotation.

Planet	*Length of One Rotation*
Mercury	59 days
Venus	−243 days (rotates backward)
Earth	24 hours
Mars	24.6 hours
Jupiter	9.9 hours

32. Which planet spins most rapidly on its axis?

33. Which planet's day is almost equal to an Earth day?

34. In what direction does Earth rotate?

---

35. On August 15, Jesse saw a group of starlike objects in the night sky. A month later he noticed that one of the objects had moved (see the diagram). What inference can you make about the object that moved?

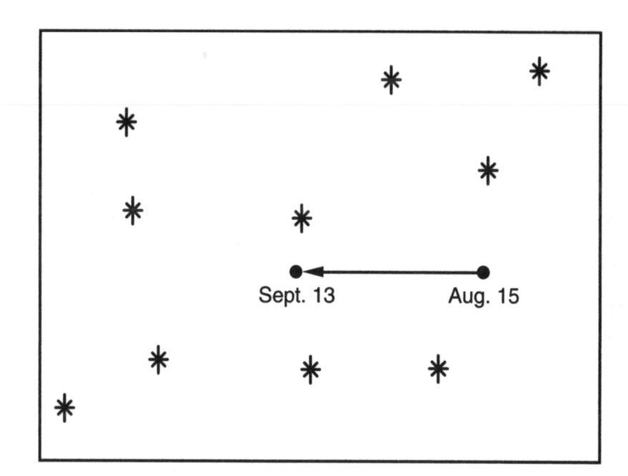

Our Place in the Universe

The Universe

The universe is composed of space and all matter and energy scattered throughout space. There are many more stars in the universe than there are grains of sand on all the beaches on Earth. However, the distances between stars are so great that most of the universe is empty space.

Stars are not distributed evenly in space, but are clustered in large groups called ***galaxies***. The universe contains billions of galaxies, and each galaxy contains billions of stars. Galaxies are separated by great distances in space. Our sun is a star in the galaxy called the *Milky Way*. From Earth on a clear night, we can see only a few thousand nearby stars in our galaxy and some nearby galaxies. Astronomers must use telescopes to see stars in other galaxies.

Distances in Space

Distances in space are so huge that they are difficult to understand. For example, the distance from Earth to our nearest neighbor star (besides the sun) is about 41,000,000,000,000 (41 trillion) kilometers! Because these numbers are so large, astronomers use a unit called a ***light-year*** to describe distances to the stars. A light-year is the distance light travels in one year, about 9.5 trillion kilometers. (See Figure 12-18.) Using light-years allows us to describe distance to stars with simpler numbers. Our sun, the closest star to Earth, is a little more than 8 light-minutes from Earth. Our nearest neighbor star is 4.3 light-years away. Some galaxies and stars are billions of light-years away. This means we are seeing the light that left these stars more than a billion years ago.

Many stars are larger than our sun. However, because they are so far away, they look like points of light in the night sky.

Formation of the Solar System

Although astronomers have evidence to support their theory of the origin of the universe and our solar system, many details are still missing. The theory states that between 10 and 15 billion years ago, a violent expansion of matter and energy created the universe, producing all the material from which the stars, planets, dwarf planets, moons, and other objects formed. This is called the big bang theory. If such a dramatic expansion did occur, you would expect to find all the galaxies moving outward and away from one another. This is exactly what astronomers have found.

Scientists also think that at some point, the expanding gases and dust particles condensed to form galaxies filled with stars. About 5 billion years ago, a dense area in the center of our local cloud of rotating gases and dust became our star, the sun. (See Figure 12-19.) The sun contains about 99 percent of the original cloud material. Soon afterward, the remaining material condensed into the planets, dwarf planets, moons, asteroids, comets, and meteoroids. The oldest rocks from Earth, the moon, and meteorites are about 4.5 billion years old. This information is evidence that all these objects formed at about the same time.

Figure 12-18. *A beam of light travels about 9.5 trillion kilometers per year. The light from a star that is 47.5 trillion kilometers from Earth takes 5 years to reach us.*

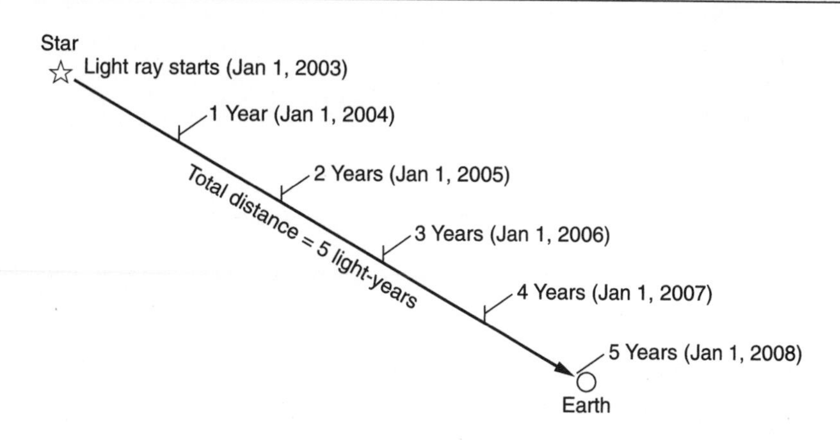

Figure 12-19. *(a) Cloud of gas and dust in space, (b) condensation of cloud, (c) early sun and planet formation, (d) present-day solar system.*

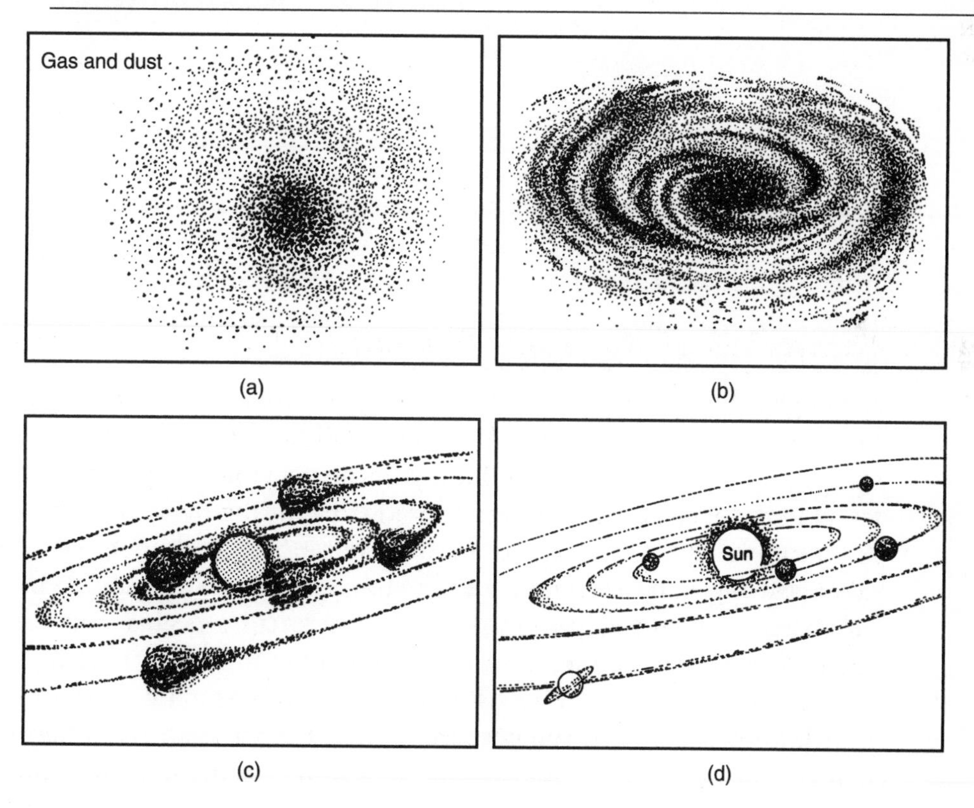

Earth's Position in the Universe

Our sun is one of more than 200 billion stars in the Milky Way galaxy, a large, disk-shaped group of stars that looks like a pinwheel. Our galaxy is

100,000 light-years across and about 20,000 light-years thick in the center. As shown in Figure 12-20, our sun, along with Earth and the other members of our solar system, is located about two-thirds out from the galaxy's center in one of its smaller spiral arms.

Figure 12-20. *The Milky Way galaxy and the location of the solar system in the galaxy.*

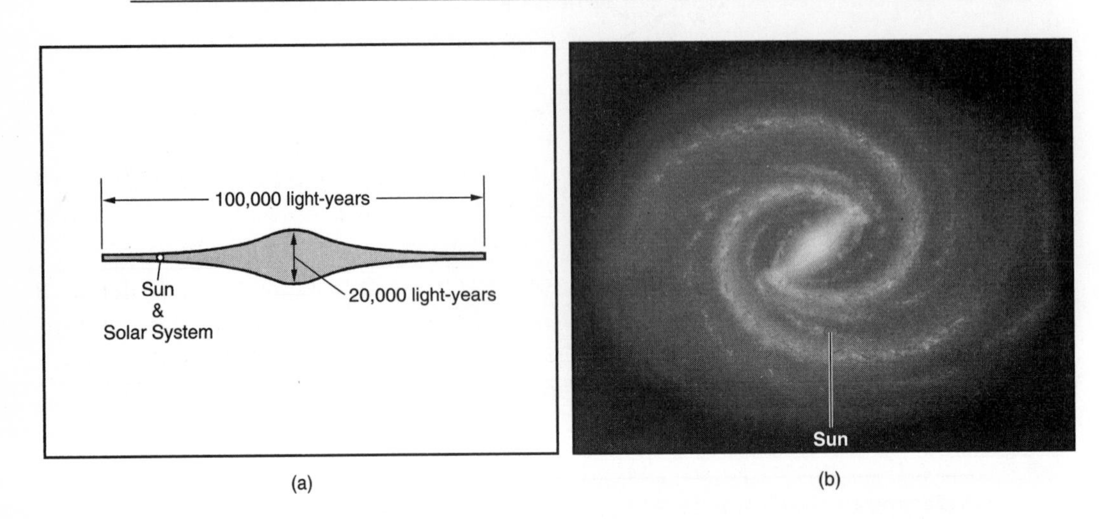

On a clear night, only a few thousand neighbor stars (all within our galaxy) are visible. If you look carefully, you can see a broad band of closely packed stars that looks like a band of haze, or cloud across the night sky. This milky patch across the sky contains many stars also in the Milky Way galaxy. Although the stars appear near each other, the distances between them are huge. Other galaxies and their stars are so distant that you need a telescope to see most of them. When you look at the night sky, you can see a very small portion of the universe that is near us.

Stars

Astronomers think stars begin as clouds of gas and dust in space. Gravity causes the material to collapse, forming the nucleus of a new star. As the material continues to collapse, a thermonuclear reaction begins. This reaction changes hydrogen into helium, releasing energy in the form of light and heat. In their early stage of development, stars appear reddish because they are relatively cool. As the star becomes hotter, it expands and its surface color may change from red, to orange, to yellow, and possibly to white. If the star is massive enough, it may become a bright blue-white star. Eventually, the star stabilizes. Our sun is an average-size yellow star in this stable stage of development. (See Figure 12-21.) Stars, like planets, are nearly spherical.

When the star's hydrogen is used up, the star collapses. Collapsing compresses the star and causes it to get hotter until it expands rapidly, becoming a cooler red giant or supergiant.

Figure 12-21. *A star undergoes many changes during a typical life cycle.*

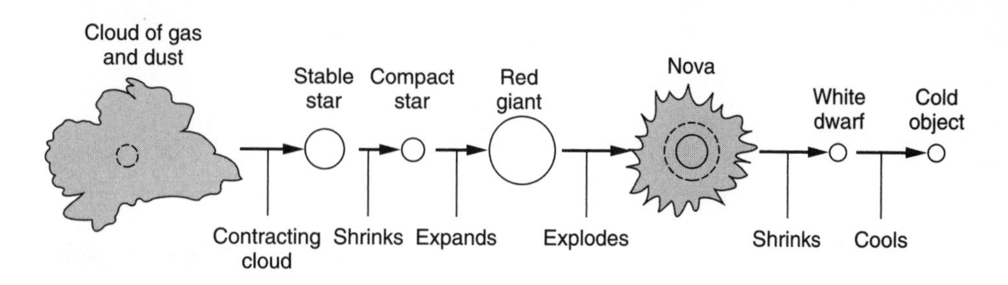

The giant star eventually explodes, losing its outer layer with a burst of energy. This bright flare-up of a star is called a *nova*. Finally, the star shrinks into a white dwarf and eventually may become a cold space object.

The amount of time it takes for a star to go through its life cycle depends on the amount of material in the original cloud and the size of the star. Our sun is believed to be 5 billion years old, and scientists think it will last another 5 billion years.

Review Questions

Part I

36. The big bang theory of the universe suggests the
    (1) universe is shrinking
    (2) universe is expanding
    (3) galaxy is expanding
    (4) solar system is expanding

37. Which fact suggests that our solar system is about 4.5 billion years old?
    (1) The planets all revolve in the same direction.
    (2) The sun is more than 4.5 billion years old.
    (3) The oldest meteorites, Earth rocks, and moon rocks are about 4.5 billion years old.
    (4) The light from some stars has traveled more than 4.5 billion light-years.

38. All the stars in our night sky are located
    (1) in the Milky Way galaxy
    (3) in the solar system
    (2) in a nearby galaxy
    (4) between galaxies

39. Where are the sun and solar system located?

(1) They are in the center of the Milky Way galaxy.

(2) They are within the spiral arms of the Milky Way galaxy.

(3) They are outside of the Milky Way galaxy.

(4) They are on the outer edge of the Milky Way galaxy.

40. Stars appear to have life spans of billions of years. During that time they

(1) never change

(2) constantly get larger

(3) constantly get smaller

(4) change in size and temperature

Base your answers to questions 41 and 42 on the table below. The table gives some data about four stars.

Star Data Table

Star	*Absolute Magnitude (Brightness)*	*Apparent Magnitude (Brightness)*	*Distance*
Sirius	1.4	−1.5	8.8
Vega	0.5	0.04	26.4
Capella	−0.6	0.05	45.6
Antares	−4.7	1.0	423.8

41. The numbers in the Distance column most likely are measured in

(1) miles

(2) kilometers

(3) light-years

(4) years

42. Apparent magnitude is how bright a star appears when viewed from Earth. The lower the magnitude, the brighter the star. Which star appears the brightest in the night sky?

(1) Sirius

(2) Vega

(3) Capella

(4) Antares

43. The Milky Way galaxy is believed to contain billions of stars, and its shape is best described as

(1) spherical, like a ball

(2) puffy, like clouds

(3) flat with arms, like a pinwheel

(4) oval, like a football

44. Which of the following objects is the smallest?

(1) sun

(2) galaxy

(3) solar system

(4) Earth

Part II

Base your answers to questions 45 through 47 on the table below and your knowledge of science. The table lists the distance from Earth to some objects.

Distances of Some Objects from Earth

Object	*Average Distance*
Sun	149,600,000 km (1 AU)
Moon	384,000 km
Neptune's orbit	4,497,000,000 km (30 AU)
Neighbor star	4.3 light-years
Neighbor galaxy	2,200,000 light-years
Distant galaxy	65,000,000 light-years

45. How does the distance between the planets in our solar system compare with the distance between stars?
46. Why do astronomers use light-years rather than kilometers or astronomical units to measure distance to stars and galaxies?
47. What units are the most convenient to use when measuring the distance to Eris?

Chapter 13 Science, Technology, and Society

Vocabulary at a Glance

global warming
science
society
system
technology

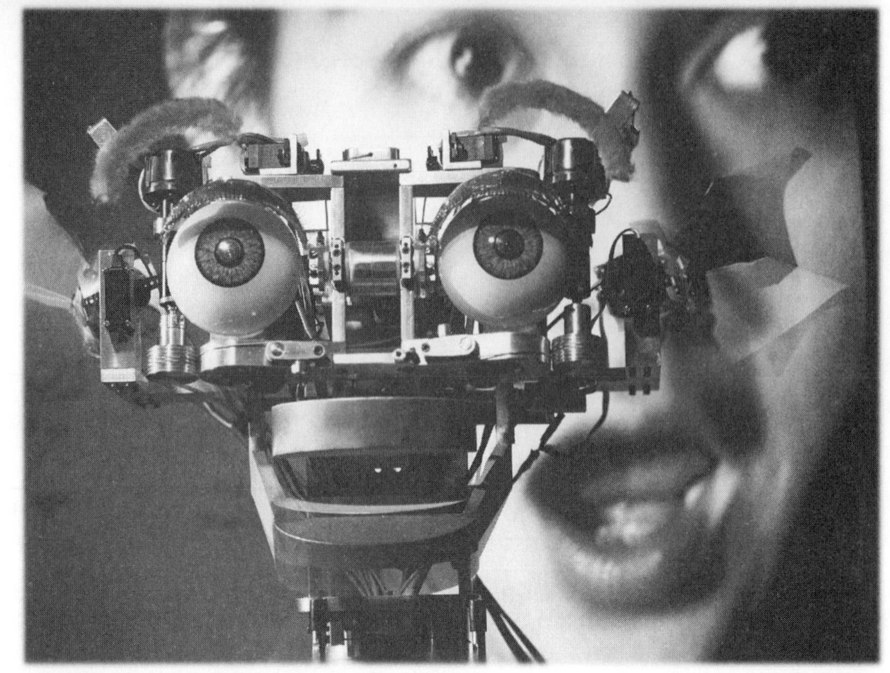

Kismet, a robot built at M.I.T.'s Artificial Intelligence lab, can produce facial expressions, such as surprise.

Major Concepts

- The purpose of science is to learn about the natural world by asking questions; the purpose of technology is to find practical ways to apply what we learn to solve problems. Science and technology often help to advance each other. The search for knowledge is a driving force for science and technology.
- We all use the products and processes of technology in almost everything we do.
- Technology can extend or improve our ability to do things; many products of technology affect the environment in some way.
- Science, technology, and society interact constantly. Often, a change in one area causes changes in the other two areas.
- Every technological product or process has advantages and disadvantages.

Science and Technology

Science and technology affect the lives of people all over the world. *Science* is the process of finding answers to questions about the natural world so that we can better understand it. By helping us understand nature, science helps us predict the outcome of physical events.

Some questions that science attempts to answer include:

- What is the nature of matter?
- How did the universe form?
- How did life evolve on Earth?
- How will Earth change in the future?

Technology finds practical ways to apply scientific knowledge to develop new products and processes that help us solve problems. Some problems that technology attempts to solve include:

- How can we improve gas mileage?
- How can we make farmland more productive?
- How can we control pollution?
- How can we replace fossil fuels?

There are major fields of science; for example, life science, Earth science, and physical science. Each field contains other, more specific sciences. (See Figure 13-1.) Biologists, chemists, and geologists are types of scientists. Engineers, computer programmers, and medical technicians are examples of people who work with technology.

Figure 13-1. *The major fields of science and some related sciences.*

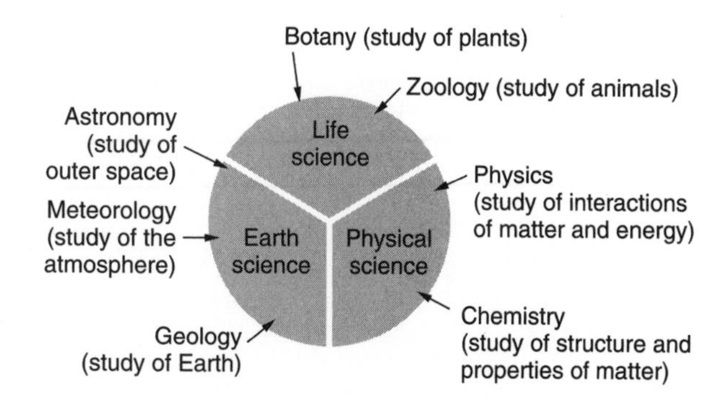

Science and Technology Advance Each Other

Science advances technology and technology advances science. Scientific discoveries often lead to the development of new or improved devices and processes. Some of these new devices and processes may lead to new discoveries in science or to a better understanding of scientific principles. For example, scientists discovered how light refracts or bends when it passes through different types of lenses. The discovery of this property of light led to the invention of the telescope, microscope, and eyeglasses. Using these devices, scientists made many more discoveries about the natural world. Every technological advance is based in some way on scientific principles, as the examples in Table 13-1 show.

Table 13-1. *Relationship of Scientific Principles and Technology*

Scientific Principle	*Technological Device/Process*
Low temperatures kill or reduce growth of microorganisms.	Refrigerators and freezers
Sunlight contains energy.	Solar heating systems and solar cells
Splitting atoms of radioactive elements produces heat.	Nuclear power plants
Every action produces an equal and opposite reaction.	Rocket engines and jet engines

The development of most technological processes and products involves knowledge from more than one field of science. For example, the artificial heart shown in Figure 13-2 was invented by using knowledge from biological science (the structure of the human heart) and physical science (the mechanical principles of how the heart works).

Figure 13-2. *Scientific knowledge from several different fields made possible the development of the artificial heart.*

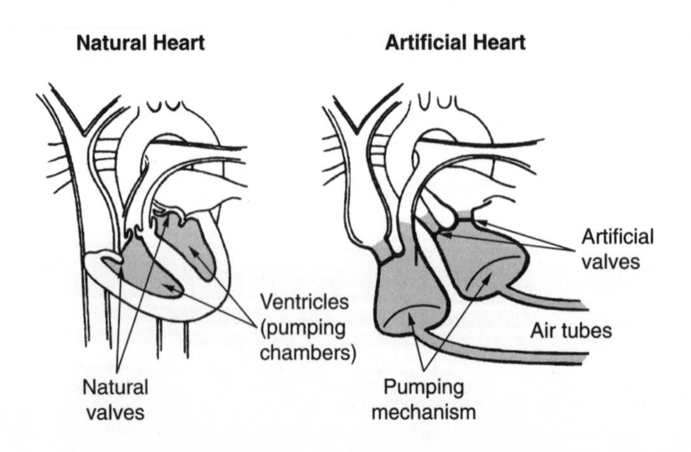

Search for Knowledge

Throughout the ages, people have wanted to learn more about Earth and the universe. The search for knowledge drives science and technology. It causes scientists to ask such questions as, who are we, and what is our place in this vast universe? It also leads to the development of new technologies to answer these questions. For example, improved and bigger telescopes help us to see farther into the universe. Table 13-2 shows that in space exploration, science and technology assist each other in the search for knowledge about the universe.

Table 13-2. *Space Exploration Highlights for Years 1957–2008*

1957	U.S.S.R. places *Sputnik 1* into orbit around Earth
1958	U.S.A. places *Explorer 1* into orbit around Earth
1959	Satellites orbit the sun and moon
1960	First weather satellite is placed in orbit
1961	First astronaut and cosmonaut sent into space
1964	*Ranger 7* takes close-up pictures of the moon
1965	First cosmonaut and astronaut make space walk; first close-up images of Mars
1966	First spacecraft to orbit the moon and first to land softly on the moon
1967	Capsule dropped to Venus's surface sends data back to Earth
1968	First manned spacecraft to orbit the moon and return
1969	First landing of astronauts on the moon; high-resolution pictures of Mars
1970	First probe to land softly on Venus
1971	First space station is placed in orbit
1972	First close-up pictures of Jupiter
1973	*Skylab* placed in orbit and maintained by crews
1974	*Mariner 10* takes close-up images of Mercury
1977–89	*Voyager* spacecrafts fly by and gather information from Jupiter, Saturn, Neptune, and Uranus; the spacecrafts continue to journey into interstellar space
1986	Spacecraft gathers information from Halley's Comet
1989	Spacecraft crashes into Jupiter and sends information back to Earth
1990	Spacecraft placed in orbit around the sun; Hubble Telescope placed in orbit around Earth
1996	Spacecraft placed in orbit around Mars
1998	Russia launches first mission to construct International Space Station
2001	Spacecraft studies Eros, a near-Earth asteroid
2003	Spacecrafts place two rovers (*Spirit* and *Opportunity*) on the surface of Mars
2004	Spacecraft orbits and photographs the rings around Saturn
2005	First soft landing on Saturn's moon Titan
2006	Stardust mission returns samples of comet to Earth
2008	Phoenix Mars Lander searches for water on Mars

Technology as a System

A *system* is a group of related parts that work together for a common purpose. Some technological processes and devices can be thought of as systems. The parts of a system act in a series of steps. These steps are input, compare and control, process, output, and feedback.

You can apply these steps to a home-heating system that uses a furnace and a thermostat. (See Figure 13-3.) You apply *input* to the system by setting the thermostat to the desired temperature. The thermostat *compares* the actual room temperature to the set temperature. The thermostat *controls* the furnace, turning it on if the temperature is too low. The *process* is the burning of fuel in the furnace, which produces the *output*—heat. The changing room temperature provides *feedback* to the thermostat, which turns the furnace off when the desired temperature is reached. In this way, the system maintains a selected temperature.

Figure 13-3. *A home-heating system is a technological example of a system.*

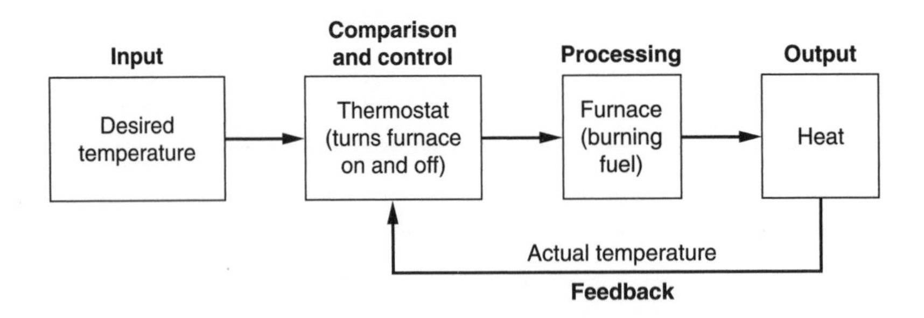

Using Technology

In almost everything you do—wearing clothing, sleeping on a bed, watching television, eating with a knife and fork, riding a school bus—you are using the products of technology. People use technology for many different reasons. (See Figure 13-4.) We use radios and telephones, calculators and computers, binoculars and telescopes, cell phones, and iPods, and other devices to extend or improve our ability to do things. Machines and appliances help us do work that requires more than human strength, and at speeds faster than humanly possible. To overcome physical limitations, people use devices like eyeglasses, hearing aids, and heart pacemakers.

We use many products of technology to change our environment. For example, we use electric lights so we can remain active after dark. Every technological process or device affects the environment in some way. Some of these effects may be harmful. Lightbulbs and other appliances use electricity. Many electric power plants consume nonrenewable natural resources and cause air and water pollution. However, technology can also protect the en-

Figure 13-4. *Some technological devices are used to extend human abilities (telescope), do work faster (sewing machine), and overcome limitations (wheelchair).*

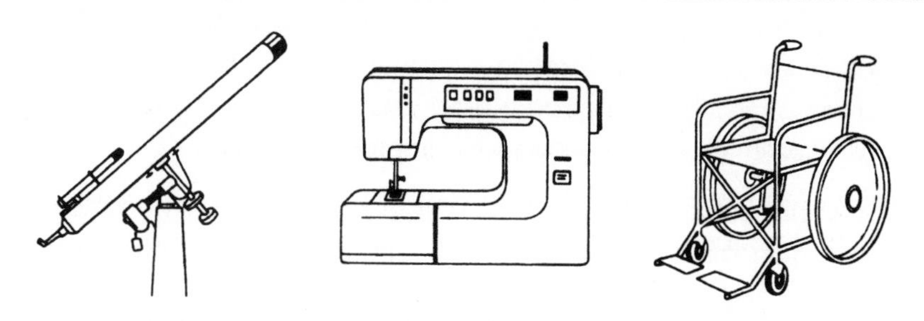

vironment. Sewage-treatment plants remove harmful materials from wastewater and pollution-control devices in cars and factories help to reduce air pollution. Table 13-3 gives some examples of helpful and harmful effects of technology on the environment.

Table 13-3. *Technology and Our Environment*

	Technology	How It Affects the Environment
Helpful effects	Electric lights	Extend light into the night
	Dams	Store water, generate electricity, and create lakes for recreation
	Sewage-treatment plants	Reduce pollution of rivers and lakes
Harmful effects	Electric generators	Consume nonrenewable natural resources and cause pollution
	Cars, boats, and planes	Consume nonrenewable resources and cause pollution

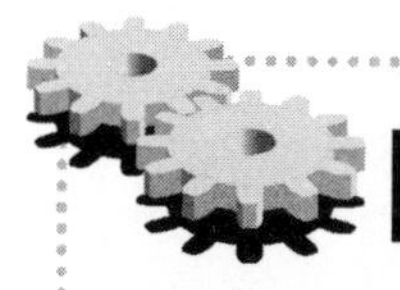

Process Skill

Using a Radio Communication Time Line

Before 1896, all telegraph signals were sent over wires. When wires were damaged, communication stopped until the wires were repaired. Guglielmo Marconi (1874–1937), an Italian physicist also known as the "father of radio," accomplished the first wireless transmission and reception of a telegraph signal in 1896. His accomplishment soon led to the development of radio transmission. Many previous scientific discoveries and technological devices helped Marconi achieve his success.

Many important scientific breakthroughs occurred during the 150-year period before Marconi's discovery. Scientists had to discover and understand electricity, electromagnetic waves, and magnetism. In addition, technological devices such as the telegraph, microphone, and transmitter had to be invented before wireless telegraph transmission and reception could occur. After Marconi's accomplishment, further scientific and technological advancements led to the radio broadcasts we enjoy today. These advancements include vacuum tubes, transistors, and improved circuitry.

The time line shows some of the scientific and technological advancements that led to wireless communication. Carefully examine the time line and answer the questions that follow:

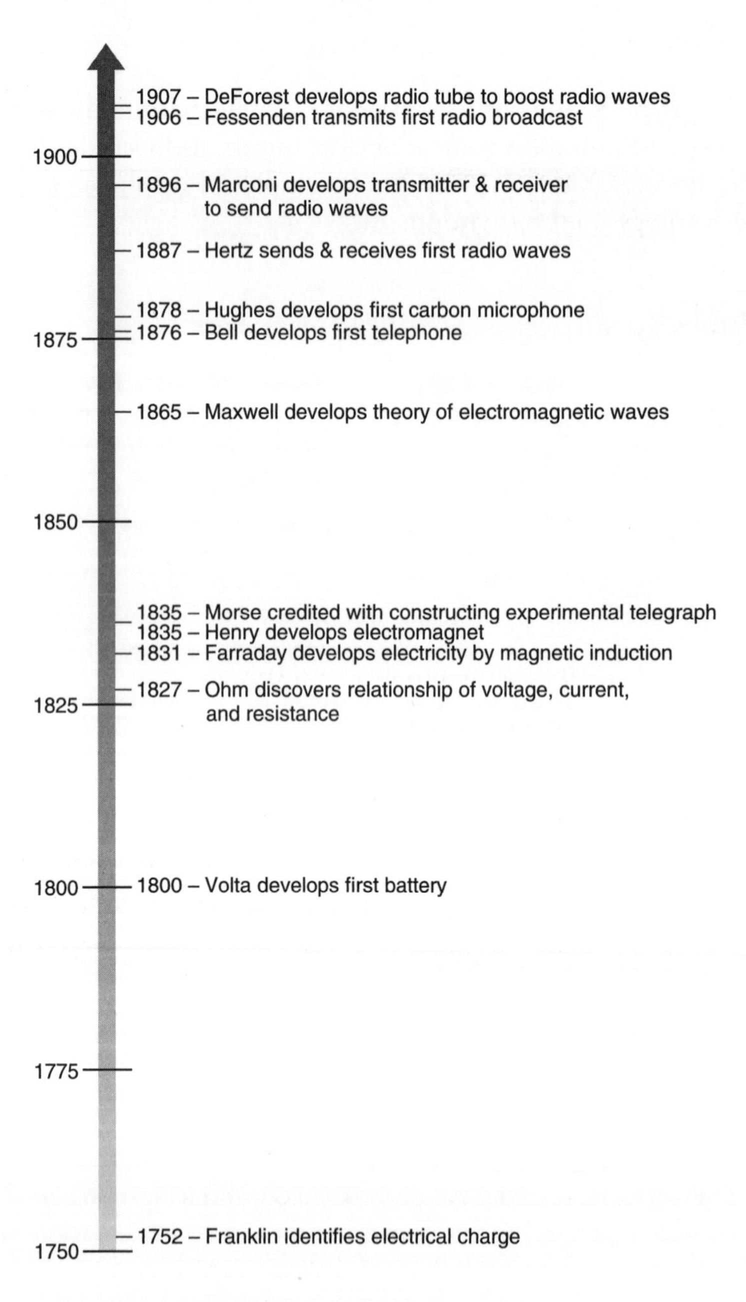

Questions

1. List three scientists who contributed scientific discoveries that led to wireless communication. List three scientists who contributed technological devices that led to wireless communication.

Answer the following questions based on this additional data: In 1883, Thomas Edison discovered that a hot filament emitted electrons, and in 1904, John Fleming developed the radio tube from Edison's discovery.

2. Was Edison's achievement a scientific discovery or a technological device?
3. Was Fleming's achievement a scientific discovery or a technological device?

Review Questions

Part I

1. The process of asking questions and finding answers to help us understand the natural world is called
   (1) science
   (2) industry
   (3) technology
   (4) renewing resources

2. Using scientific knowledge about magnetism to build a compass is
   (1) a scientific discovery
   (2) a technological development
   (3) predicting future physical events
   (4) observing the natural world

3. In 2005, engineers built a space probe and landed it on Titan, one of Saturn's moons. The probe sent pictures back to Earth, adding to our scientific knowledge. Which statement does this best demonstrate?
   (1) New technology sometimes builds on past technology.
   (2) Advances in technology cause some devices to become obsolete.
   (3) Technology affects our environment.
   (4) Technology helps advance our scientific knowledge.

4. Which statement bests describes humans and technology?
   (1) All people interact with the products of technology.
   (2) Some people interact with the products of technology.
   (3) Only a few people interact with the products of technology.
   (4) Only engineers and scientists interact with the products of technology.

5. Which of these technological devices affects the environment in some way?

(1) electric coffeepot
(2) car
(3) air conditioner
(4) all of these

6. In this air-conditioning system, the input is the

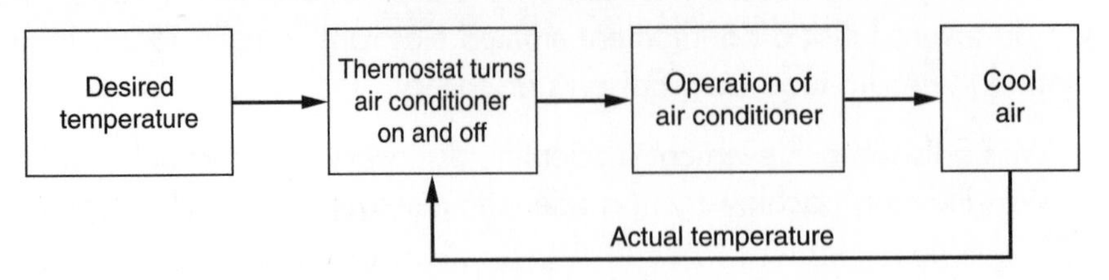

(1) desired temperature
(2) cool air
(3) actual temperature
(4) thermostat

7. A moving car is an example of a system. The following diagram shows how the components of a car system work together. The feedback in this system is provided by the

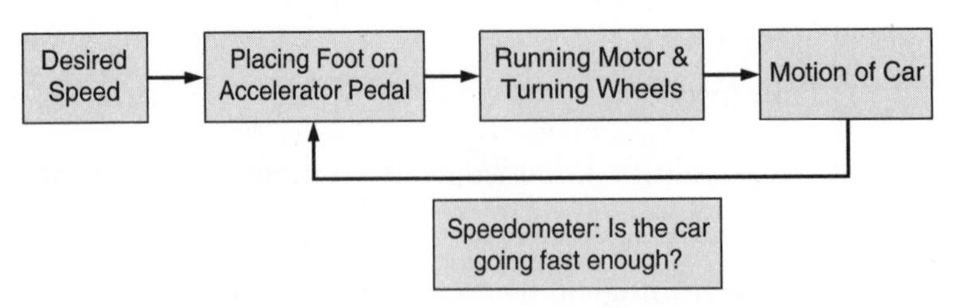

(1) accelerator
(2) motor
(3) wheels
(4) speedometer

8. What are three technological devices that extend our abilities?

(1) heart pacemaker, furnace, traffic light
(2) binoculars, telephone, calculator
(3) hammer, furnace, DVD recorder
(4) airplane, plumbing, coal mining

9. A technological device or process may become outdated, or obsolete, when a new device or process is developed that does a better job. An example of a technological device that has been replaced by a new device is

(1) buses replaced cars
(2) DVDs replaced videotapes
(3) eyeglasses replaced contact lenses
(4) shoes replaced sneakers

10. Which statement describes how space technology has helped us learn about the universe?

    (1) Scientists observe the moon from Earth.

    (2) Pluto has been reclassified as a dwarf planet.

    (3) Mars appears to have water-formed features.

    (4) Space probes have been sent to other planets.

Part II

Base your answers to questions 11 through 13 on the table below. The table lists the number of different word-producing technologies (machines) used in a television newsroom from 1955 to 2005. These technologies are listed in five-year intervals.

Numbers of Different Word-Producing Machines in a Newsroom, 1955–2005

	Word-Producing Technologies		
Year	Manual Typewriters	Electric Typewriters	Word Processors
1955	45	0	0
1960	35	10	0
1965	25	20	0
1970	10	40	0
1975	5	50	0
1980	0	40	10
1985	0	20	40
1990	0	5	50
1995	0	5	52
2000	0	1	57
2005	0	0	73

11. What trend do you see in the number of electric typewriters used in the newsroom between 1975 and 2005? What trend do you see in the number of word processors used in the newsroom during the same period?

12. What word-producing technology is most likely used in the television newsroom today?

13. Describe what happens to old technology as newer and better technology is produced.

Base your answers to questions 14 and 15 on the diagrams below and your knowledge of science. The diagrams show two home-heating systems.

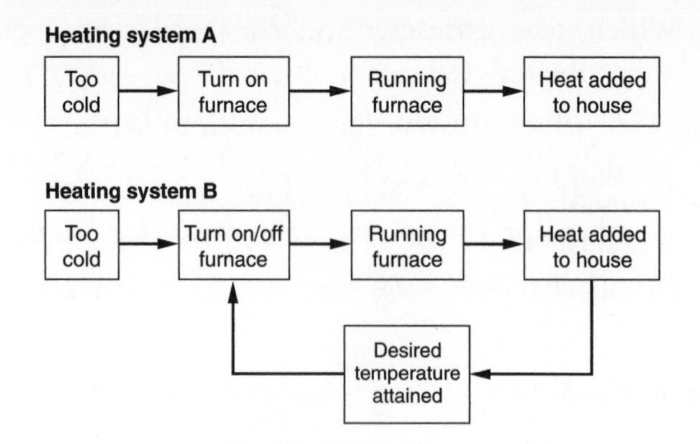

14. Which part of the system is missing from heating system A?
15. What would happen to the furnace in heating system B if the input were changed to "too warm"?

Interaction of Science, Technology, and Society

Effects of Science and Technology on Society

A *society* is a group of individuals who generally think and interact in similar ways. Science, technology, and society interact with one another constantly. (See Figure 13-5.) Often a change in one area will affect the other two. For example, scientific discoveries about the structure of matter led to many technological developments, including the production of microprocessors on tiny silicon chips. These "microchips" made possible many products that have affected society by improving health care, communications, and transportation.

Figure 13-5. *Science, technology, and society are constantly interacting and affecting one another.*

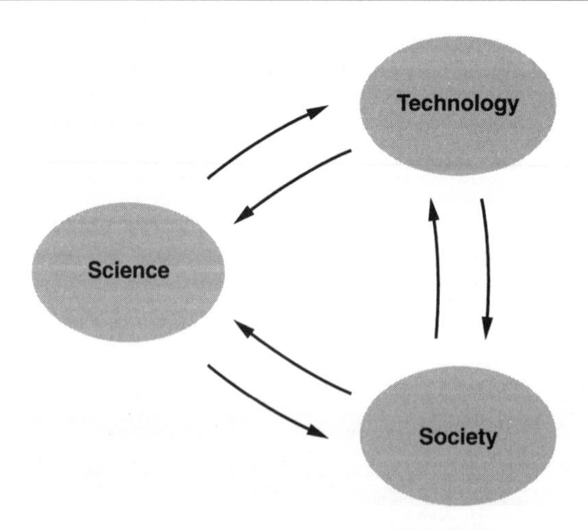

Our culture, economy, and social systems are affected by new developments in science and technology. During the 1800s, the United States changed from a farming society to a highly industrialized society. This period of cultural, economic, and social change was caused by technological developments such as new machines and new ways to power them. Some other examples of how science and technology have affected society are shown in Table 13-4.

Table 13-4. *Some Ways Science and Technology Have Affected Society*

Raised Our Standard of Living
Health care products allow us to live longer, healthier lives.
Work-saving home appliances provide more leisure time.
Created New Businesses and Industries
Still photography led to the motion picture industry.
Music recordings changed from phonograph records to tapes to CDs and MP3s.
Eliminated Businesses and Industries
Refrigerators and freezers eliminated the need for the ice-delivery industry.
Cars and buses eliminated the need for horse-drawn wagons, carriages, and stagecoaches.
Eliminated, Modified, and Created Career Choices and Job Opportunities
Computers replaced telephone operators.
Typists had to learn to use computers for word processing.
Computer systems need computer analysts to keep them running.
Solved Society's Problems
Vaccines eliminated certain diseases.
Communications systems, such as cell phones and the Internet, allow people to work together although they may be separated by great distances.

While science and technology have solved many problems, they have also created problems. Science and technology contribute to environmental pollution, garbage, and hazardous waste. (See Figure 13-6 on page 374.) In 2005, we threw away more than 206 million pieces of computer equipment. About 15 percent were recycled, but most of the other 85 percent went to landfills, causing environmental problems. Solving such problems requires the help of consumers, along with government, industry, science, and technology.

Figure 13-6. *Garbage disposal is a social problem. Much garbage consists of products of science and technology.*

Effects of Society on Science and Technology

Society affects science and technology in many ways. The needs of society often promote the development of new technologies. The need to help people overcome diseases and disabilities led to the development of new medical procedures, such as chemotherapy and laser eye surgery, and new devices, like artificial organs, limbs, and joints.

The attitude of a society may shift the direction of scientific research and technological development. Public opinion has pushed science to find a cure for AIDS. In contrast, public attitudes have largely discouraged the use of animals to test the safety of new cosmetics.

The public's attitude determines if an existing technology is accepted. Nuclear energy is a good example. Most people agree that nuclear energy has advantages and disadvantages. However, people disagree about whether its advantages outweigh its disadvantages. Public attitude against nuclear energy has led some countries to ban its use. Yet other countries generate most of their electricity with nuclear energy. Public opinion will certainly influence the future of nuclear energy. (See Figure 13-7.)

Figure 13-7. *People's attitudes toward nuclear power will affect the acceptance and use of this technology.*

Historical, Political, and Social Factors Affect Science

Historical, political, and social factors also influence science. Sometimes these factors have caused science to surge forward. At other times, these factors have caused scientific progress to lag and become a controversial social issue. Some examples will show how each of these factors can affect science.

Historical Factor. Galileo Galilei (1564–1642) (see Figure 13-8), an Italian scientist, built one of the first telescopes. Using his telescope, Galileo studied astronomical objects such as the sun, moon, and stars. He recorded his observations and made numerous calculations of the positions and motions of the sun, moon, and stars. Based on his observations, Galileo concluded that the sun was located in the center of the solar system and did not move around Earth. He also stated that Earth traveled around the sun. These conclusions supported an unpopular theory (heliocentric theory). However, church officials at that time did not support the heliocentric theory and condemned Galileo, saying that his conclusions were absurd. Galileo was imprisoned in his home for the rest of his life.

Church officials supported the historical belief that Earth was the center of the universe and that the sun, moon, stars, and other astronomical objects revolved around Earth (geocentric theory). Galileo's evidence was unacceptable because of the historical beliefs of the church.

Figure 13-8. *Galileo Galilei, an Italian scientist, was the first to use a telescope to study the night sky.*

Political Factor. On October 4, 1957, the Soviet Union (Russia) placed a satellite called *Sputnik 1* into Earth orbit. This event caused fear within the United States. Politicians saw Russia's achievement as a threat to our military defense and a sign of a decline in our science capabilities. In January 1958, the United States launched *Explorer 1*, showing that it also could orbit a satellite around Earth. These events marked the beginning of "the space race" between the United States and the Soviet Union.

The space race remained highly competitive for the next three decades. The launching of *Sputnik 1* caused the United States to recognize that in 1957 there was a technological gap between it and the Soviet Union. This resulted in a major push to improve science education in the United States. The competition between the two countries did much to promote science and technology in this country. Since the breakup of the Soviet Union, the competition has decreased. In fact, the United States and Russia are working together on the *International Space Station*.

Social Factor. Genetic engineering changes the genetic makeup of plants and animals to improve some of their traits. Genetically changing corn and wheat plants to increase their yield would help eliminate hunger in the world. Using genetic engineering (biotechnology) to replace disease-producing genes with healthy genes in humans could wipe out certain diseases and increase life expectancy.

However, issues regarding who decides what crops are improved and what diseases are eliminated could produce social unrest. Countries with this ability would possess power over countries without this ability. In recent years, genetic engineering has become a controversial issue for those who recognize the pros and cons of its use.

Global Effects of Technology

Technology used in one country may have an international or global influence. In 1985, an accident at the Chernobyl nuclear power plant in Ukraine in Eastern Europe released radiation that affected several neighboring countries. The radiation contaminated livestock, crops, and water. Acid rain is another example of how the use of technology in one area can affect other regions. Industries in the Midwest sometimes create air pollution that drifts eastward with the prevailing winds. This causes acid rain to fall in New York, New England, and parts of Canada.

Greenhouse gases (see Chapter 11), such as carbon dioxide and methane, trap heat close to Earth instead of letting it radiate into space. The burning of fossil fuels releases large amounts of carbon dioxide into the atmosphere. The amount of carbon dioxide released by industry and transportation has increased almost 25 percent in the last 25 years. ***Global warming*** is an increase in temperature believed to be caused by an increase in greenhouse gas emissions. If global warming continues, scientists expect a 3°C (5.4°F) increase during the 21st century. This increase will cause glaciers to melt and sea levels to rise. Rising sea levels along the coastlines will stress humans and other living things that live there.

On the positive side, technological advances allow people around the world to interact more frequently. Communication satellites let us make phone calls or send E-mail and pictures by computer and cell phone to people around the world. The Internet quickly gives us access from our home or office to a wide variety of information. With the help of satellites, television lets us watch events as they happen. (See Figure 13-9.) Technology helps us predict natural disasters such as the approach of hurricanes and tsunami waves. In the future, technology may make it possible to predict earthquakes. When these events occur, sensors on the ground and in the sky inform us immediately.

Figure 13-9. *Communication satellites let us watch televised events from all over the world.*

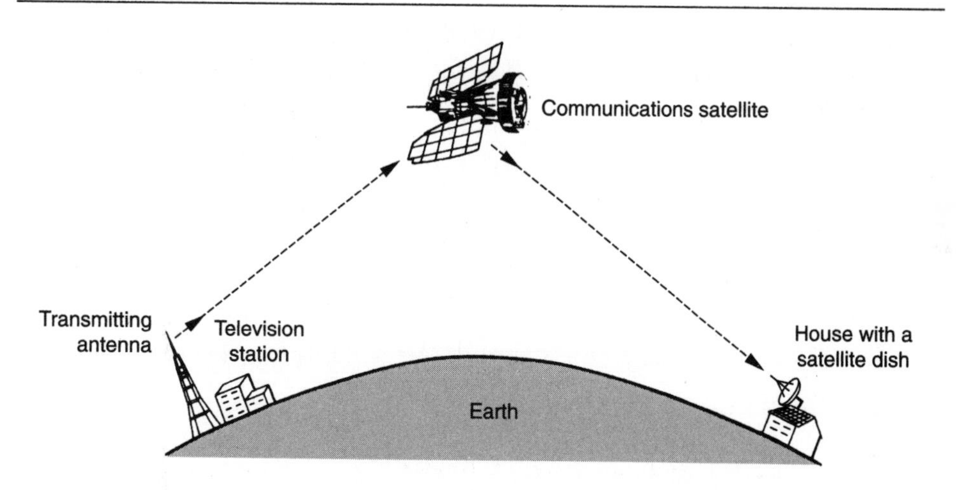

Review Questions

Part I

16. Society affects technology by
    (1) creating problems that need to be solved
    (2) providing funds for research and development
    (3) its attitudes toward new research or products
    (4) all of the above

17. An example of a job that has been created in response to a new technology is
    (1) construction worker
    (2) computer analyst
    (3) schoolteacher
    (4) police officer

18. An example of science and technology affecting society is the

(1) discovery of a vaccine to prevent polio

(2) discovery of bacteria

(3) discovery of the atom

(4) discovery of global warming

19. Carbon dioxide released into the atmosphere by industry and motor vehicles appears to be causing

(1) an increase in technology

(2) a negative attitude toward technology

(3) the creation of new industries

(4) global warming

20. In the past, secretaries used typewriters to write letters. Today they use computers and word-processing software. This example demonstrates

(1) technology has created new secretarial jobs

(2) technology has eliminated jobs

(3) technology has changed secretarial jobs

(4) technology has not affected secretarial jobs

21. The launching of an Earth-orbiting satellite by the Soviet Union in 1957 caused concern in the United States. This is an example of a

(1) historical factor affecting our need to improve science knowledge

(2) political factor affecting our need to improve science knowledge

(3) social factor affecting our need to improve science knowledge

(4) science factor affecting technology

22. Which choice best shows the relationship illustrated in the diagram?

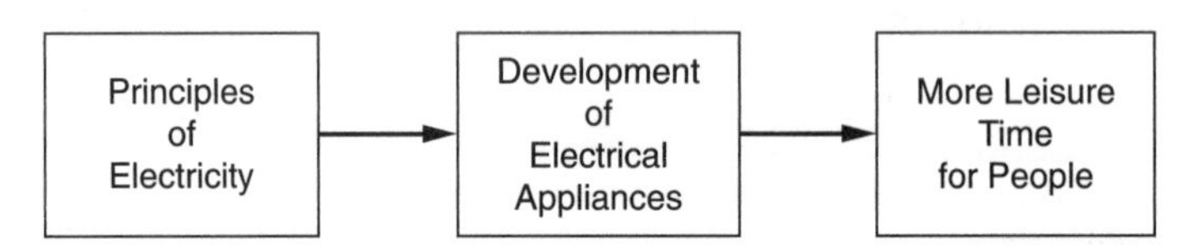

(1) Science affects technology and technology affects society.

(2) Science affects society and society affects technology.

(3) Technology affects society and society affects science.

(4) Society affects technology and technology affects science.

23. In Europe, many people use irradiated milk (milk treated with radiation to kill microorganisms). Unopened containers of irradiated milk can be stored at room temperature for a long time. In the United States, many people are suspicious of irradiated milk and are slow to accept it.

The above paragraph illustrates that

(1) technology has affected society by raising our standard of living

(2) people's attitudes can affect acceptance and use of technological processes

(3) products of technology may have an international effect

(4) technology has caused some products to become obsolete

24. The following graph shows how average human life expectancy has changed over time. This change is most likely a result of advances in

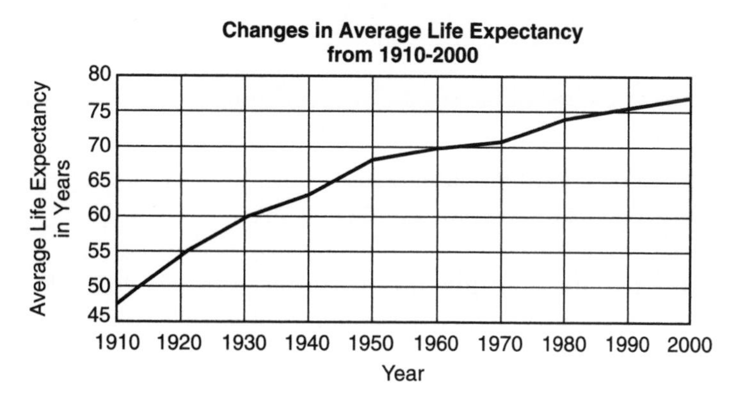

(1) medical technology

(2) environmental technology

(3) educational technology

(4) communication technology

25. The presidential election of 2000 demonstrated a problem in our voting process. The problem was caused by the method used to cast a ballot. The problem demonstrates a relationship between

(1) science and society

(2) science and technology

(3) society and technology

(4) long-term and short-term effects of technology

Part II

Base your answers to questions 26 through 28 on the table below and your knowledge of science. The table shows how the percentage of the workforce in three job areas has changed from 1800 to 2000.

Percent of Workforce in Three Fields, from 1800 to 2000

Year	*Agriculture*	*Information*	*Industry*
1800	75%	5%	20%
1850	60%	5%	35%
1900	40%	5%	55%
1950	10%	20%	70%
2000	5%	75%	20%

26. Copy the circle below. Draw and label a pie chart that shows the total workforce in the year 2000.

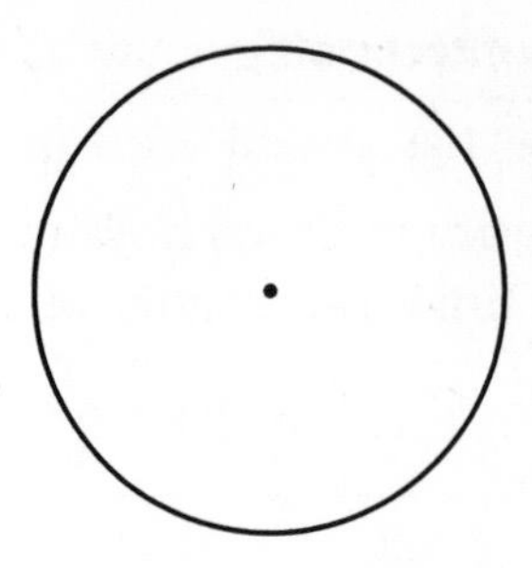

27. Describe how changes in information technologies affected the industry workforce from 1950 to 2000.

28. Give one example of how new technology changed the information workforce between 1950 and 2000.

Base your answers to questions 29 through 31 on the graph below and your knowledge of science. The graph shows atmospheric concentration of carbon dioxide from 1900 to 2006 in parts per million (ppm).

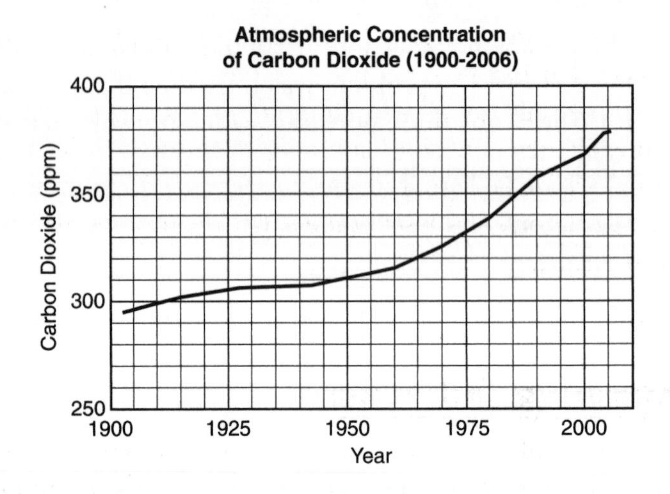

29. What caused the increase in atmospheric carbon dioxide?

30. How is an increase in carbon dioxide affecting the atmosphere? What is this process called?

31. Predict one way the surface of Earth will change if the amount of carbon dioxide in the atmosphere continues to increase throughout the 21st century.

Technology Increases Our Choices

People have more choices in their everyday lives because of the products of technology. For example, cable and satellite television, videocassette recorders, and CD, DVD, and MP3 players have increased our choices in home entertainment. Figure 13-10 shows some leisure activities that are available because of technology.

Figure 13-10. *Many leisure-time activities are avaialbe because of technology.*

Technology gives people more transportation and home appliance choices. For instance, people may travel from New York City to Chicago by car, bus, train, or plane. Shoppers may choose from a wide selection of home, kitchen, and shop appliances produced by technology. Some of these items include DVD players, microwave ovens, computers, cell phones, graphing calculators, and laser guide saws. Technology also gives us the choice to shop online.

Assessing Technology

Every technological process or device provides benefits and burdens for people and the environment. For instance, the automobile has given people greater mobility and contributed to our nation's wealth. However, cars, buses, and trucks contribute to air pollution and cause deaths and injuries in traffic accidents. Table 13-5 on page 382 lists benefits and burdens of some technological devices and processes.

Our society monitors the effects of many technological devices and processes, including medical treatments, food additives, industrial chemicals, and processes for generating electricity. Government agencies and public-interest groups perform this task.

Table 13-5. *Benefits and Burdens of Technology*

Technological Process or Device	Benefits	Burdens
Nuclear energy	Nonpolluting electricity	Risk of accidents; radioactive wastes
Painkilling drugs	Relieve pain	Addiction through abuse
Computers	Increased ability to process data	Loss of jobs; wrist and eye problems
Space travel	Increased knowledge	High cost
Life-sustaining medical devices	Keep people alive	Decisions about when to use them or stop using them
Chemical fertilizers	Increased crop yields	Upset the ecology of lakes and streams; possible harm to humans
Artificial sweeteners	Convenience for diabetics and dieters	Possible increased risk of cancer
Refrigerants	Storage and preservation of food	Released into atmosphere, they destroy ozone layer
Herbicides and pesticides	Destroy unwanted plants and insects	Upset the natural food chain of some animals; possible harm to humans
Cell phones	Improve communications	Possibly cause cancer

Technology and Decision Making

We make decisions about the use of technology almost all the time. To make these decisions wisely, we consider both short-term and long-term effects. Sometimes the short-term benefits of a technology outweigh its long-term burdens. For example, dentists agree that the benefits of using x-rays to find cavities in your teeth outweigh the possible long-term dangers of brief exposures to the radiation. (See Figure 13-11.)

Figure 13-11. *"Weighing the benefits and burdens of technology."*

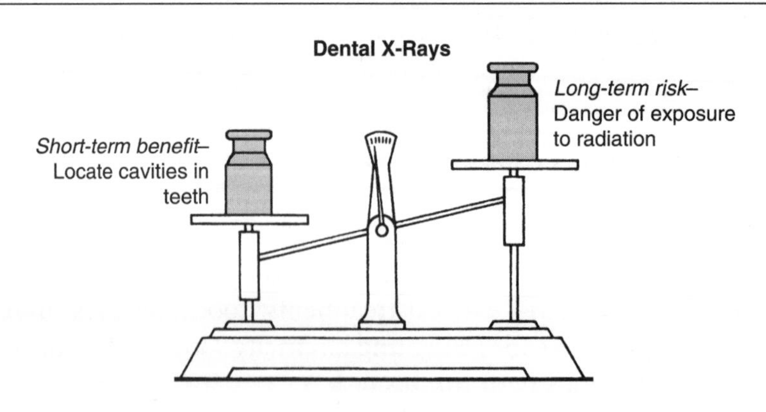

In other cases, long-term benefits may outweigh short-term burdens. Wearing a seat belt in a car may be a momentary discomfort. Over time, however, the use of seat belts reduces deaths and injuries from car accidents. Society's consideration of short-term and long-term effects has led to using unleaded gasoline for cars (which causes less pollution) and the recycling of cans, bottles, and newspapers (to reduce waste and conserve resources).

Review Questions

Part I

32. Which lists both a benefit and a burden of using the Internet?
    (1) offers useful products, and provides up-to-date news
    (2) provides entertainment, and decreases healthy physical activity
    (3) causes eye strain, and discourages reading books
    (4) provides a useful educational tool, and offers games

33. Most oil tankers have a single hull that may break open easily if the ship runs aground, causing oil spills. However, some tanker ships are now being built with a double hull that is less likely to break. This best illustrates that
    (1) technology has increased our choices in life
    (2) technology may be modified to reduce or eliminate its disadvantages
    (3) government should constantly monitor technology to determine possible harmful effects
    (4) decisions about technology often involve trade-offs between benefits and burdens

34. People with diabetes, who should not eat sugar, can safely use the artificial sweetener saccharin. Saccharin also has fewer calories than sugar and causes less tooth decay. However, some experiments have shown that using saccharin may increase the risk of getting cancer. A burden for people who use saccharin is
    (1) it is an artificial sweetener
    (2) it takes the place of sugar
    (3) it increases the risk of getting cancer
    (4) it has fewer calories than sugar

35. Big cars get poor gas mileage, but they are safer than small cars in a traffic accident. Small cars get good gas mileage but offer little protection in an accident. For these reasons, many people buy medium-sized

cars, which get fairly good gas mileage and offer some protection in accidents. This is an example of

(1) a burden of technology on the environment

(2) monitoring short-term and long-term effects of technology

(3) a decision about technology that involves a trade-off between benefits and burdens

(4) modifying technology to reduce or eliminate its drawbacks

36. Although the pesticide DDT was effective in killing insects that damage crops, its use was banned when it was found to be harmful to humans and wildlife. This shows that

(1) technological products have only disadvantages

(2) technological products have increased our choices

(3) the use of a technological product may be stopped if its disadvantages outweigh its advantages

(4) the use of a technological product may depend on people's attitudes

37. A burden to the widespread use of digital cameras, instead of film cameras, is

(1) the loss of film processing jobs

(2) you know instantly how your picture looks

(3) you never run out of film

(4) pictures are more expensive to produce

38. A few years ago people used CDs to record and store music. Today people use MP3 players for the same purpose. This is an example of

(1) people's attitudes causing a change in technology

(2) a new technology replacing an old technology

(3) a technology burden becoming a technology benefit

(4) responding to technology harming the environment

Part II

Base your answers to questions 39 through 41 on the information below and your knowledge of science.

Some social problems can be solved or helped by laws that control the use of technology. These laws provide benefits but may also burden us. Some laws may deny our right to privacy or freedom of choice. Other laws may be too costly to put into action and enforce. Listed below are some problems caused by technology. Beside each problem is the law that controls the problem and the benefits and burdens of the law.

Problem	*Law*	*Benefit*	*Burden*
Water pollution from industries	Effluent controls law	Cleaner water	Higher industrial costs
Head injuries from motorcycle accidents	Mandatory motorcycle helmet law	Reduces freedom of choice; uncomfortable	Fewer head injuries in motorcycle
Crimes committed with easily obtained handguns	Handgun control law	Fewer crimes; harder for criminals to get guns	Harder for people to get guns for protection or hunting
Bottle and can litter	Bottle deposit law	Cleaner streets and sidewalks	Inconvenience of returning bottles and cans

39. Which two laws deal with environmental problems?

40. Which law has its benefit and burden given in the wrong order?

41. Most states have a seat belt law. How is the seat belt law beneficial? How is the seat belt law a burden?

Base your answers to questions 42 and 43 on the information below and on your knowledge of science.

In the late 1900s, the bottle deposit bill was passed. This law requires people to pay an extra five cents for every beverage bottle or can. The deposit is refunded when the empty bottles and cans are returned. The graph below shows the number of bottles collected on a roadside, 1980 to 2007.

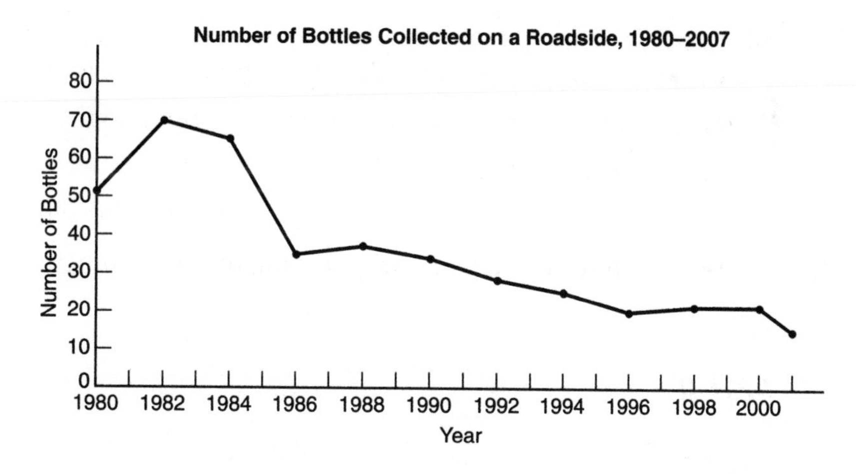

42. In what year did the bottle deposit bill most likely become law? How do you know?

43. Has the bottle deposit bill been successful? Explain.

Practice Test 1

Part I

1. Fossil shell fragments in a rock indicate the rock is most likely

   (1) igneous
   (2) sedimentary
   (3) metamorphic
   (4) volcanic

2. Rocks are classified as igneous, sedimentary, or metamorphic based on

   (1) size
   (2) shape
   (3) color
   (4) how they formed

3. Which statement best describes the distribution of heat inside Earth?

   (1) The core is cool and the crust is hot.
   (2) The temperature is the same throughout Earth.
   (3) The core is hot and the crust is cool.
   (4) The mantle is hot and the core is cool.

4. The diagram below shows how mountains change over long periods of time. Which process causes mountains to change as shown in the diagram?

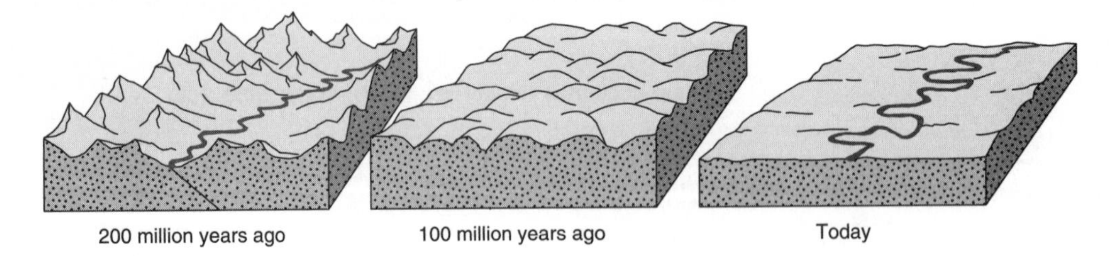

   (1) erosion
   (2) earthquakes
   (3) convection
   (4) deposition

5. The diagram below shows ice forming in the cracks of rocks. This process causes rocks to break into smaller pieces. The reason ice breaks rocks into smaller pieces is

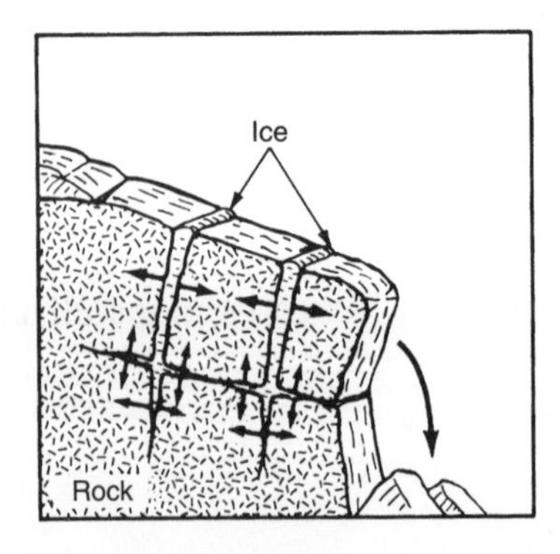

(1) water contracts when it freezes
(2) water expands when it freezes
(3) rocks get hard when they freeze
(4) rocks absorb water

6. Movement of Earth's crustal plates is caused by

(1) ocean currents
(2) convection currents in Earth's mantle
(3) the gravitational pull of the moon
(4) convection currents in Earth's atmosphere

7. What evidence suggests the continents were once connected?

(1) Some continents are the same size.
(2) All the continents are moving in the same direction.
(3) Some continents fit together like pieces of a jigsaw puzzle.
(4) All the continents have mountains and glaciers.

Base your answers to questions 8 through 10 on the table below and your knowledge of science. The table shows various data collected for Buffalo, New York, on February 29, 2008.

Updated: 4:21 P.M. EST on February 29, 2008

Observed at Buffalo, New York	
Temperature	28°F
Windchill	17°F
Humidity	90%
Dew point	26°F
Wind	SE at 14 mph
Air pressure	29.96 falling
Conditions	Precipitation
Visibility	0.7 mile

8. The information in the table describes Buffalo's

(1) climate conditions
(2) weather conditions
(3) weather forecast
(4) seasonal change

9. What type of precipitation is most likely occurring in Buffalo?

(1) rain
(2) snow
(3) hail
(4) sleet

10. The air pressure reading is stated in what units?

(1) inches of mercury
(2) millibars
(3) isobars
(4) degrees

11. An air mass that forms over the Gulf of Mexico and moves into Alabama is usually

(1) cold and humid
(2) cold and dry
(3) warm and humid
(4) warm and dry

12. All matter is composed of tiny particles called

(1) atoms
(2) chromosomes
(3) cells
(4) metals

13. Which instrument might be used to determine the volume of an irregularly shaped solid?

(1) graduated cylinder
(2) thermometer
(3) ruler
(4) funnel

14. Hydrogen peroxide is a compound often used to clean contact lenses. When exposed to light, hydrogen peroxide changes into water and oxygen gas. This change is an example of

(1) a physical change
(2) a chemical reaction
(3) photosynthesis
(4) evaporation

The diagram below shows the inner portion of the solar system.

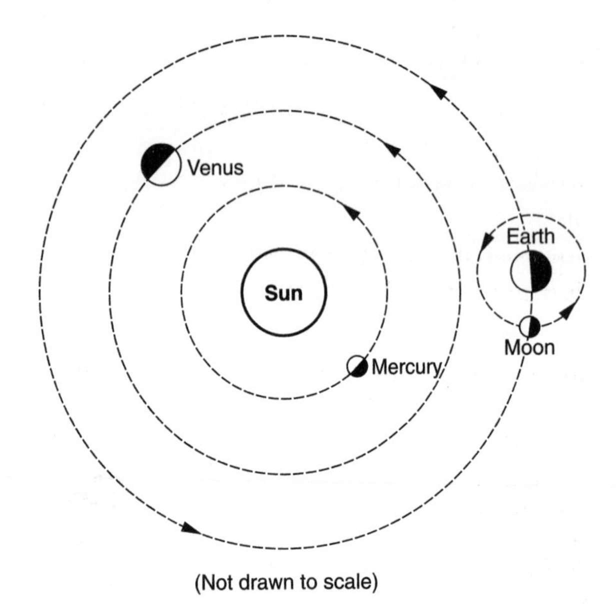

15. From where are we viewing the inner portion of the solar system?

(1) from above the North Pole
(2) from above the South Pole
(3) from the sun
(4) from Mars

16. During the night, the stars appear to rise in the east and set in the west. The reason for this is

(1) Earth rotates from west to east
(2) Earth rotates from east to west
(3) Earth rotates from north to south
(4) Earth rotates from south to north

17. The graph below shows the changing tides during a 30-hour period of time.

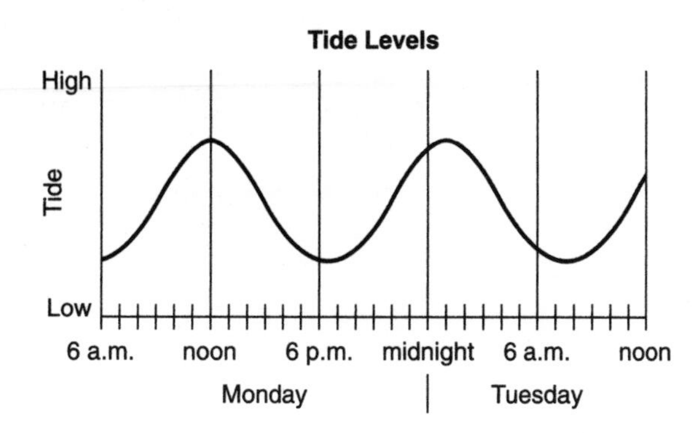

Approximately how much time is there between high tide and the next low tide?

(1) 24 hours
(2) 12 hours
(3) 6 hours
(4) 3 hours

18. A person standing on a bus suddenly stumbles toward the back of the bus when the bus begins moving. Which statement best summarizes the scientific concept that caused this reaction?

(1) A falling object accelerates at a constant speed.
(2) A body at rest resists a change in its motion.
(3) The force of friction is greater than the force of a moving bus.
(4) The force of friction slows a moving body.

19. The diagram below shows two children pulling on a rope with constant but unequal forces.

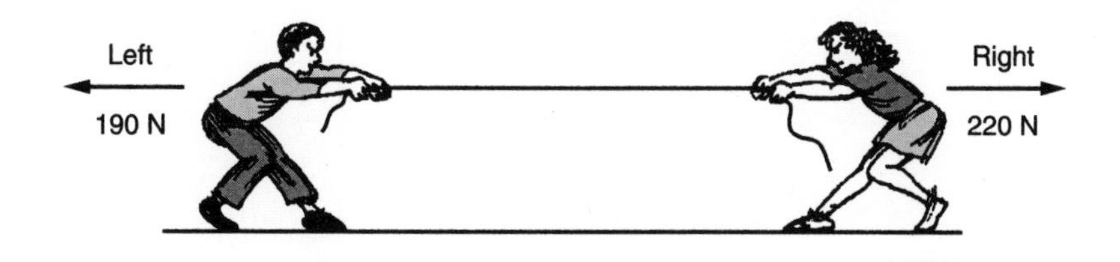

What will be the result of the action of the two children?

(1) They will move to the left.
(2) They will move to the right.
(3) They will not move.
(4) They will move to the right and then to the left.

20. The Periodic Table of the Elements is used to determine the

(1) properties of elements
(2) properties of molecules
(3) age of elements
(4) age of molecules

21. What simple machine is being used to open the can of paint in the diagram below?

(1) wheel and axle
(2) inclined plane
(3) pulley
(4) lever

22. The problem of global warming, if it continues, will eventually affect life on Earth. Solving the global warming problem presents a challenge to

(1) society and science
(2) society and technology
(3) science and technology
(4) society, science, and technology

23. The diagram below shows two identical balls (A and B) on shelves at different heights.

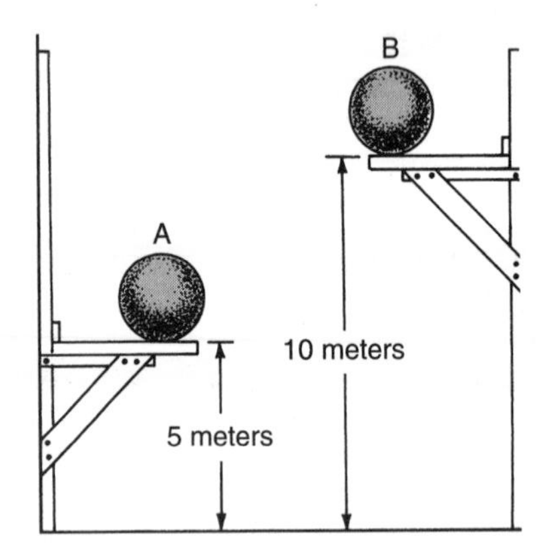

Which statement is true?

(1) Ball A has the greater potential energy.
(2) Ball B has the greater potential energy.
(3) Ball A has the greater kinetic energy.
(4) Ball B has the greater kinetic energy.

24. In the diagram below, the man is watching a fish in a pond. The fish is at position B, but the man sees the fish at position *A*.

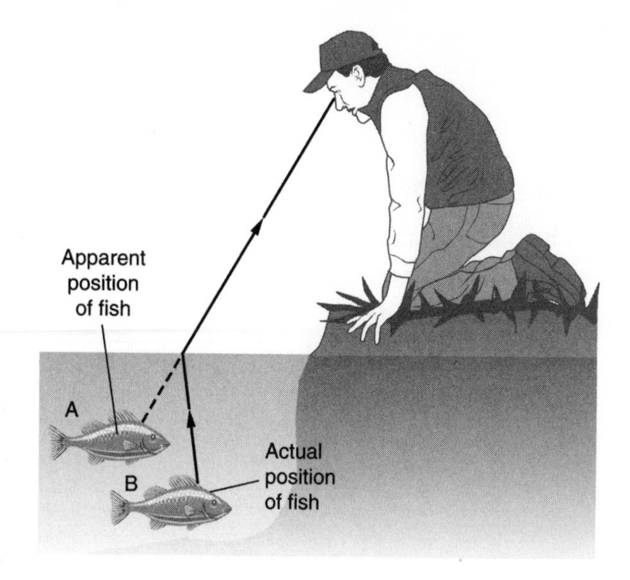

What property of light causes the fish to appear at position A?

(1) refraction
(2) reflection
(3) absorption
(4) transmission

25. A lightbulb changes electrical energy to light energy. In the process, some electrical energy is

(1) changed to heat energy
(2) changed to sound energy
(3) created
(4) destroyed

26. Which energy source is not renewable?

(1) coal
(2) moving water
(3) solar power
(4) wind

27. As the temperature of a metal pot on the stove increases from 50°C to 80°C, the motion of the molecules in the pot

(1) increases
(2) decreases
(3) remains the same
(4) increases then decreases

28. Organs are to tissues as tissues are to

(1) organ systems
(2) nuclei
(3) cells
(4) genes

29. When we perspire, water is removed from our bodies by a process called

(1) excretion
(2) respiration
(3) locomotion
(4) transport

30. Oxygen enters the body through the respiratory system, and is sent to the cells by the

(1) excretory system
(2) reproductive system
(3) circulatory system
(4) endocrine glands

31. The term "metamorphosis" describes how a

(1) caterpillar changes into a butterfly
(2) lion cub becomes a mature lion
(3) single bacterium divides to form two new bacteria
(4) pollen grain is transferred from the stamen to the pistil

Base your answers to questions 32 through 34 on the food web below and your knowledge of science.

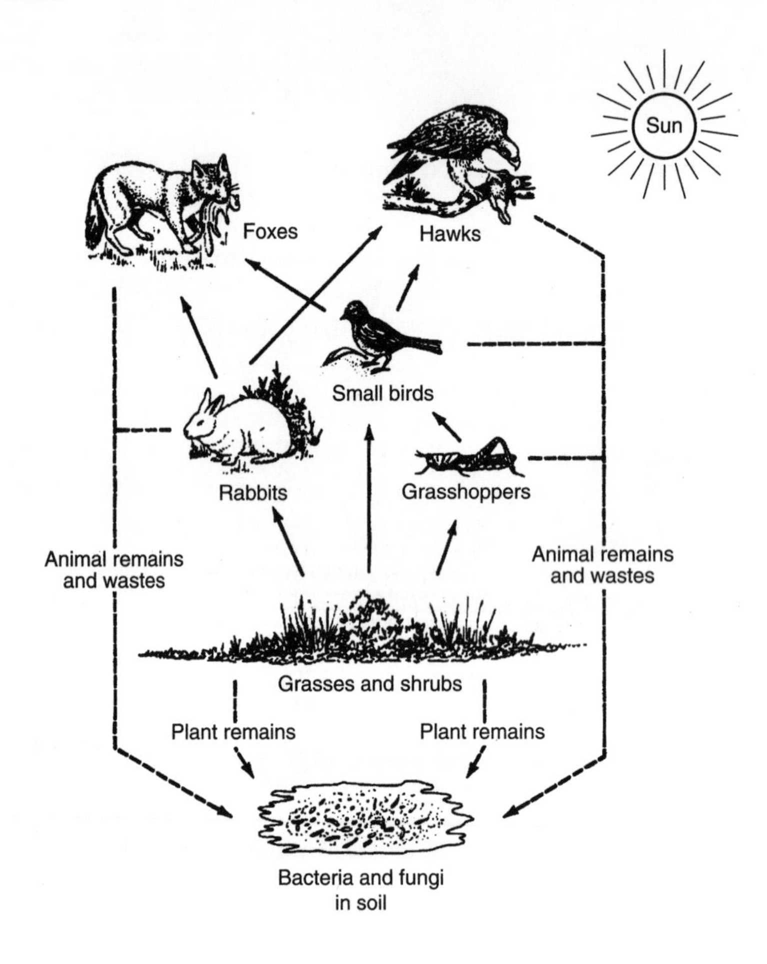

32. What is the role of hawks?

(1) herbivore
(2) producer
(3) decomposer
(4) carnivore

33. The *best* description of the role of the small birds in this food web is

(1) omnivore
(2) herbivore
(3) carnivore
(4) producer

34. The *original* source of all of the energy in this food web is the

(1) bacteria and fungi
(2) foxes
(3) grasses and shrubs
(4) sun

35. Which body system is most directly responsible for responses to stimuli?

(1) nervous
(2) respiratory
(3) excretory
(4) digestive

36. Which organism shown below consists of only one cell?

(1) paramecium
(2) fruit fly
(3) tree
(4) mushroom

37. What is the main function of the skeletal system?

(1) supports and protects the body
(2) moves materials to and from the cells
(3) removes waste products from the body
(4) nourishes the cells

38. Infectious diseases may be caused by

(1) mutations in humans
(2) poor diet
(3) bacteria
(4) allergies

Base your answers to questions 39 through 41 on the drawings of leaves and the dichotomous key below.

A Key to Identifying Some Trees Common New York Based upon Their Leaves		
Couplet	Description	
1a 1b	Broad leaves Needlelike leaves	Go to 2 Go to 3
2a 2b	Leaves opposite each other Leaves alternating	Maple Go to 4
3a 3b	Long pointed leaves Short flattened leaves	Pine Hemlock
4a 4b	Compound leaf (leaf consists of several smaller leaflets) Simple leaf	Walnut Oak

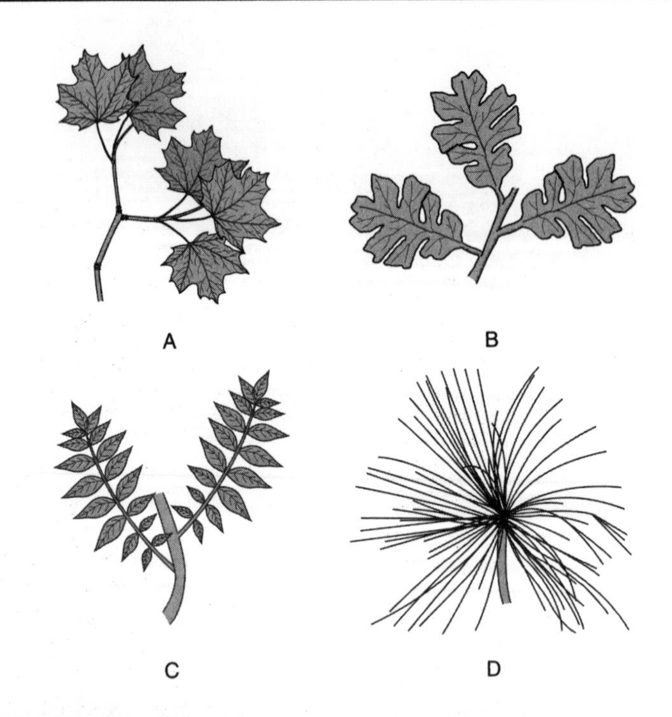

39. Diagram A most likely shows leaves from

(1) a pine
(2) an oak
(3) a maple
(4) a walnut

40. What is a common feature of both the walnut and the oak?

(1) needlelike leaves
(2) leaves alternating
(3) leaves opposite each other
(4) short flattened leaves

41. Which tree is *not* shown in the diagrams?

(1) walnut
(2) oak
(3) hemlock
(4) maple

42. Newer varieties of corn have been produced that are resistant to insect damage. The new corn is produced by changing some of the genes within the cells of the plant. This process is called

(1) selective breeding
(2) genetic engineering
(3) natural selection
(4) cross fertilization

43. A student maintains a small terrarium—a container that holds several living things. The terrarium contains worms, frogs, crickets, and green plants. What do all of these living things have in common?

(1) They reproduce asexually.
(2) They are composed of cells.
(3) They are able to produce their own food.
(4) They have the same type of circulatory system.

44. When someone claps his hands in front of your face, you blink. This reflex action most directly involves

(1) the respiratory system and the excretory system
(2) the nervous system and the muscular system
(3) the reproductive system and the skeletal system
(4) the digestive system and the nervous system

45. What process is illustrated by the following formula?

egg + sperm → zygote

(1) pollination
(2) cleavage
(3) fertilization
(4) mitosis

Part II

Base your answers to questions 46 through 48 on the map below and your knowledge of science. The map shows earthquake locations in the United States. Each dot represents a major earthquake.

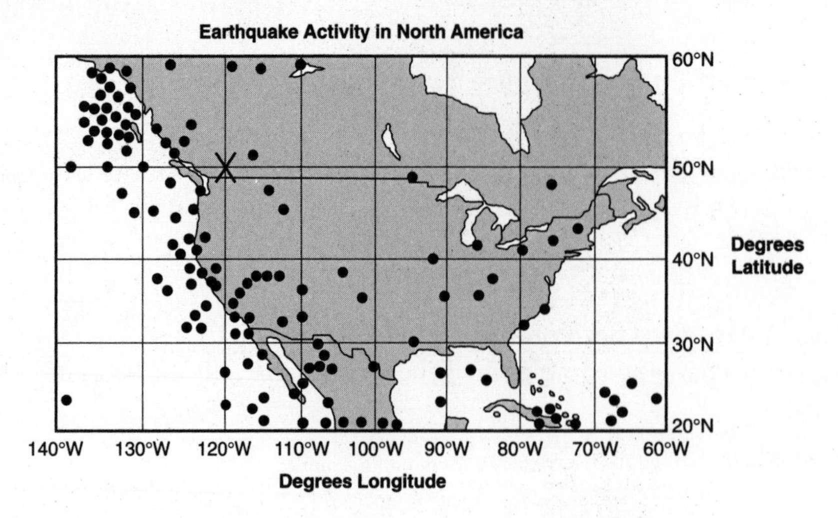

46. Why are there more earthquakes along the Pacific coastline than anywhere else in the United States? [1]

47. What are the latitude and longitude of location X? (Include a value, unit, and direction in your answers.) [1]

48. Name two items a resident of California should have in an earthquake emergency kit. [2]

---

49. State two precautions people should take when a hurricane warning is issued. [2]

Base your answers to questions 50 through 52 on the diagram below and your knowledge of science. The diagram shows the position of Earth in relation to the sun at the beginning of each season.

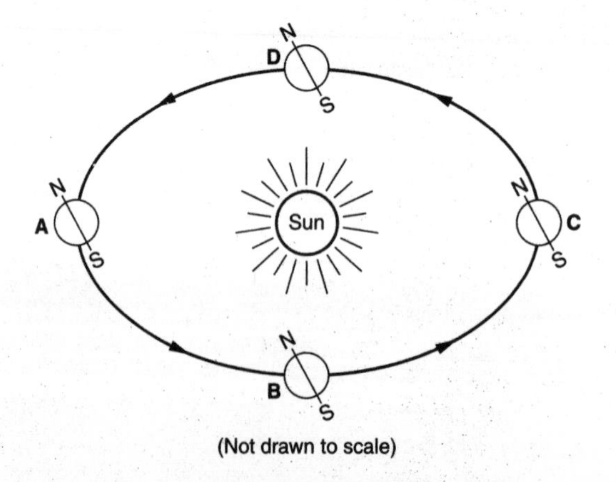

50. How much time does it take Earth to move from position A to position B? [1]

51. Which season would begin in the Northern Hemisphere when Earth is at position C? [1]

52. Describe the relationship between seasonal change in the Northern Hemisphere and the position of the sun's vertical rays striking Earth. [2]

Base your answers to questions 53 through 55 on the information in the travel line below and your knowledge of science. The travel line shows the time and the distance traveled by a car in 0.5-hour intervals.

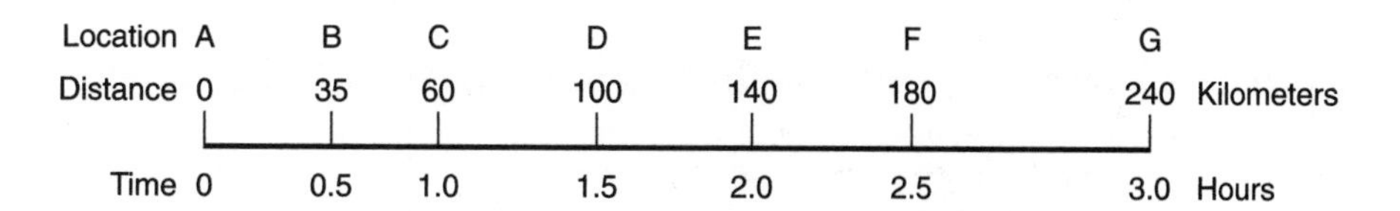

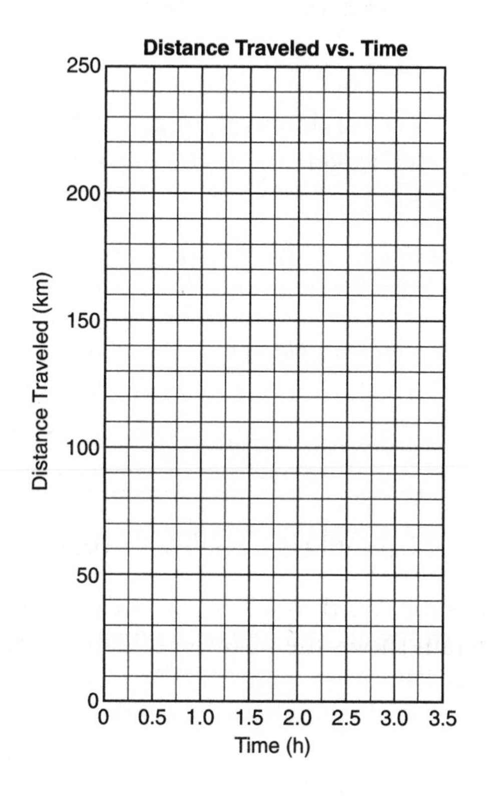

53. Copy the grid onto your answer paper and draw an X on the grid to represent the travel time and distance traveled by the car for each 0.5-hour interval. Connect the X's to produce a graph that represents the travel line. [2]

54. What was the average speed of the car for the 3 hours it traveled? Show your work. [2]

55. (a) Between what two locations did the car travel at the fastest speed? [1]
(b) How fast did the car travel between the two locations? [1]

Base your answers to questions 56 and 57 on the diagram below and your knowledge of science. The diagram shows two bar magnets. The north pole (N) is labeled on one magnet. The pattern of iron filings between the two magnets is also shown in the diagram.

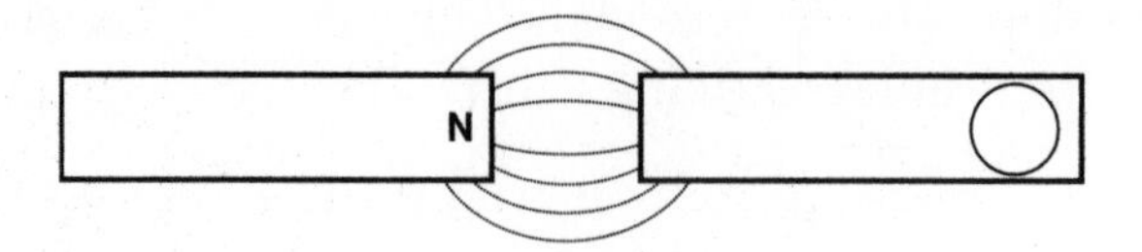

56. What is the polarity of the circled position on the second magnet? [1]

57. Describe what will happen if you push the two magnets closer together. [1]

Base your answers to questions 58 through 61 on the information below and your knowledge of science.

Copper filings, sugar, oil, and water are mixed in a beaker. After a few minutes, the sugar seems to have disappeared. The oil floats on the top of the water, while the copper filings rest on the bottom of the beaker.

58. What property of the oil causes it to float on top of the water? [1]

59. Why does the sugar appear to disappear? [1]

60. What could be done to make the sugar disappear *faster*? [1]

61. Identify one method that can be used to remove the copper from the mixture. [1]

Base your answers to questions 62 and 63 on the diagram below and your knowledge of science. The diagram illustrates what happens when ice cubes are added to two glasses, one containing water and the other containing alcohol. (Assume that the ice cubes in both glasses are identical.)

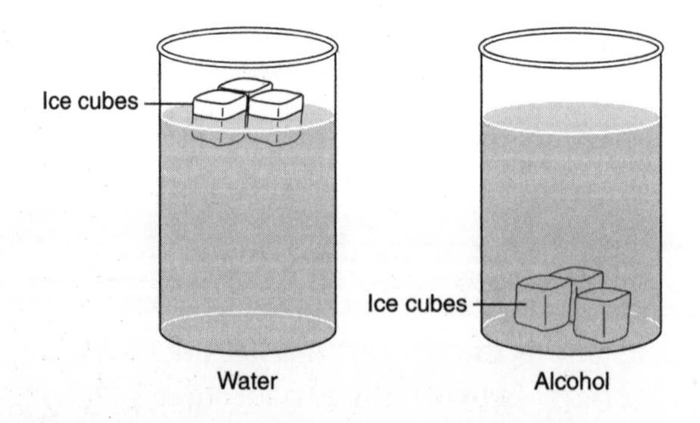

62. Which liquid has a greater density, alcohol or water? Explain your answer. [2]

63. After the ice cubes have melted completely, which glass will contain a *mixture*?

Explain your answer. [2]

Base your answers to questions 64 and 65 on the reading passage below and your knowledge of science.

Global warming is a long-term change in Earth's climate. Many scientists believe that by the end of this century people will no longer be able to live in many of the areas they inhabit now.

64. Most scientists believe that we contribute to global warming when we burn fossil fuels. Which gas produced in the burning of fossil fuels is believed to contribute to global warming? [1]

65. What change in the environment might occur if the temperature rises high enough to melt all of Earth's glaciers? [1]

Base your answers to questions 66 and 67 on the reading passage below and your knowledge of science.

Lions do *not* eat plants. Yet if the plants were to disappear from their environment, the lions would starve.

66. Explain why lions can survive only where there is adequate plant life. [2]

67. Identify the role of the plant and the role of the lion in a food chain. [2]

---

68. The table below lists some common structures of a plant cell and their functions. In the table, there are three blank spaces, marked *a*, *b*, and *c*. Fill in the spaces by writing the name of the cell structure that performs the function, or the name of the function performed by the given cell structure. [3]

Structure and Function of Plant Cell Structures

Plant Cell Structure	*Function*
a	Allows materials to enter and leave the cell
Chloroplast	*b*
Mitochondria	Produce energy for the cell
c	Contains the chromosomes and DNA needed for reproduction
Cell wall	Protects the cell and maintains its shape

Base your answers to questions 69 through 71 on the following reading passage and your knowledge of science.

The Republic of China built an enormous dam on the Yangtze River. Hydroelectric power plants along the Three Gorges Dam will eventually produce as much energy as fifteen nuclear plants. It is hoped that the new power plants will reduce China's reliance on fossil fuels, which are now used to produce most of China's electricity. However, building the dam has created a huge lake, flooding an area larger than the city of New York, forcing more than one million people to leave their homes.

69. Give an example of a fossil fuel. [1]

70. Why might hydroelectric plants be better for the environment than power plants that burn fossil fuels? [1]

71. What is one argument *against* building an enormous dam, like the Three Gorges Dam? [1]

Base your answers to questions 72 and 73 on the reading passage below and your knowledge of science.

Many organisms must maintain a constant body temperature to survive. These organisms have many different body structures and behaviors that help them maintain their body temperature in many different kinds of environments.

72. Seals and sea lions have a thick layer of fat under their skins. How does this layer of fat help them to maintain a constant body temperature? [1]

73. On a hot day, humans sweat. How does sweating help humans maintain their body temperature? [1]

Practice Test 2

Part I

1. Which statement correctly describes *all* rocks?

(1) Rocks contain fossils.
(2) Rocks form underwater.
(3) Rocks contain minerals.
(4) Rocks form from a hot liquid solution.

2. When viewed through a microscope, a rock sample contains grains of sand and pieces of shells. What type of rock is it?

(1) igneous
(2) sedimentary
(3) metamorphic
(4) volcanic

3. The hydrosphere includes

(1) mountains
(2) air
(3) rocks
(4) oceans

4. Letter *A* in the diagram is pointing to Earth's

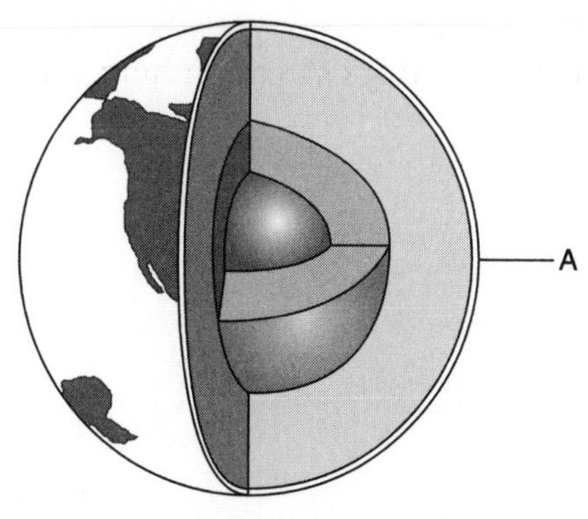

(Not drawn to scale)

(1) crust
(2) mantle
(3) outer core
(4) inner core

5. The primary force causing erosion of Earth's surface is

(1) the sun
(2) earthquake activity
(3) gravity
(4) animal activity

6. The diagram below shows the shape of a small rock before it was thrown into a stream.

Which statement best describes how the rock will change over time?

(1) The rock will become smaller and more irregularly shaped.
(2) The rock will become larger and more irregularly shaped.
(3) The rock will become smaller and smoother.
(4) The rock will be unchanged.

7. The table below lists the distribution of solar radiation that reaches Earth.

Distribution of Solar Radiation That Reaches Earth

Reflected by atmosphere	30%
Absorbed by atmosphere	20%
Absorbed by Earth's surface	50%

Which pie graph best represents this information?

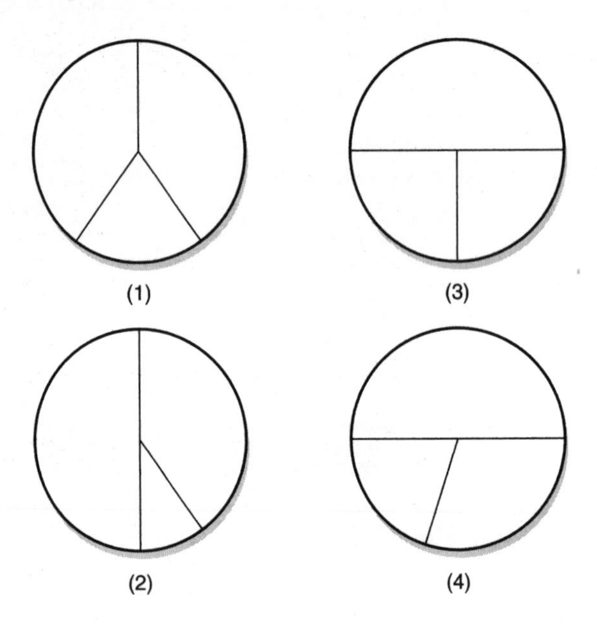

(1) 1 (2) 2 (3) 3 (4) 4

8. Water left in a dish on the kitchen table evaporates slowly. What could be done to make the water evaporate more quickly?

(1) Cover the dish.
(2) Place the dish in a dark area.

(3) Place the dish high on a shelf.
(4) Place the dish in sunlight.

9. The diagram below shows the relative position of the sun and Earth on June 21.

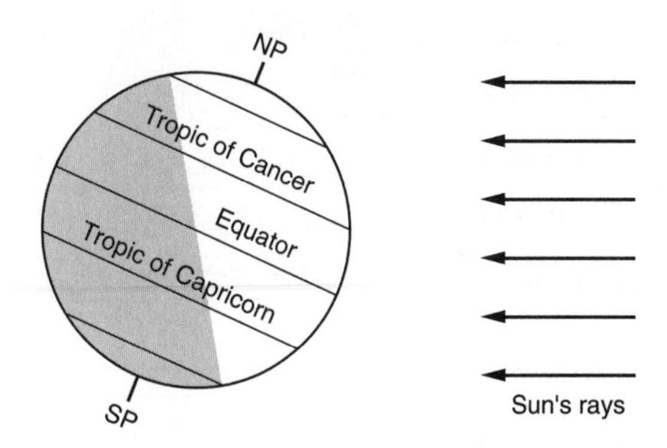

At what location are the sun's vertical rays striking Earth?

(1) North Pole (NP)
(2) Tropic of Cancer
(3) Equator
(4) Tropic of Capricorn

10. Maxine saw a bright star over a tree at 8:00 P.M. (position *A* in the diagram). An hour later, the star was at position *B*. What caused the star to change position?

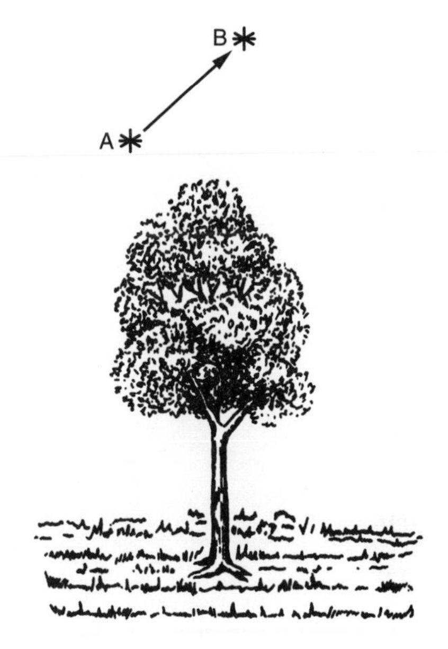

(1) the rotation of Earth
(2) the revolution of Earth
(3) the movement of the star
(4) the tilt of Earth's axis

11. The diagram below shows the relationship of Earth and the sun's rays in space.

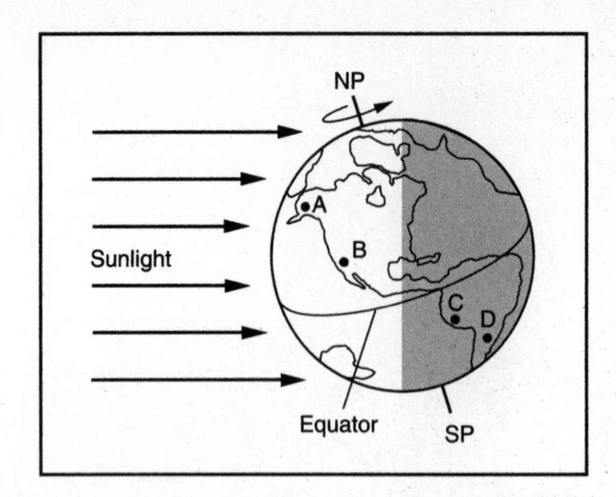

Which location is having the greatest number of daylight hours each day?

(1) location *A*
(2) location *B*
(3) location *C*
(4) location *D*

12. Which statement is true about *all* objects that revolve around the sun?

(1) They produce their own light.
(2) They reflect sunlight.
(3) They travel at the same speed.
(4) They are spherically shaped.

13. Oil is used in a car engine to decrease

(1) work output
(2) efficiency
(3) kinetic energy
(4) friction

14. Which object represents a *simple* machine?

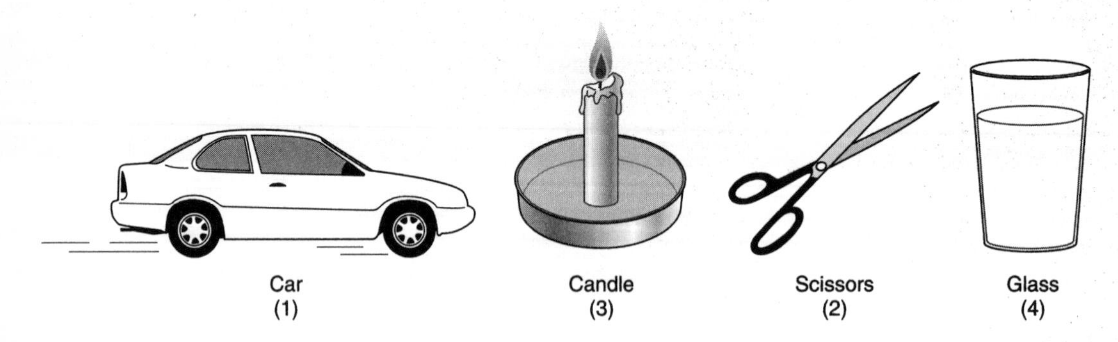

(1) car
(2) candle
(3) scissors
(4) glass

15. What property of light is illustrated in the diagram below?

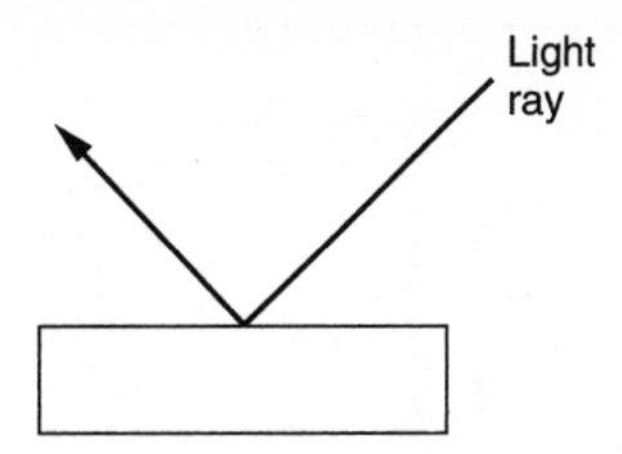

(1) refraction
(2) reflection
(3) absorption
(4) transmission

Base your answers to questions 16 and 17 on the sequence of events shown below and your knowledge of science.

Burning coal → Heating water to produce steam → Turning generator to produce electricity → Lights in your home

16. The sequence of events describes

(1) energy recycling
(2) how energy increases
(3) energy reduction
(4) energy transformation

17. Heated water produces steam that is used to turn a generator. What type of energy causes the generator to turn?

(1) heat energy
(2) chemical energy
(3) mechanical energy
(4) sound energy

18. What process transfers energy from the sun to Earth?

(1) reflection
(2) conduction
(3) convection
(4) radiation

19. Which item is the best conductor of electricity?

(1) copper wire
(2) glass stirring rod
(3) rubber rod
(4) plastic comb

Base your answer to question 20 on the demonstration below and your knowledge of science. The demonstration shows a bell inside a glass jar.

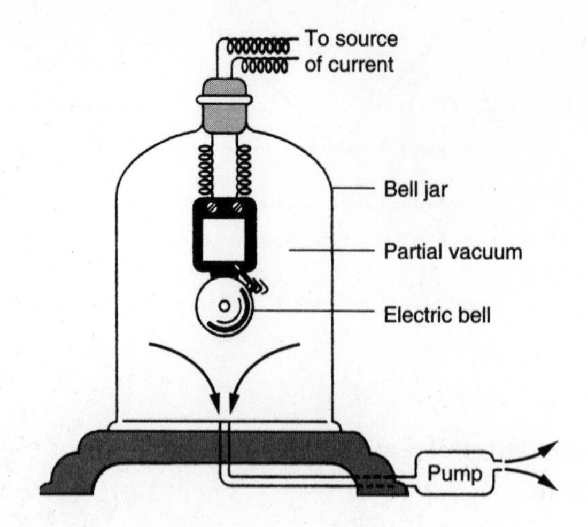

20. Before the air is removed from the jar, you could hear the bell ringing, but after the air is removed, creating a vacuum, you could *not* hear the bell ringing. This demonstration shows

(1) sound travels in a straight line
(2) sound does not travel through glass
(3) sound cannot travel in a vacuum
(4) the speed of sound decreases inside a glass jar

21. A skier racing downhill has what state of energy?

(1) potential
(2) kinetic
(3) light
(4) chemical

22. The diagram below shows Earth orbiting the sun and the moon orbiting Earth.

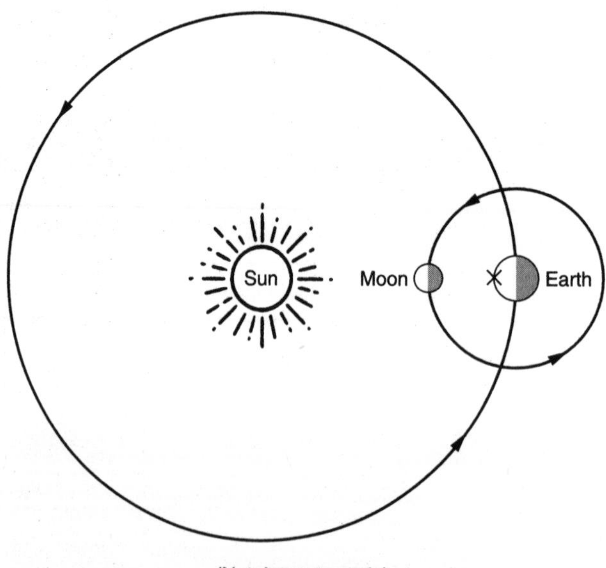

(Not drawn to scale)

What moon phase would a person observe from position *X* on Earth?

(1) new moon
(2) first quarter
(3) full moon
(4) last quarter

23. On a hot, sunny day, which land surface will absorb the greatest amount of heat energy from sunlight?

(1) a square meter of beach sand
(2) a square meter of blacktop road
(3) a square meter of grass lawn
(4) a square meter of water

24. In humans, what two functions are performed by the skeletal system?

(1) absorbs and transports oxygen
(2) transports and removes wastes
(3) protects and supports the body
(4) controls all body activities

25. Which body system breaks down food for absorption into the blood?

(1) digestive
(2) nervous
(3) excretory
(4) respiratory

26. A boy touches a hot object and immediately pulls his arm away from the object. Which three systems work together to produce this response?

(1) endocrine, circulatory, excretory
(2) respiratory, circulatory, digestive
(3) nervous, muscular, skeletal
(4) muscular, endocrine, respiratory

Base your answers to questions 27 through 29 on the diagram below and your knowledge of science. The diagram shows a model of inheritance in pea plants.

	T	*T*
t	*Tt*	*Tt*
t	*Tt*	*Tt*

Key:
T = tall
t = short

27. This diagram is called a

(1) Periodic Table
(2) pedigree chart
(3) flowchart
(4) Punnett square

28. The parent with the genes *tt* is a

(1) hybrid tall pea plant
(2) pure tall pea plant
(3) hybrid short pea plant
(4) pure short pea plant

29. The plants produced by crossing the parent plants shown in the diagram are

(1) 100% tall
(2) 100% short
(3) 50% tall, 50% short
(4) 25% short, 75% tall

30. A Calorie is a unit of

(1) mass
(2) volume
(3) temperature
(4) energy

31. It begins to rain, so you open your umbrella. A change in the environment that causes an organism to respond is called a

(1) stimulus
(2) response
(3) reflex
(4) instinct

32. Which process removes carbon dioxide from the air?

(1) photosynthesis
(2) respiration
(3) excretion
(4) digestion

33. Which is a renewable source of energy?

(1) gasoline (2) coal (3) fuel oil (4) wind

34. Charcoal is made mostly of carbon and can be used to grill food. When the charcoal burns, it produces carbon dioxide gas and releases heat. This is an example of

(1) a physical change
(2) a chemical reaction
(3) condensation
(4) melting

35. A student buys a 1-liter container of orange juice. What property of the juice is measured in liters?

(1) volume
(2) density
(3) mass
(4) solubility

36. A merlin is a hawk that preys on small birds. If the population of small birds in an environment were to decrease, we would expect the merlins in that environment to

(1) increase in number
(2) produce their own food
(3) decrease in number
(4) reproduce faster

37. In multicellular organisms, which process most directly contributes to growth and repair?

(1) digestion
(2) cell division
(3) excretion
(4) locomotion

38. Which is the best example of ecological succession?

(1) Robins fly south for the winter.
(2) The population of sparrows in New York remains constant.
(3) Over time, grasses and small shrubs are replaced by large trees.
(4) Hawks and snakes compete for the same prey.

39. Which is an indication that photosynthesis is taking place?

(1) the white color of root hairs
(2) the green color of a leaf
(3) the red color of a flower
(4) the orange color of a sweet potato

40. In the water cycle, water moves from the oceans into the air, and from the air back into the oceans. Which processes help return water to the ocean?

(1) evaporation and boiling
(2) condensation and precipitation
(3) condensation and boiling
(4) precipitation and freezing

Base your answers to questions 41 and 42 on the reading passage below and your knowledge of science.

Mosquitoes lay their eggs on water. A small, wormlike organism called a wriggler emerges from the egg and remains near the water's surface. Eventually, the wriggler stops moving and forms a pupa. The pupa floats on the water while the adult mosquito develops inside it. The adult mosquito emerges from the pupa after about four days.

41. The reading passage describes

(1) the life cycle of a mosquito
(2) the breeding habits of mosquitoes
(3) how mosquitoes mutate
(4) how mosquitoes spread disease

42. The process of turning from a wriggler to an adult mosquito is called

(1) mutation
(2) germination
(3) metamorphosis
(4) germination

43. You can "catch" a cold, the flu, or measles from a friend, but you cannot catch heart disease or diabetes. This is because cold, flu, and measles are caused by

(1) mutations
(2) poor diet
(3) microorganisms
(4) allergies

44. Nutrients enter the blood through tiny structures in the small intestine called villi.

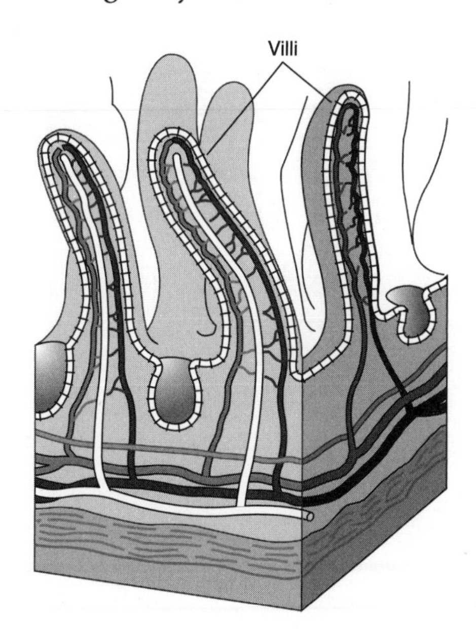

Which two body systems work together in the small intestine?

(1) respiratory and circulatory
(2) digestive and circulatory
(3) respiratory and excretory
(4) skeletal and digestive

45. Which illustrates a behavioral adaptation?

(1) Some flies resemble wasps.
(2) Seals have thick layers of fat that help maintain their body temperature.
(3) A penguin's wing is suitable for swimming rather than for flying.
(4) Herds of zebra migrate across East Africa as the weather turns dry.

Part II

46. The table below lists the three types of rocks. From what type of material did each rock type form (a–c)? [3]

Rock	*Formed From*
Igneous	(a)
Sedimentary	(b)
Metamorphic	(c)

Base your answers to questions 47 through 50 on the cross section below and your knowledge of science. The cross section shows the mechanism that causes crustal plates to move.

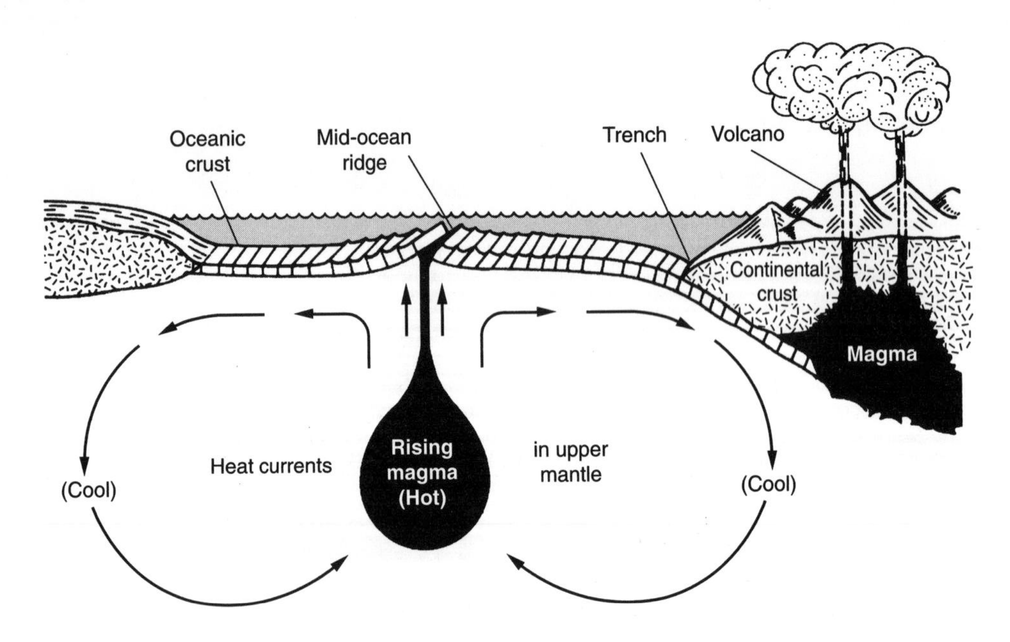

47. What ocean floor feature is found where crust is produced? [1]

48. What ocean floor feature is found where the ocean crust slides under the continental crust? [1]

49. Describe two differences between ocean floor crust and continental crust. [2]

50. What method of heat transfer is occurring in the upper mantle? [1]

Base your answers to questions 51 through 53 on the diagram below and your knowledge of science. The diagram represents the water cycle. Copy the diagram on your answer paper and do the following:

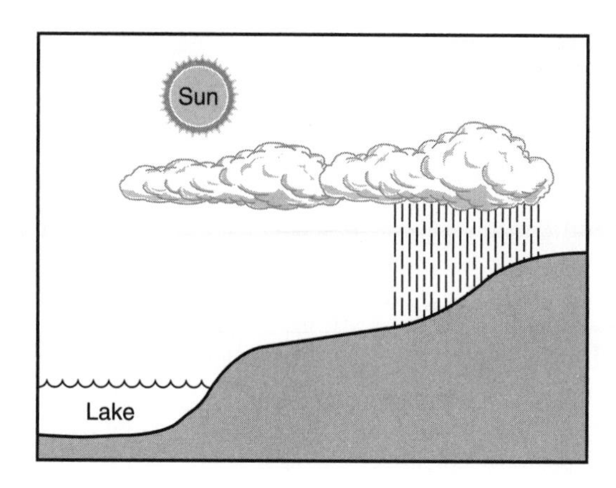

51. Draw and label three arrows in the diagram to show the location of (A) evaporation, (B) precipitation, and (C) runoff. [3]

52. Place an *X* at the location where water gains heat energy. [1]

53. What is the energy source for the water cycle? [1]

Base your answers to questions 54 through 56 on the diagram below and your knowledge of science. The diagram shows a spring scale pulling a block of wood across a tabletop.

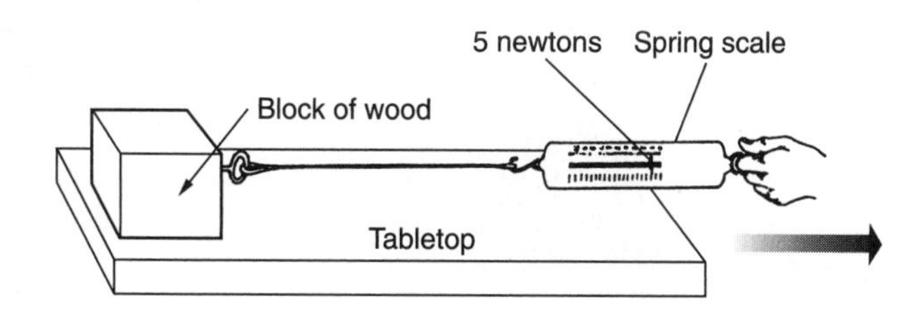

54. What force is resisting the movement of the block of wood? [1]

55. What change could be made in the setup to *decrease* the amount of force necessary to move the block of wood? [1]

56. If a same-size block of wood with less mass were used in the setup, how would the force most likely change? [1]

Base your answers to questions 57 and 58 on the diagram below and on your knowledge of science. The diagram shows a water phase change represented by the arrow in the diagram.

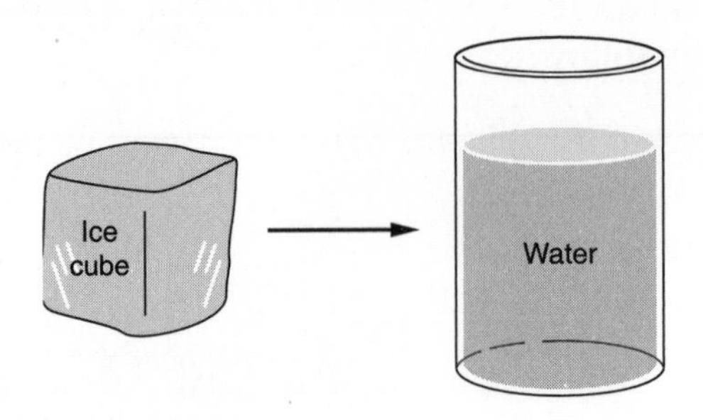

57. What type of phase change is shown in the diagram? [1]

58. Which has a greater volume, 100 grams of water or 100 grams of ice? Explain why. [1]

Base your answers to questions 59 through 61 on the graph below and your knowledge of science. The graph shows the solubility of three solid solutes, salt *A*, salt *B*, and salt *C*.

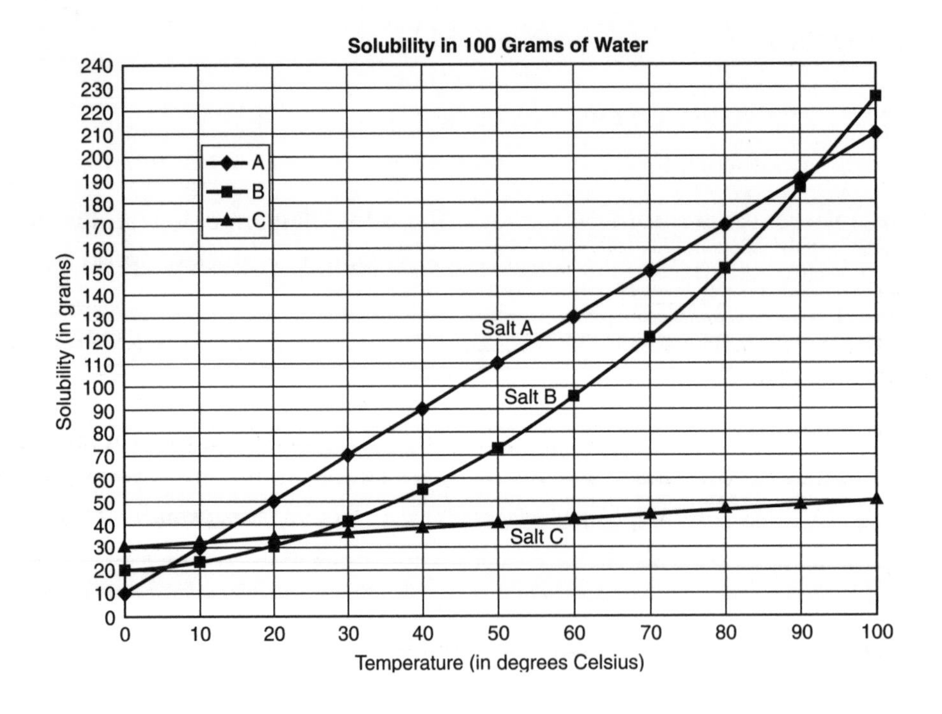

59. Which of these solutes is most soluble in water at a temperature of 5.0°C? [1]

60. Between 20°C and 30°C, which solute shows the greatest *increase* in solubility? [1]

61. A student tests the solubility of an unknown salt. The student found that at room temperature, 25°C, the maximum amount of the salt that can dissolve in 100 grams of water is 60 grams. Which of the three salts on the graph might be the same substance as the unknown salt? [1]

Base your answers to questions 62 through 64 on the diagram below and your knowledge of science. The diagram shows a green plant. Three of the plant's structures are labeled *A*, *B*, and *C*.

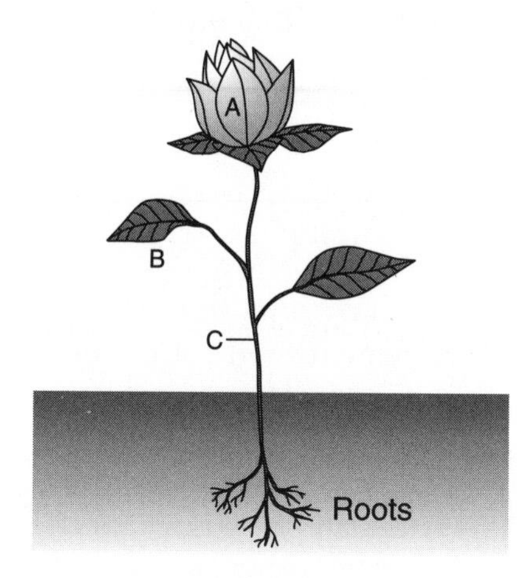

62. What are two functions of the roots? [2]

63. Write the letter that corresponds to the plant structure where reproduction occurs. [1]

64. What is the main function of structure B? [1]

Base your answers to questions 65 through 67 on the reading passage below and your knowledge of science.

The crude oil that is used to make gasoline and other fuels is carried on large ships called oil tankers. If one of these ships becomes damaged, it could spill oil into the ocean. Because the oil floats on the surface of the water, it is especially dangerous to sea birds and mammals, such as seals and otters, that depend on the sea for food.

65. What *physical* property of the oil causes it to float on the water? [1]

66. Some of the chemicals that are produced from crude oil are burned in power plants that produce electricity. What gas is added to the environment when oil is burned? [1]

67. Identify one *renewable* energy source that can be used to produce electricity. [1]

Base your answers to questions 68 and 69 on the following information and your knowledge of science. The chemical equation below shows a process that takes place in green plants and other organisms containing chlorophyll.

energy + carbon dioxide + water → sugar + oxygen

68. What is this process called? [1]

69. What form of energy is absorbed during this process? [1]

Base your answers to questions 70 and 71 on the information given about four-o'clock plants and your knowledge of science.

When four o'clock plants with red flowers are crossed with four-o'clock plants with white flowers, the resulting plants have pink flowers. In these plants, neither white nor red is dominant. When both the gene for red color and the gene for white color are present, the flowers are pink.

The Punnett square below represents a cross of four-o'clock plants. Use the key and Punnett square below to answer the questions.

	R	*W*
R	*RR*	*RW*
W	*RW*	*WW*

Key:
R = gene for red flowers
W = gene for white flowers

70. What percentage of the next generation of plants, as shown in the Punnett square, would have red flowers? [1]

71. When four-o'clock plants with red flowers are crossed with other four o'clock plants with pink flowers, how many different colors appear in the next generation? Explain how you arrived at your answer. Construct a Punnett square in your explanation. [2]

---

72. Identify two structures that are found in plant cells but not in animal cells. [2]

Use the diagram below and your knowledge of science to answer questions 73 through 75. The diagram shows a food web for a community.

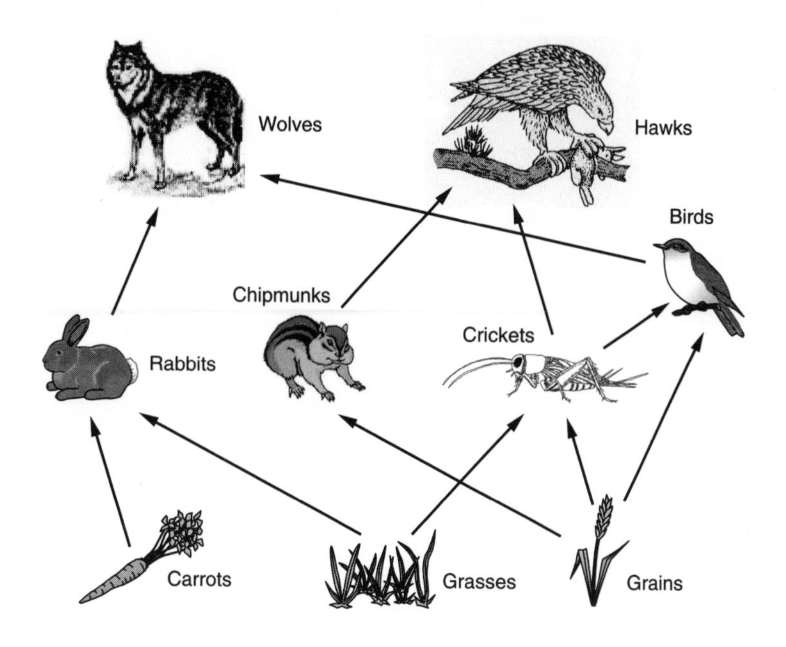

73. Which animal shown in this food web relies on only one source of food? [1]

74. List *two* animals that compete with the crickets for food. [2]

75. Which *two* animals in this food web are carnivores? [2]

Glossary

A

acceleration The rate of change in velocity.

acid rain Rain that has been made more acidic than normal by pollutants in the atmosphere.

adaption A characteristic that helps an organism survive in its habitat.

air mass A large body of air that has roughly uniform temperature and humidity throughout.

air pressure The force with which air presses down on Earth's surface.

air temperature A measurement of the amount of heat energy in the atmosphere.

altitude The height above sea level.

amplitude The height of the crest or the depth of the trough of a wave measured from the undisturbed surface.

arteries Blood vessels that carry blood away from the heart.

asexual reproduction Reproduction that involves only one parent, producing offspring that are genetically identical to the parent.

asteroids Pieces of rock of many different sizes and shapes that revolve around the sun.

astronomical unit (AU) The average distance from Earth to the sun.

atmosphere The gases that surround Earth.

atom The smallest particle of an element that has properties of that element.

axis of rotation An imaginary line through an object, around which the object spins.

B

bacteria of decay Microorganisms that break down dead organisms and return nutrients to the environment.

bedrock The unbroken, solid rock portion of Earth's crust.

behavioral adaptations Special behaviors that help organisms adapt to changes in their environment.

biodiversity Describes the great variety of lifeforms on Earth or within a habitat.

blood A liquid tissue that contains red and white blood cells, and platelets, and also carries dissolved gases, nutrients, hormones, and wastes.

blood vessels Tubes through which the blood flows.

boiling The rapid change in phase from liquid to gas, during which bubbles of gas form within the liquid.

boiling point The temperature at which a substance changes rapidly from a liquid to a gas.

bone The hard parts of the skeleton.

brain The organ, located within the skull, that controls thinking and body activities.

bronchi The two tubes that branch off from the lower end of the trachea, connecting it to the lungs.

buoyancy The tendency of an object to float.

buoyant Able to float.

C

calorie A unit used to measure and compare the amount of heat energy contained in substances such as fuels.

cancer A disease that is the result of abnormal cell division.

capillaries Tiny blood vessels, connecting arteries to veins, through which materials are exchanged between the blood and the body's cells.

carnivore A meat-eating animal.

cartilage A flexible tissue that acts as a cushion between bones and provides flexibility at the ends of bones.

cell The basic unit of all living things.

cell division The process by which cells reproduce, wherein a parent cell splits into two new daughter cells.

cell membrane The outer covering, or "skin," of a cell, which controls the flow of materials into and out of the cell.

cellular respiration A life process that occurs in all cells, in which nutrients from digested food are combined with oxygen to release energy and produce the wastes carbon dioxide and water.

chemical bond The link that joins one atom to another in a molecule.

chemical change A change that results in the formation of one or more new substances; a chemical reaction.

chemical energy A form of potential energy stored in the bonds of atoms and molecules.

chemical property A characteristic that a substance displays when it undergoes a change to a new substance or substances.

chlorophyll A green substance found in the leaves of plants, which is necessary for photosynthesis to take place.

chloroplast The organelle in plant cells that contains chlorophyll.

circuit breaker A device that prevents overloading of an electric circuit by interrupting the flow of electricity when it reaches a dangerous level.

climate The long-term average weather over a large area.

cloud A mass of tiny water droplets or ice crystals floating high in the atmosphere.

coal A black rock formed from the remains of ancient swamp plants; it is a fossil fuel.

coarse-adjustment knob The part of a microscope used to focus under low power.

cold front The boundary formed when a cool air mass pushes into and under a warm air mass.

comet A relatively small object made up of rock, ice, and dust that orbits the sun in a long stretched-out elliptical orbit.

commensalism A symbiotic relationship in which one organism benefits while the other is not affected.

community All the different populations that live within a habitat.

compound A substance that is formed when two or more different elements combine chemically.

compound microscope A microscope that uses two lenses.

concentrated A strong solution.

concentration The strength of a solution.

condensation The changing of water vapor into droplets of liquid water; more generally, the change in phase from gas to liquid.

conduction The transfer of heat by direct molecular contact.

conductor A material through which electricity can flow.

conservation The saving of natural resources through wise use.

constellation A pattern formed by stars in the night sky, which reminded people of animals or characters in ancient myths.

consumer An organism that obtains nutrients by eating other organisms.

contour line A line on a topographic map that connects points of equal elevation.

core Earth's center, which is made up of an outer and inner zone.

corrosion The chemical wearing away of a metal.

crust The outermost rock layer of Earth, which contains all Earth's surface features.

cytoplasm The watery substance that fills the cell, where most life processes occur.

D

decomposer An organism that breaks down the remains and wastes of other living things.

density The quantity that compares the mass of an object to its volume.

$$density = \frac{mass}{volume}$$

digestion The breaking down of the nutrients into a useable form.

dilute A weak solution.

Doppler effect The apparent change in the frequency of a wave (light or sound) that occurs when the source and/or the observer are in motion relative to each other.

dormancy A state in which an organism is inactive while it awaits more favorable conditions in its environment.

dwarf planet A celestial body that orbits around a star (but is not a star or a satellite), has enough mass that its gravity makes it spherically shaped, and has not cleared rocky objects from its orbit.

E

earthquake A shaking or vibrating of Earth's crust, usually caused by the sudden movement of rocks sliding along a fault.

eclipse An event during which the moon's shadow falls on Earth or Earth's shadow falls on the moon.

ecological succession The natural process by which one community of living things is replaced by another community in an orderly, predictable sequence, until a stable climax community appears.

ecosystem The living members of a community, plus the nonliving elements of their environment.

effort The force applied to a simple machine to overcome the resistance.

electric circuit A complete path for the flow of electricity.

electric energy A form of energy produced by the flow of electrons from one point to another point.

electromagnetic spectrum A continuous band of electromagnetic energy waves that include: radio waves, microwaves, infrared waves, visible light, ultraviolet light, x-rays, and gamma rays.

element One of the basic substances that form the building blocks of matter.

elimination The removal of undigested materials from the body (egestion).

endangered species A species that is in danger of extinction.

energy The ability to do work.

environment The surroundings in which an organism lives, including both living and nonliving things.

erosion The process whereby rock material at Earth's surface is removed and carried away.

evaporation The changing of liquid water into water vapor (gaseous water); more generally, the change in phase from liquid to gas.

evolution The process by which a species gradually becomes a different species.

excretion The process of removing the waste products of cellular respiration from the organism.

extinct Species that no longer exist.

F

faulting The process in which internal forces cause Earth's crust to break and slide along a fracture called a fault.

fertilization The joining of an egg cell and a sperm cell, during sexual reproduction, to begin the development of a new individual.

field of view The area you see through the microscope.

fine-adjustment knob The part of a microscope used to focus under high power.

first law of motion An object at rest will remain at rest and an object in motion will remain in motion unless an outside force acts on the object.

folding The process whereby rock layers in Earth's crust are squeezed into wavelike patterns called folds.

food chain A sequence of organisms through which nutrients are passed along in an ecosytem.

food web A number of interconnected food chains.

force A push or a pull.

fossil The remains or traces of an ancient organism.

fossil fuel A fuel that was formed from the remains of ancient plants or animals; examples include oil, coal, and natural gas.

freezing (1) The change in phase from liquid to solid. (2) The storing of food at temperatures below 0°C (32°F), to slow the growth of microorganisms that can spoil food.

freezing point The temperature at which a substance changes from a liquid to a solid.

frequency The number of waves that pass by a fixed point in a given amount of time.

front The boundary between two different air masses.

fulcrum The point around which a lever can turn.

full moon The phase of the moon that occurs when Earth is between the sun and the moon, so that all of the moon's lighted side can be seen from Earth.

fuse A device, used in an electric circuit, containing a thin metal strip that melts to interrupt the flow of electricity when the circuit becomes overheated.

G

galaxy A large cluster of billions of stars.

gene A piece of genetic information that influences a trait.

genetic engineering The process of changing the genetic material of organisms.

gland An organ that makes and secretes (releases) chemicals called hormones.

global warming An increase in temperature believed to be caused by an increase in greenhouse gas emissions.

greenhouse effect The trapping of heat in Earth's atmosphere by carbon dioxide, can lead to global warming.

growth The increase in size of an organism.

H

habitat The particular environment in which an organism lives.

heart An organ, made mostly of muscle, that contracts (beats) regularly to pump blood throughout the body.

heat energy A form of kinetic energy associated with vibrating molecules.

herbivore A plant-eating animal.

hibernate To enter a sleeplike state of reduced body activity; how some animals survive the winter.

hibernation A sleeplike state of reduced body activity that some animals enter to survive the winter.

high-pressure system An area in the atmosphere where air is sinking, causing high surface air pressure; also called a high.

hormone A chemical "messenger" secreted by a gland into the bloodstream, which carries the hormone to an organ that responds in some way.

humidity The moisture (water vapor) present in the atmosphere.

hurricane A huge, rotating storm that forms over the ocean in the tropics, with strong winds and heavy rains.

hydroelectric energy Electricity produced by using the energy of flowing water to turn the turbines of a generator.

hydrosphere The liquid water part of Earth.

I

igneous rock A volcanic rock formed by the cooling and hardening of hot, liquid rock material.

inclined plane A simple machine that consists of a flat surface with one end higher than the other, such as a loading ramp.

inertia The tendency of an object at rest to remain at rest or an object in motion to remain in motion.

infectious disease A disease caused by microorganisms that can be transmitted from one individual to another.

ingestion The process of taking in food.

insoluble Not able to dissolve in a given solvent.

insulation Material used to reduce or slow the flow of heat from one area to another.

insulator A material through which electricity cannot flow.

involuntary muscles Muscles that we do not consciously control.

J

joint A place where one bone is connected to another one.

K

kidneys A pair of organs that filter wastes from the blood and help control the water and mineral balance of the body.

kinetic energy Energy that an object has because of its motion.

L

landform A large land feature defined by its height, steepness of slope, and type of bedrock present.

latitude The distance measured in degrees, north and south of the equator.

Law of Conservation Energy Energy can neither be created nor destroyed.

Law of Conservation of Matter Matter can neither be created nor destroyed in a chemical reaction.

lens A piece of transparent glass or plastic with curved surfaces that bend light rays to form an image.

lever A simple machine consisting of a bar or rod that can turn around a point called the fulcrum.

life cycle The changes that an organism undergoes as it develops and produces offspring.

light A visible form of radiant energy that moves in waves.

light-year The distance that light travels in one year, about 9.46 trillion kilometers.

lithosphere The solid part of Earth.

liver An organ that produces urea from excess amino acids, removes harmful substances from the blood, and secretes bile, a digestive juice.

locomotion The movement of the body from place to place.

longitude The distance measured in degrees east and west of the prime meridian.

longitudinal wave (L-wave) An earthquake wave that travels along Earth's surface.

low-pressure system An area in the atmosphere where air is rising, causing low surface air pressure; also called a low.

lungs A pair of organs, located in the chest, that contain millions of tiny air sacs, in which the exchange of respiratory gases between the blood and the environment takes place.

lymph A fluid that bathes all body cells and acts as a go-between in the exchange of materials between the blood and the cells.

lymph vessels Tubes in which waste-laden lymph is collected and returned to the bloodstream.

M

machine A device that transfers mechanical energy from one object to another object.

magnification The number of times the image of a specimen is enlarged.

mammary glands The female breasts, which produce milk to nourish newborn offspring.

mantle The layer of Earth below the crust.

map A flat model that shows a portion of Earth's surface.

mass The amount of matter in an object.

matter Anything that has mass and takes up space.

mechanical energy A form of kinetic energy in a moving object that is doing work.

melting The change in phase from solid to liquid.

melting point The temperature at which a substance changes from a solid to a liquid.

metabolism The sum of all the chemical reactions that take place in the body.

metal A shiny solid that conducts electricity. Metals are found to the left on the Periodic Table of the Elements.

metamorphic rock A rock produced when existing igneous or sedimentary rock undergoes a change in form caused by great heat, pressure, or both.

metamorphosis The process of a complete change in body form during development from juvenile to adult stage.

meteor A rock fragment from space that enters Earth's atmosphere and burns up, producing a bright streak of light.

meteorite A meteor fragment that reaches Earth's surface.

meteoroid A tiny piece of rock, some as small as dust, that travels through space, believed to have come from a comet.

methane A gas produced by decaying plant matter and animal wastes, the main component of natural gas. Methane produced in this way is a renewable resource; however methane is also a greenhouse gas.

microorganism An organism that is very small, usually too small to be seen with the unaided eye.

microscope A tool used by scientists to magnify tiny objects such as cells.

migration Moving from one environment to another, where conditions are more favorable.

mineral A naturally occurring, solid inorganic (nonliving) substance with characteristic physical and chemical properties.

molecule The smallest particle of a compound.

moon A ball of rock that revolves around a planet.

moon phases The apparent changes in the shape of the moon produced by the relative positions of the sun, moon, and Earth.

motion A change in the position of an object relative to another object, which is assumed to be at rest.

mountain A feature on Earth's surface that rises relatively high above the surrounding landscape.

muscles Masses of tissue that contract to move bones or organs.

mutation A change in the genetic material.

mutualism A symbiotic relationship in which both organisms benefit.

N

natural gas A gaseous fossil fuel found trapped deep underground, often with oil deposits.

natural selection The process that favors those organisms that are best able to survive and reproduce.

nerves Thin strands of tissue, composed of neurons, that carry impulses throughout the body.

neurons Cells that make up the nervous system, which receive and transmit information in the form of impulses.

new moon The phase of the moon that occurs when the moon is between Earth and the sun, so that the moon cannot be seen from Earth.

noble gases A group of gaseous elements that seldom react with other elements, which are placed in the extreme right column of the Periodic Table of the Elements.

noninfectious disease A disease that cannot be transmitted from one individual to another.

nonmetal Solids and gases found at the right of the Periodic Table of the Elements, which are poor conductors of electricity

nonrenewable resource A resource that is not replenished by nature within the time span of human history.

nuclear energy The energy stored within the nucleus of an atom, used by nuclear power plants to produce electricity.

nuclear waste The poisonous, radioactive remains of the materials used to fuel nuclear power plants.

nucleus (1) The structure within the cell that controls cell activities and contains genetic material. (2) The center of an atom, which contains the protons and neutrons.

nutrients Food substances that supply an organism with energy and with materials for growth and repair.

nutrition The process that includes ingestion, digestion and elimination.

O

oil A thick, black, liquid fossil fuel, found trapped underground; also called petroleum.

omnivore A consumer that can eat both plants and other animals.

orbit The path of an object in space that is revolving around another object.

organ A group of tissues that act together to perform a function.

organism A living thing.

organ system A group of organs that act together to carry out a life process.

ovaries The female reproductive organs that produce egg cells.

oviducts Tubes that connect the ovaries to the uterus.

oxidation or *"burning"* The reaction in which a substance combines with oxygen and releases energy.

P

parasitism A symbiotic relationship in which one organism benefits while the other is harmed.

parallel circuit An electric circuit that has two or more paths for the electricity to flow through.

phases (1) The changing apparent shape of the moon, as seen from Earth. (2) The three forms, or states, of matter—solid, liquid, and gas.

photosynthesis The process by which green plants produce food, using sunlight, carbon dioxide, and water; oxygen is given off as a by-product.

physical adaptation A physical characteristic that enables an organism to survive under a given set of conditions.

physical change A change in the appearance of a substance that does not change the chemical makeup of the substance.

physical property A characteristic of a substance that can be determined without changing the identity of the substance.

plain A broad, flat landscape region at a low elevation, usually made of layered sedimentary rocks.

planet A celestial body that orbits around a star, is spherically shaped, and has no rocky objects in its orbit.

plateau A large area of Earth's surface made up of horizontally layered rocks, found at a relatively high elevation.

plate tectonics The theory that Earth's crust is broken up into a number of large pieces, or plates, that move and interact, producing many of Earth's surface features.

plutoids A sub-class of dwarf planets that orbit the sun beyond Neptune

pollutant A harmful substance that contaminates the environment, often produced by human activities.

population All the members of a particular species that live within a habitat.

potential energy Stored energy that an object has because of its position or chemical makeup.

precipitation Water, in the form of rain, snow, sleet, or hail, falling from clouds in the sky.

prevailing winds Global winds that commonly blow in the same direction at a given latitude.

primary wave (P-wave) An earthquake wave that can travel through liquids and solids.

producer An organism that makes its own food. Most producers are green plants.

profile The side view of a landform projected from a straight line on a contour map.

pulley A simple machine consisting of a rope or chain wraped around a wheel with a grooved rim. It is used to change the direction of force or decrease the force used to move a heavy object attached to it.

Punnett square A diagram used to predict the probability of an organism inheriting a given trait.

R

radiation The transfer of energy in the form of waves.

regulation The process that helps an organism maintain a constant internal environment.

renewable resource A resource that is replenished by nature within a relatively short time span.

reproduction The life process by which organisms produce new individuals, or offspring.

resistance The force a machine has to overcome.

respiration (1) The process of taking in oxygen from the environment and releasing carbon dioxide and water vapor. (2) See also cellular respiration.

respiratory system The system that brings oxygen into the body and removes carbon dioxide.

response The reaction of a living thing to a change in its environment.

revolution The movement of an object in space around another object, such as the revolution of the moon around Earth.

rock A natural, stony material composed of one or more minerals.

rotation The spinning of an object around its axis, such as Earth around its axis.

S

science The study of the natural world.

screw An inclined plane that is wrapped around a wedge or cylinder.

seafloor spreading New rock material upwelling along the mid-ocean ridge and moving east and west away from the ridge in a conveyor beltlike fashion, pushes the seafloor out in opposite directions.

secondary wave (S-wave) An earthquake wave that can travel only through solids.

second law of motion The relationship among force, mass, and acceleration: $F = m \times a$.

sedimentary rock A rock formed from layers of particles, called sediments, that are cemented together under pressure.

selective breeding A process in which individuals with the most desirable traits are crossed or allowed to mate with the hope that their offspring will show the desired traits.

sense organs Organs that receive information from the environment. The sense organs include the eyes, ears, nose, tongue, and skin.

series circuit An electric circuit that has a single path of electricity through the devices attached to it.

sexual reproduction Reproduction that involves two parents, producing offspring that are not identical with either parent.

simple machine A device that makes work easier by decreasing the effort needed while increasing the distance the effort is applied.

skin The organ that covers and protects the body, and excretes wastes by perspiring.

smog A haze in the atmosphere produced by the reaction of sunlight with pollutants from cars and factories.

society A group of individuals who generally think and interact in similar ways.

soil A mixture of small rock fragments, water, air, and decayed organic material that covers much of Earth's land surface.

solar energy Energy from the sun.

solar system The sun and all the objects that revolve around it, including the planets and their moons, dwarf planets, asteroids, comets, and meteors.

solubility The maximum amount of solute that can dissolve in a given amount of solvent at a given temperature.

soluble Able to dissolve in a given solvent.

solute The substance that dissolves in the solvent.

solution A mixture in which the components remain evenly distributed.

solvent The substance that dissolves the solute.

sound A form of mechanical energy produced by a vibrating object.

sound waves Alternating layers of compressed and expanded air particles that spread out in all directions from a vibrating object.

species A group of organisms of the same kind that can produce fertile offspring. The subgroups of a genus.

specimen The object to be viewed.

speed The distance traveled per unit of time, for example, meters per second.

sperm ducts Tubes through which sperm cells pass upon leaving the testes.

spinal cord The thick cord of nerve tissue that extends from the brain down through the spinal column.

stain A material used to color certain cell structures.

stimulus A change in the environment that causes an organism to react in some way.

storm A natural disturbance in the atmosphere that involves low air pressure, clouds, precipitation, and strong winds.

succession See ecological succession.

symbiosis A relationship in which two or more different organisms live in close association with one another; that is, when one organism lives on or inside another one.

system A group of related elements or parts that work together for a common purpose.

T

technology The application of scientific knowledge and other resources to develop new products and processes.

testes The male reproductive organs that produce sperm cells.

thermal pollution An increase in the temperature of a body of water, caused by human activities, that may be harmful to living things in that environment.

third law of motion For every action there is an equal and opposite reaction.

thunderstorm A brief, intense rainstorm that affects a small area and is accompanied by thunder and lightning.

tides The rise and fall in the level of the ocean's waters that take place twice each day.

tissue A group of similar cells that act together to perform a function.

topographic map A map that uses contour lines to show the form and shape of the land's physical features.

tornado A violent whirling wind, sometimes visible as a funnel-shaped cloud.

trachea The tube that connects the nose and mouth to the bronchi, which lead to the lungs; also called the windpipe.

transport The process of moving materials throughout an organism.

Tropic of Cancer The latitude 23.5° north of the equator where the vertical ray of the sun strikes on June 21.

Tropic of Capricorn The latitude 23.5° south of the equator where the vertical ray of the sun strikes on December 21.

troposphere The lowest layer of the atmosphere in which all weather occurs.

U

uranium A radioactive element found in certain rocks and used as a fuel for nuclear power plants.

uterus The organ of the female reproductive systems within which the offspring develop; also called the womb.

V

vagina The birth canal.

veins Blood vessels that return blood to the heart.

velocity The speed of an object in a certain direction.

virus A nonliving, disease-causing core of genetic material surrounded by a protein coat.

volcano (1) An opening in Earth's surface through which hot, liquid rock flows from deep underground. (2) A mountain formed by a series of volcanic eruptions.

volume The amount of space an object occupies.

voluntary muscles Muscles that we consciously control.

W

warm front The boundary formed when a warm air mass slides up and over a cool air mass.

water cycle The process in which water moves back and forth between Earth's surface and the atmosphere by means of evaporation, condensation, and precipitation.

watt A unit that measures the rate at which energy is used.

wavelength The distance from one point on a wave to the corresponding point in the next wave.

weather The short-term condition of the atmosphere over a small area.

weather forecasting An attempt to make accurate predictions of future weather.

weathering The breaking down of rocks into smaller pieces.

wedge A double-sided inclined plane.

wheel and axle A modified lever that consists of a large wheel and a smaller wheel, or axle, at its center.

wind The movement of air over Earth's surface.

wind direction The direction from which the wind is blowing.

winter storms Blizzards and ice storms.

work The moving of an object over a distance by a force.

Index

Index

U

V

W

X

Z

Photo Credits

Cover: M101 Spiral Galaxy, © STScI/NASA/Corbis
Page 27: Student and teacher, Jupiter Images
Page 61: Roller coaster, Jupiter Images/© John Fox Images
Page 101: Nuclear power plant and wind farm, Comstock Images
Page 123: U.S. Air Force Thunderbirds, The Stocktrek Corp.
Page 138: Boys playing baseball, Getty Images/Rob Atkins
Page 152: Tropical fish: Image Source Black
Page 175: Lion family: Comstock Images
Page 196: Blood vessels of the hand, © Frans Lanting/ Corbis
Page 221: Meerkats, iStockphotoExclusive/© Peter Malsbury
Page 243: Rock formation in desert, Creatas Images/Don Hebert
Page 272: Receding glacier, ©Paul Souders/Corbis
Page 276: The Grand Canyon, Jupiter Images/Radius Images
Page 280: The Grand Tetons, Fotosearch/Corbis
Page 295: Wintry countryside, © Thierry Brésillon/Godong/Corbis
Page 326: Tornado, Courtesy of NOAA
Page 330: M101 Spiral Galaxy, © STScI/NASA/Corbis
Page 346: Moon, Courtesy of NASA
Page 358: Milky Way, NASA/JPL-Caltech
Page 362: Expressive robot, © George Steinmertz/Corbis
Page 375: Galileo Galilei, © Bettmann/Corbis